Lecture Notes in Computer Science

Commenced Publication in 1973
Founding and Former Series Editors:
Gerhard Goos, Juris Hartmanis, and Jan van Leeuwen

Wenyin Liu   Josep Lladós
Jean-Marc Ogier (Eds.)

Graphics Recognition

Recent Advances
and New Opportunities

7th International Workshop, GREC 2007
Curitiba, Brazil, September 20-21, 2007
Selected Papers

 Springer

Volume Editors

Wenyin Liu
City University of Hong Kong
Department of Computer Science
Hong Kong, China
E-mail: csliuwy@cityu.edu.hk

Josep Lladós
Universitat Autònoma de Barcelona
Dept. Ciències de la Computació
08193, Bellaterra, Spain
E-mail: josep@cvc.uab.es

Jean-Marc Ogier
Université de La Rochelle
Pôle Sciences et Technologie
7042, La Rochelle Cédex 1, France
E-mail: jmogier@univ-lr.fr

Library of Congress Control Number: Applied for

CR Subject Classification (1998): I.4, I.7.5, I.4.6, D.2.2

LNCS Sublibrary: SL 6 – Image Processing, Computer Vision, Pattern Recognition, and Graphics

ISSN   0302-9743
ISBN 978-3-540-88184-1 Springer Berlin Heidelberg New York

Springer is a part of Springer Science+Business Media

springer.com

© Springer-Verlag Berlin Heidelberg 2008

Typesetting: Camera-ready by author, data conversion by Scientific Publishing Services, Chennai, India
Printed on acid-free paper   SPIN: 12463305   06/3180   5 4 3 2 1 0

Preface

This book contains refereed and improved papers presented at the Seventh IAPR Workshop on Graphics Recognition (GREC2007), held in Curitiba, Brazil, September 20-21, 2007. The GREC workshops provide an excellent opportunity for researchers and practitioners at all levels of experience to meet colleagues and to share new ideas and knowledge about graphics recognition methods. Graphics recognition is a subfield of document image analysis that deals with graphical entities in engineering drawings, sketches, maps, architectural plans, musical scores, mathematical notation, tables, diagrams, etc. GREC2007 continued the tradition of past workshops held at Penn State University, USA (GREC 1995, LNCS Volume 1072, Springer, 1996); Nancy, France (GREC 1997, LNCS Volume 1389, Springer, 1998); Jaipur, India (GREC 1999, LNCS Volume 1941, Springer, 2000); Kingston, Canada (GREC 2001, LNCS Volume 2390, Springer, 2002); Barcelona, Spain (GREC 2003, LNCS Volume 3088, Springer, 2004); and Hong Kong, China (GREC 2005, LNCS Volume 3926, Springer, 2006). GREC2007 was also the first edition of a GREC workshop held at the same location of the ICDAR conference and it facilitated people to attend to both events.

The program of GREC2007 was organized in a single-track 2-day workshop. It comprised several sessions dedicated to specific topics. For each session, there was an invited presentation describing the state of the art and stating the open questions for the session's topic, followed by a number of short presentations that contributed by proposing solutions to some of the questions or presenting results of the speaker's work. Each session was then concluded by a panel discussion. Session topics included technical documents, maps and diagrams understanding, symbol and shape description and recognition, information retrieval, indexing and spotting, sketching interfaces and on-line processing, feature and primitive analysis and segmentation, performance evaluation and ground truthing. In addition, a panel discussion on the state of the art and new challenges was organized as the concluding session of GREC2007.

Continuing with the tradition of past GREC workshops, the program of GREC2007 also included two graphics recognition contests: a symbol recognition contest, organized by Philippe Dosch and Ernest Valveny, and an arc segmentation contest, organized by Daniel Keysers and Faisal Shafait. In these contests, test images and ground truths are prepared in order for contestants to have objective performance evaluation conclusions on their methods.

After the workshop, all the authors were invited to submit enhanced versions of their papers for this edited volume. The authors were encouraged to include ideas and suggestions that arose in the panel discussions of the workshop. Every paper was evaluated by two or three reviewers. At least one reviewer was assigned from the attendees to the workshop. Papers appearing in this volume were selected, and most of them were thoroughly revised and improved based on the reviewers' comments. This volume is organized in seven sections, reflecting the workshop session topics.

We want to thank all paper authors and reviewers, contest organizers and participants, and workshop attendees for their contributions to the workshop and this volume.

Specially, we gratefully acknowledge Karl Tombre for leading the panel discussion and Luiz Eduardo S. Oliveira for his great help in the local arrangement of the workshop.

The Eighth IAPR Workshop on Graphics Recognition (GREC2009) is planned to be held at La Rochelle, France.

April 2008

Liu Wenyin
Josep Llados
Jean-Marc Ogier

Table of Contents

Information Retrieval, Indexing and Spotting

Sketching Interfaces and On-Line Processing

Feature and Primitive Analysis and Segmentation

Performance Evaluation and Ground Truthing

Panel Discussion

Automatically Making Origami Diagrams

Jien Kato, Hiroshi Shimanuki, and Toyohide Watanabe

Department of Systems and Social Informatics
Graduate School of Information Science, Nagoya University
Furo-cho, 464-8603, Chikusa-ku, Nagoya, Japan
jien@is.nagoya-u.ac.jp

Abstract. A traditional way to explain the folding process of origami works is usually using a sequence of diagrams (called origami illustrations). Since making origami illustrations is a heavy burden for origami designers, new original works are difficult to be published in time. To solve this problem, we propose a unique framework for automatically making origami illustrations based on 2-D crease patterns, and moreover give the key techniques to fully support proposed framework.

Keywords: Origami diagram making, graph recognition, transform form 2-D crease pattern to 3-D origami model.

1 Introduction

Origami is popular art of folding paper. In its long history, not only a large number of traditional works have been accumulated, but many original works are designed and published into origami drill books every year. For most of origami lovers, the origami drill books [1], in which a folding process for a piece of work is explained by a sequence of diagrams (called origami illustrations), are probably the most convenient means for learning and enjoying origami making. Recent years, accompanied with the Internet spreads, considerable sum of original origami works come to announce via web pages, but there is no change in the point to explain folding process by a series of origami illustrations (e.g. [2], [3]).

However, making origami illustrations requires a large amount of labor. Since it is a heavy burden for designers, new origami works are usually difficult to be published in time. To solve this problem, this paper proposes an interactive supporting framework that makes it possible to realize automatically making origami illustrations. Although many researches on origami simulation, design, etc. have been conducted recently, such a framework has never been reported.

2 Basic Idea

A straightforward approach to automatically making origami illustrations is probably to simulate the folding process of origami works. That means the system needs to always keep the 3-D models of origami works, and moreover to update the models every time folding operations are applied. Obviously, that is a too tough task to be achieved.

W. Liu, J. Lladós, and J.-M. Ogier (Eds.): GREC 2007, LNCS 5046, pp. 1–8, 2008.

To avoid keeping and manipulating the 3-D models all the time, our approach is originally based on 2-D crease patterns. The basic idea is to utilize the 2-D crease patterns as much as possible, and transform the crease patterns into 3-D models only in unavoidable instances. Therefore, as shown in Fig.1, the concept of our system can be explained by a circular model where information is enhanced from 2-D (crease patterns) to 3-D (3-D origami model); than the 3-D information is further reduced into 2.5-D (origami illustrations) and finally back to 2-D (crease patterns). A big advantage of our approach is that when a folding operation is applied, instead of updating the 3-D origami model, we need just to add some creases on the 2-D crease pattern so that it can correspond to origami's updated state. Obviously, the latter is much easier than the former.

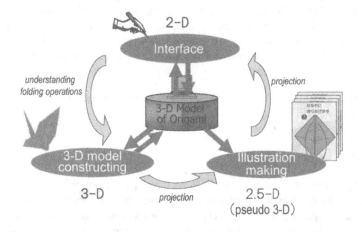

Fig. 1. Circular model for information processing and transformation

Some key techniques to support the basic idea have been developed. First, we have constructed an interface which allows the designers to communicate with the system by freehand drawing. That is, origami designers can input origami illustrations by pen (initially) or draw symbols on existing illustrations to tell the system which operation will be done next. The interface can understand the users' intention. Second, we have also developed algorithms to reflect the users' intention into a foldable crease pattern (sometimes, multiple crease patterns). Third, we have established a method to transform creases patterns into 3-D origami models using computer graphic techniques. To make origami illustrations, we just need to project the 3-D origami models from an appropriate perspective.

3   Interactive User Interface

We use the notation of origami illustrations proposed by Yoshizawa [4] that has already become a kind of international standard. There are five kinds of oft-used folding operations. They are divided into basic folding ("mountain", "valley" folding) and complicated folding ("tucking in", "covering", and "opening"). Fig. 2 shows

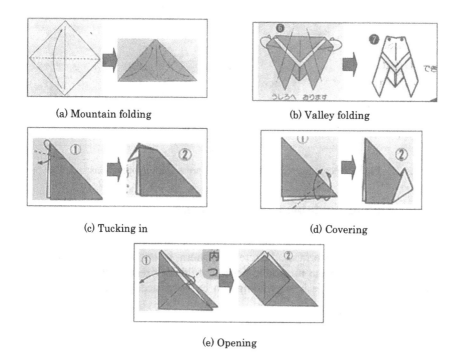

(a) Mountain folding

(b) Valley folding

(c) Tucking in

(d) Covering

(e) Opening

Fig. 2. Five kinds of frequently used origami operations and their notations in origami dill books

some illustrations of these operations. From Fig.2 (a)-(e), we can find that recognition of these operations mainly depends on if we can recognize some handing-written straight lines and curves. Concretely, it is necessary to recognize two types of straight lines (solid/broken), arrows (solid/hollow), arcs (C/S-type) and also necessary to make use of the stroke orders of theses symbols.

Our methods to recognize these symbols are very simple. Given a hand-written curve, we first draw an axis that links the beginning and ending of the curve with the direction from beginning toward the ending. Then, we integrate the curve along the axis. Let the parts of positive and negative integration be d^+ and d^-. For example, the C/S-type arcs and straight lines can be distinguished in the way described as follows.

1. The curve is recognized as a C-type arc (Fig.3 (a)), which is used for describing operations excluding "tucking in", if they meet: $d^+ > TH_{arc\_right}$
2. The curve is recognized as straight line (Fig.3(b)), if they meet: $\left|d^+\right| + \left|d^-\right| < \varepsilon$.
3. The curve is recognized as S-type arc (Fig.3 (c)), which is used for describing "tucking in", if they meet: $d^+ + d^- < \varepsilon$.

Here, TH_{arc_right} and ε are a large and small thresholds. The folding operations are finally determined based on individual symbols and their spatiotemporal relations.

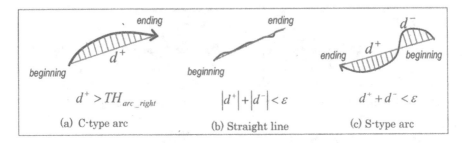

$$d^+ > TH_{arc\_right} \qquad \left|d^+\right| + \left|d^-\right| < \varepsilon \qquad d^+ + d^- < \varepsilon$$

(a) C-type arc   (b) Straight line   (c) S-type arc

Fig. 3. Recognition of symbols that express folding operations

4   Generation of Foldable Crease Patterns

According to information about the type and location of the folding operation obtained via the user interface, the crease pattern is updated to meet the uses' intention. We summarize our method to generate foldable crease patterns as follows.

The creases caused by a basic folding can be generated based on the following features:

(1) **Connectivity:** the terminal points of a crease locate either on an outside-edge of the square (virtual origami sheet) or share an inner point with another crease caused by the same fold.

(2) **Symmetry:** when a terminal point of a crease (A) reaches to an existing one (B), a new crease (C) extends from the connective point in the direction that makes creases A and C symmetrical against crease B. Creases A and C have opposite attributes (mountain/valley).

Fig.4 shows an example of generating creases made by a basic folding applied to a simple origami model. As to origami models in complicated states, the creases can be generated by repeatedly making use of feature (2) until a terminal point reaches to an outside-edge of the square or no more new creases can be generated (stop conditions).

We generate creases caused by a complicated folding using their own features and foldablity theorem called Local Flatness Conditions [5], [6] to limit the number of the folds we need to take into account. "Tucking-in" and "covering" have some identical/different features with each other that are described as follows.

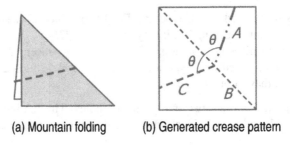

(a) Mountain folding   (b) Generated crease pattern

Fig. 4. Calculation of crease patterns for basic folding

(1) **Identical Features:**
- Generated creases (A and B) are symmetrical to an existing crease (called symmetrical axis).
- The symmetrical axis is separated into two segments, and the attribute of one piece (C) is reversed.

(2) **Different Features:**
- Generated creases have the same (opposite) attribute with (from) the symmetrical axis for tucking-in (covering).
- The creases generated by tucking-in (covering) form an acute (obtuse) angle with the symmetrical axis.

The creases caused by "tucking-in" and "covering" are simply generated based on above features in two steps. First, the new creases are made on the position given by a fold-line, using the method to make creases for basic folding. Second, the attribute of a part of the symmetrical axis is reversed. The way to make creases for "opening" can be referred in [7]. Some examples can be seen in Figs.5.

5 Construction of 3-D Origami Model

We construct the 3-D origami model based a foldable crease pattern using Affine transformation. Affine transformation relies on the neighboring relationships among faces (polygons in crease patterns) and the information attached to creases. A face-crease graph (CP-graph), where the nodes and links indicate the faces and creases, is thus introduced to easily grasp these face relationships and crease information. In Fig.6, we show a crease pattern (left), the face-crease graph (middle) and expression of a crease (right).

(1) Formalization
From the viewpoint of easily constructing 3-D origami models, a crease is specified by both of its position(coordinates) and the angle θ $(-\pi \le \theta \le \pi)$ between two faces that share the crease. The crease $C_{i,j}$ is specified by the unit vector $\hat{c}_{i,j}$ (clockwise to F_i), and the vector $t_{i,j}$ that links the origin and the beginning of $C_{i,j}$ with the direction from the former to the latter.

(2) Affine Transformation
An arbitrary face F_0 in the crease pattern is fixed on x-y plane. We calculate the transformation matrix for other faces (notated by F_p) with two steps. First, a path, the shortest one is desirable, is found out from F_0 to F_p . Let the path be

$$F_0, \cdots, F_q, \cdots, F_p . \tag{1}$$

So, the creases on this path can be indicated by

$$C_{0,1}, \cdots, C_{q-1,q}, \cdots, C_{p-1,p} . \tag{2}$$

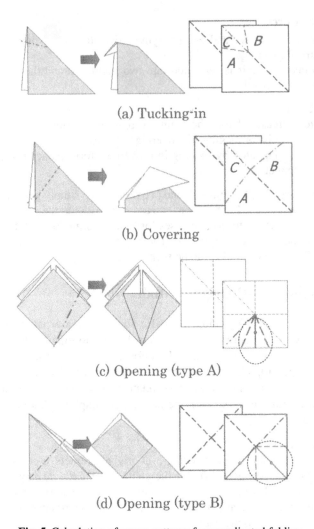

(a) Tucking-in

(b) Covering

(c) Opening (type A)

(d) Opening (type B)

Fig. 5. Calculation of crease patterns for complicated folding

Then, the transformation matrix based on the sequence of creases shown in above is calculated. Let rotation matrix in association with angle θ of $\hat{c}_{q-1,q}$ be R_q, and translation matrix in association with vector $t_{q-1,q}$ be T_q. The Affine transformation matrix for crease $C_{q-1,q}$ can be written as

$$X_q = T_q R_q T_q^{-1}.$$ (3)

So, the 3-D Affine transformation matrix for face F_p along the path we have found out can be calculated by

$$Z_q(x, y, 0, 1) = X_1 \ldots X_q \ldots X_p(x, y, 0, 1)^T \text{ for } (x, y) \in F_p.$$ (4)

By such a transformation for each face in the crease pattern, it is possible to constitute the 3-D origami model. We show an example in Fig.7.

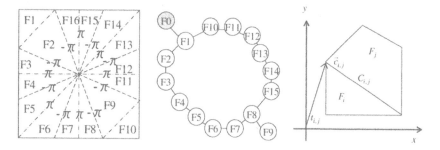

Fig. 6. An example of crease patterns (left), FC-graph (middle) and formalization of a crease

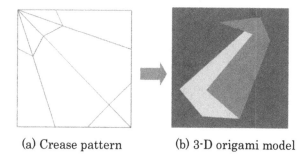

(a) Crease pattern   (b) 3-D origami model

Fig. 7. An example of constructing the 3-D origami model based on a 2-D crease pattern

6 Conclusions

In this paper, we have proposed a framework for automatically making origami illustrations with the viewpoint to accelerate publication of origami works. We also described the key techniques that fully support the proposed framework. The experiments for testing individual units of the system have shown very good results. As the future work, the integration of different units has to be completed.

References

1. Yamaguchi, M.: Delightful Complete Works of Origami. Shufu To Sekatsusha (1997)
2. Barber, A.: Origami.com, `http://www.origami.com/indx.html`
3. City Plan Co., Ltd: Origami club, `http://www.origami-club.com/index.html`
4. Magazine: Origami Tanteidan. No. 92 (2005)
5. Belcastro, S., Hull, T.: Modeling the Folding of Paper into Three Dimensions Using Affine Transformations. Linear Algebra and Applications 348, 273–283 (2002)

6. Kawasaki, T.: R(γ)=1. In: Proc. of the 2nd International Meeting on Origami Science and Scientific Origami, pp. 32–40 (1994)
7. Kato, J., Shimanuki, H., Watanabe, T.: Understanding and Reconstruction of Folding Process Explained by a Sequence of Origami Illustrations. In: The Fourth International Conference on Origami in Science, Mathematics, and Education, Pasadena, California, USA, September 8–10 (2006)

An Adaptative Recognition System Using a Table Description Language for Hierarchical Table Structures in Archival Documents

Isaac Martinat, Bertrand Coüasnon, and Jean Camillerapp

IRISA / INSA, Campus universitaire de Beaulieu, F-35042 Rennes Cedex, France
Isaac.Martinat@irisa.fr, Bertrand.Couasnon@irisa.fr,
Jean.Camillerapp@insa-rennes.fr

Abstract. Archival documents are difficult to recognize because they are often damaged. Moreover, variations between documents are important even for documents having a priori the same structure. A recognition system to overcome these difficulties requires external knowledge. Therefore we present a recognition system using a user description. To use table descriptions in analyzing the image, our system uses the intersections of two rulings with a close extremity of one or each of these two rulings. We present some results to show how our system can recognize tables with a general description and how it can deal with noise with a more precise description.

Keywords: archival documents, table structure analysis, knowledge specification.

1 Introduction

Many works were carried out on table recognition [1], but very few have been carried out on tables from archival documents. These documents are difficult to analyze because they are often damaged due to their age and conservation. The rulings can be broken and skewed or curved. Another difficulty is that paper is thin, ink bleeds through the paper, thus rulings of flip side can be visible. That is why these tables are very difficult to recognize. We want also to recognize sets of documents with a same logical structure whose physical structure can change from one page to the next. To overcome these difficulties, a recognition system needs to have a priori knowledge. Therefore we propose a recognition system using a user description from a language. This language allows to define the logical and physical structures of the tables. The advantage of our language is to describe in the same specification a logical structure with important variations in physical structures (figures 1 and 5). In this paper, we will first present the related work on table representations and on archival document recognition. In section 3 we propose a language to describe tables. Section 4 explains how our recognition system works and uses table descriptions. Before concluding our work, we will show with some results that our system can recognize very different kinds of tables with a same general description and can also recognize noisy and very damaged tables with a more precise description and we validated our system on 7783 images.

W. Liu, J. Lladós, and J.-M. Ogier (Eds.): GREC 2007, LNCS 5046, pp. 9–20, 2008.

2   Related Work

2.1   Table Representations for Editing

Wang [2] proposed a model for editing tables, which is composed of a logical part and a physical part. The logical part contains row and column hierarchy. In the physical part, for a cell or a set of cells, the user specifies the separator type, size, content display (like font size, alignment), ... When a user edits a table, the number of columns and rows must be known. Many other descriptions for tables exist in different languages such as XML but they are for generating tables. For editing, a description must be complete for data, the number of cells is fixed. For table recognition we need to have one description for a set of tables that can have variations between them like the number of columns and rows or the hierarchy.

2.2   Archival Document Recognition

Many works were carried out on table recognition [1] but very few on damaged tables in archival documents. The analysis of these documents is difficult because they are quite damaged. For the recognition of tables with rulings, Tubbs et al. analyzed 1910 U.S. census tables [3] but coordinates for each cell of the tables are given by manually typing an input of 1,451 file lines. The drawback of this method is the long time spent by the user to define this description and the recognized documents can not have physical variations. Nielson et al. [4] recognized tables whose rows and columns are separated by rulings. Projection profiles are used to identify rulings. For each document a mesh is created, and individual meshes are combined to form a template with a single mesh. Individual meshes must be almost identical to be combined, so this method can not recognize documents with important variations between them.

For other archival document recognition methods, a graphical interface is used to recognize archive biological cards [5], lists of Word War II [6]. Esposito et al. [7] designed a document processing system *WISDOM++* for some specific archival documents (articles, registration card) and the result of this analysis can be modified by the user. Training observations are generated from these user operations. All these methods are used for a very specific type of document and the information given by the user is very precise. To help the archival document recognition, systems use a user description [3,8], a graphical interface [5,6], information of other documents of the same type [4] or user corrections [7]. All these methods use external knowledge. However, the table definition process is often quite long and too precise, so these systems do not allow important variations between documents.

We presented in [8] a specific description system for military forms of the 19th Century. We also showed that a general system was not able to recognize these archival documents. This specific description took a long time to write, therefore it is necessary to have a faster way of describing tables. In [9] we presented a table recognition system using a short user description but this system was limited, the row and column numbers were fixed. Furthermore the row and column hierarchies could not be described. Therefore, we propose a general table recognition system for tables using a table description language which can be adapted to damaged archival tables with the introduction of a more precise description.

3   Table Language for Table Recognition

A table is a set of cells organized with columns and rows. We want to recognize the organization of a table, which means locating the cells of a table and labeling each cell with the name and the hierarchy of its column and the name and the hierarchy of its row. We also have to detect table structures from very damaged documents. To solve these two difficulties, we need to use a user description.

3.1   Specification Precision Levels

The language we propose is composed of two parts. The first one is a logical part which describes the row and column hierarchy. The second one is a physical part which allows to specify the row and column separators, and optionally also allows to define the number and/or size of columns and/or rows. The advantage of this language is to describe tables with different levels of precision. On the one hand, the description can be very genesral. In this case, documents with important differences can be recognized with the same description but documents to recognize can not contain noise. For example, for a general description, a multi-row hierarchy can be described without specifying the number of rows for each level. On the other hand, the description given by the user can be very precise. In this case, very damaged documents can be recognized but for the same description, variations between documents can not be important. For example, for a precise description, the numbers of rows and columns can be specified. For a more precise description, sizes can also be given for some columns and some rows. The user can change easily and quickly from a general description to a precise one by adding or modifying some specifications with different precisions as in figure 1. This language also allows to specify a general and precise description, for example the description can be precise for the columns where the number is fixed, and general for the rows where the number is unfixed.

3.2   Table Language Definition

We will now use the term *element* rather than column or row. We propose a language like Wang's model, composed of two parts, a logical one and a physical one. The main differences between our language and Wang's model is that our language allows to specify for a table an unfixed number of columns and rows. In the logical part, the user describes element hierarchy (COL, ROW) and the relationship between columns and rows (COLS_IN_ROW). The physical part is optional, it allows to specify the number of repetition times for an element (REPEAT, REPEAT+ if the number is unfixed), the size of an element, the separator types (SEPCOL, SEPROW). The user can also describe specific separators for some cells (SEPCELL).

3.3   Language Examples

These examples (figure 1) show that a description is easy and fast to write. The words in capital letters are reserved words of the language. To modify the general description to a more precise description, REPEAT+(1,info) is replaced by REPEAT(7,info)

//logical part
//column hierarchy
    COL cols = designation, numbers, info .
    COL designation = area, street .
    COL numbers = houseNb, householdNb, personNb.

//row hierarchy
    ROW rows = boxHead, house.
    ROW boxHead = head1, head2.
    ROW house = household.
    ROW household = person.

//Logical relationships between cols and rows and
//specification for each row the columns that are contained in.
    COLS_IN_ROW head1 = designation, numbers.
    COLS_IN_ROW head2 = area, street , house , household , person.
    COLS_IN_ROW house = houseNb, householdNb, personNb, info.
    COLS_IN_ROW household = householdNb, personNb, info.
    COLS_IN_ROW person = person, info.
    COLS_IN_ROW _ = designation, numbers, info.

//Physical part
//columns separated by rulings represented by |
//rows separated by rulings represented by -(---------)
    SEPCOL cols = | designation | numbers | **REPEAT+(1,info |)** .
    SEPCOL designation = area | street .
    SEPCOL numbers = houseNb | householdNb | personNb.

    SEPROW rows =
     -(---------------------------)
          boxHead
     -(---------------------------)
          REPEAT+(1,
            house
     -(---------------------------)   )   )

    SEPROW boxHead =
          head1
     -(-------------------)
          head2 .

    SEPROW house =
          REPEAT+(1,
            household
     -(-----------------------)))

    SEPROW household =
          REPEAT+(1,
            person
     -(---------------------)))

(a) General Description : It is used to recognize very different tables (figures 4 and 5) with an unfixed number of columns and rows at each level of hierarchy.

//logical part
//column hierarchy
    COL cols = designation, numbers, info
    COL designation = area, street .
    COL numbers = houseNb, householdNb, personNb

//row hierarchy
    ROW rows = boxHead, house.
    ROW boxHead = head1, head2.
    ROW house = household.
    ROW household = person

//Logical relationships between cols and rows and
//specification for each row the columns that are contained in.
    COLS_IN_ROW head1 = designation, numbers.
    COLS_IN_ROW head2 = area, street , house , household , person
    COLS_IN_ROW house = houseNb, householdNb, personNb, info
    COLS_IN_ROW household = householdNb, personNb, info.
    COLS_IN_ROW person = person, info.
    COLS_IN_ROW _ = designation, numbers, info

//Physical part
//columns separated by rulings represented by |
//rows separated by rulings represented by -(---------)
    SEPCOL cols = | designation | numbers | **REPEAT(7,info |)** .
    SEPCOL designation = area | street .
    SEPCOL numbers = houseNb | householdNb | personNb.

    SEPROW rows =
     -(-------------------------)
          boxHead
     -(-------------------------)
          REPEAT+(1,
            house
     -(-------------------------)   ).

    SEPROW boxHead =
          head1
     -(-----------------)
          head2

    SEPROW house =
          REPEAT+(1,
            household
     -(---------------------)))

    SEPROW household =
          REPEAT(31,
            person
     -(-------------------)))

(b) Precise Description : It is used to recognize very damaged documents (figure 6).

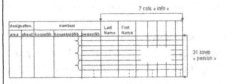

(c) Example of a table which can be recognized with the general description.

(d) Example of a damaged table which can be recognized with the precise description.

Fig. 1. Example of a general description to recognize very different tables wich can be easily modified to a precise description to recognize very damaged documents

and REPEAT+(1,person) by REPEAT(31,person). With the general description, the recognition system can recognize tables with important variations between them (figures 4 and 5). Indeed the number of columns is unfixed in this description and for the rows, at each level of hierarchy, the number of rows is unknown. The precise description allows the recognition system to recognize very damaged documents (figure 6). For documents where reverse side rulings are visible, the user can give again a more precise description in giving the approximative sizes for rows and columns. The sizes help the system to avoid detecting reverse side rulings.

4   From the Description to the Image

4.1   Final Intersections

From the image, we extract a set of line segments. Our goal is to match the image information with column and row information given by the language. Therefore we need to associate each line segment with a row or a column separator. We also need to have an intermediate level with common elements to match the image information and the user description. These elements must also be stable. To detect row and column separator within a hierarchy, we need to use line segment extremities. We need to use elements that can be derived from a user description, and these elements must easily be extracted from the image. We propose to define a specific type of intersection, called a *final intersection* which is an intersection involving at least one line segment extremity. From the user description, we can derive the *final intersections* that must be found in the image, and from the image we can extract the final intersections. More specifically, we

Fig. 2. Examples of *final intersections*, double arrows represent the intersection tolerance

define *final intersection* (figure 2) as an intersection of two rulings with the extremity of one or both of these rulings in close proximity to the other ruling. This definition includes the possibility that the two rulings may not intersect each other. In this case we define *intersection tolerance* as the distance between the two rulings. These *final intersections* allow to detect beginning and end of separators or specific changes in separator types. We do not use cross intersections because these intersections are too ambiguous. The final intersections have stronger dependencies: these intersections are typed and can be differentiated. For example some intersections can be differentiated as a table corner or as the beginning of a row separator.

4.2   Recognition System Using Final Intersections

To detect the table structure, our system perfoms an in-depth analysis of rows and for each Terminal *Row*:

- detects an horizontal Separator we call *SepH*
- from the Table Description : gets the final intersections associated with this *Row* we call *DescrInterList*
- from the Image : gets the final intersections associated with the *SepH* we call ImageInterList
- matchs the *DescrInterList* and *ImageInterList* :
  - if it succeeds, the vertical separators associated with the image final inter- sections, are labeled (and detected) with the column names from the table description.
  - If this step fails, this matching is *delayed*, which means it will be run later. When the matching is released, that is to say it is run, the detection of col- umn separators during the delay can allow the matching to succeed. The search of intersections is also extended, the intersection tolerance is automatically increased to help the matching to succeed.

The matching succeeds when the final intersections from the description are found in the image. For example, from the description if the number of intersections which must be found in the image equals the number of final intersections in the image, this matching succeeds else it fails.

4.3   System Adaptation in Function of Description Level

When a table description is precise, the system can adapt to a document using the table description, so it can recognize very noisy documents. If a table description is very gen- eral, the system will search for final intersections in the image with a small intersection tolerance, the initial value. When a table description is more precise, if after the first detection the system has not detected in the image a structure matching with the table description, the delayed row detections are released. After this release, the intersection tolerance is automatically increased to help the system to find the right structure. For example, if the number of columns is fixed, when the intersection tolerance is increased, the system searches for final intersections in larger zones and can then detect the right number of column separators. When a table description is very precise, sizes for rows and/or columns are given, the system then searches rulings in image zones delimited by these sizes. It helps the system to avoid detecting false rulings, for example the reverse side rulings.

5   Results

The system takes 14 seconds on linux with a 2.0 Ghz processor to recognize an image of 2500x3800 at 256 dpi.

5.1   Example on a Noisy Document

We will show on one synthetic example how our system can recognize noisy documents by using our language. In this first example (fig. 3), the language allows to specify that the document is a table containing 3 columns (A,B and C) and 3 rows. He specifies

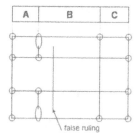

Fig. 3. Example where our system can detect a noisy document with a false ruling. Circles and ellipses represent final intersections.

also that in the second row, the separator of the column A is blank. For the analysis, a preliminary step derives from the table description the final intersections that the system must find in the image. The final intersections in this example are the circles and the ellipses on the image (fig. 3). The system starts the table recognition with the first horizontal separator, extracts from the image the final intersections of the top separator. The system extracts 3 final intersections, although with the description it would have to detect 4 final intersections, so it *delays* this matching. The system then detects the two following horizontal separators as well as the final intersections from each separator and the matching with the description succeeds. The system detects the bottom separator and as the separator for the column A is already detected, the system detects in the prolongation of this vertical separator a final intersection with a higher value of intersection tolerance. After this detection, the *delay* is released (algorithm presented in 4.2), so the system starts to detect the top separator and the final intersections associated with this separator with a higher value of intersection tolerance and as with the bottom separator, it detects the correct final intersections. This example shows how our system can recognize difficult documents. The description allows the system to eliminate false separators, and to detect separators with missing parts.

5.2   General Description

With the same general description (figure 1), the system can recognize census tables from different years (figures 4 and 5) with different structures. On figure 4, the 1881 table contains 8 columns whereas the 1911 table contains 10 columns. For a same year, the row hierarchy is different for each document, thus it is not possible to have a precise description for this hierarchy. The recognition system using this description, after the *boxHead* detection, labeled the column separators in using names from the description. For the row detection, an horizontal separator is detected, then the system gets the vertical ruling that intersects with the left extremity of the horizontal separator. From the terminal level of hierarchy, the system checks if the label of this vertical ruling matches with the specification of the row level. If this checking fails, the system tries again with the upper level of hierarchy until it finds the right level.

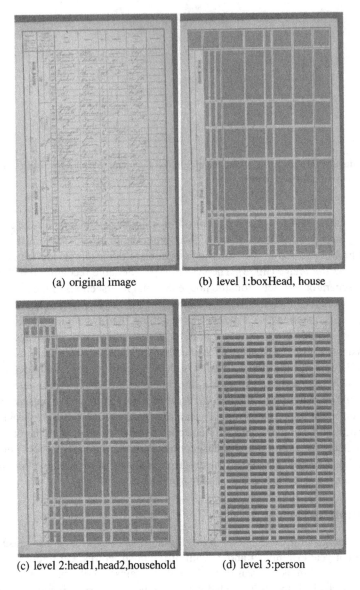

(a) original image (b) level 1:boxHead, house

(c) level 2:head1,head2,household (d) level 3:person

Fig. 4. Census Table of 1881 and the recognized structure with a general description (fig. 1), column number is unfixed like row number at each level of hierarchy

5.3 Precise Description

With a more precise description, the row and column numbers are fixed and the system can recognize the damaged archival document in figure 6. The row hierarchy is not detected but for the lowest level, the *person* rows are detected. As the system fails

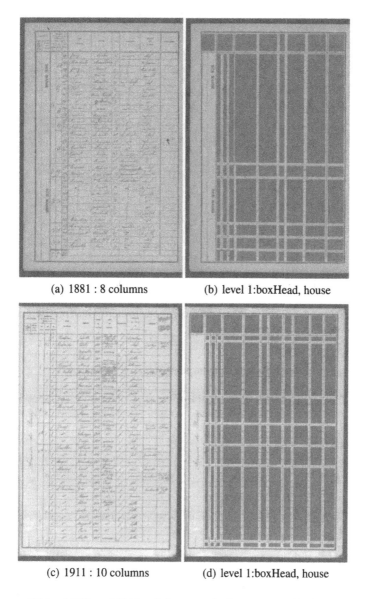

(a) 1881 : 8 columns   (b) level 1:boxHead, house

(c) 1911 : 10 columns   (d) level 1:boxHead, house

Fig. 5. Census Table of 1881 and 1911 and the recognized structures with the same general description (fig. 1), column number is unfixed like row number

to recognize structures at the first step, the intersection tolerance for ruling gaps as with final intersections is automatically increased until it recognizes the right structure. Therefore the system detects the right structure.

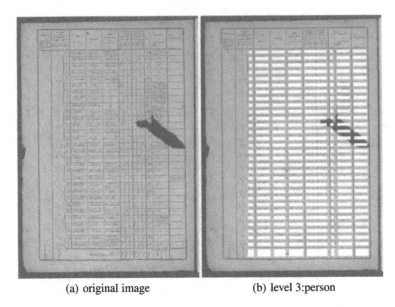

(a) original image   (b) level 3:person

Fig. 6. Damaged census table of 1876 and the recognized structure with a precise description, in which the number of rows and columns is fixed

Table 1. Results with the descriptions of the figure 1

level	number of tables	percentage of tables without any error	number of cells	number of detected cells	percentage of detected cells	rejected documents
general description						
all levels	30	66%	11,222	10,925	97,35%	0%
street	30	96%	160	140	87.5%	0%
house, fig. 4(b)	30	93%	2160	2088	96.67%	0%
household, fig. 4(c)	30	73%	2392	2264	94.64%	0%
person, fig. 4(d)	30	83%	6510	6433	98.81%	0%
precise description						
person, figure 4(d)	30	100%	6510	6510	100%	0%

5.4   Statistical Results

Hierarchical Tables. We tested 30 images with the general description of figure 1. These images each contain one table from the 1881 census. All of the code used by the system is totally presented in this figure to which we added an upper level in row hierarchy, a *street* level. Therefore the description specifies now the following row levels : *street, house, household* and *person*. Indeed for each table (figure 4), the data are presented by street, a street containing several houses, a house containing several households and a household containing serveral persons. These documents can be quite damaged. 10 images are detected with errors but there are few errors on each image. Therefore, to correctly recognize hierarchical structures, the documents must not be very damaged. However, the lowest levels of hierarchies in damaged documents can be correctly recognized with a more precise description. Table 1 shows the results with the precise description of figure 1 and all the images are well recognized. With this precise description, we have only the lowest level of hierarchy but we get 100% recognition for this level. The precision of the description allows the system to recognize more damaged documents. We did not have time to test on a larger dataset for this kind of table since we must build the groundtruth data manually. In a future work, we will enlarge this dataset.

However, we can test on a larger number of documents with precise and constant descriptions. These descriptions are very precise and contain enough information to assume that if a table is detected, then all of the cells are correctly recognized.

Large quantity of documents. For our first results on a large number of documents we ran our system with precise descriptions on tables without hierarchy. We would have wanted to test our system with fixed hierarchical tables but we did not have this kind of table. Therefore, we detect only the person level on these tables. These results are on two different sets of documents with a precise description for each set. Each set corresponds to one year of census. Table 5.4 shows the results for two years, 1831 and 1836. The description for each year contains the numbers of rows and columns as well as the sizes for each column and each row. In a future work we will study why the system can not recognize the rejected documents. One of the reasons can be the weak contrast value of certain images and the system can fail to detect rulings with weak contrast.

Table 2. Results on a large quantity of documents

year	number of tables	percentage of tables without any error	number of cells	number of detected cells	percentage of detected cells	rejected documents
1831	2722	92.32%	359,304	331,716	92.32%	7.68%
1836	5031	92.72%	950,859	881,685	92.72%	7.28%

6 Conclusion

We presented a language to describe tables. With the same language, table descriptions can be very precise for damaged document recognition as well as very general to detect tables with important variations between them. Moreover, these descriptions can be written quickly. To match table description and image information, we have shown the interest in using some specific intersections which we defined as final intersections. Finally, we have shown through our results how our system can detect a multi-level row hierarchy table with a general description. With this description an important number of different structures can be recognized. If documents are too damaged to be recognized with this description, the user can easily and quickly add or modify some specifications to get a more precise description. The system can then detect very damaged documents, which is important for the automatic processing of archival documents. We validated our system on 7753 images with a precise description and we got 92.52% recognition. In a future work, we will test on a larger dataset with a general description and we will try to decrease the number of rejected documents using a precise description.

Acknowledgments

This work has been done in cooperation with the *Archives départementales des Yvelines* in France, with the support of the *Conseil Général des Yvelines*.

References

1. Zanibbi, R., Blostein, D., Cordy, J.R.: A survey of table recognition. International Journal of Document Analysis and Recognition (IJDAR) 7(1), 1–16 (2004)
2. Wang, X.: Tabular abstraction, editing, and formatting. PhD thesis, University of Waterloo (1996)
3. Tubbs, K., Embley, D.: Recognizing records from the extracted cells of microfilm tables. In: ACM Symposium on Document Engineering, pp. 149–156 (2002)
4. Nielson, H., Barrett, W.: Consensus-based table form recognition. In: 7th International Conference on Document Analysis and Recognition (ICDAR 2003), Edinburgh, UK, pp. 906–910 (August 2003)
5. He, J., Downton, A.C.: User-assisted archive document image analysis for digital library construction. In: 7th International Conference on Document Analysis and Recognition (ICDAR 2003), Edinburgh, UK, pp. 498–502 (August 2003)
6. Antonacopoulos, A., Karatzas, D.: Document image analysis for world war 2 personal records. In: 1st International Workshop on Document Image Analysis for Libraries (DIAL 2004), Palo Alto, CA, USA, pp. 336–341 (January 2004)
7. Esposito, F., Malerba, D., Semeraro, G., Ferilli, S., Altamura, O., Basile, T.M.A., Berardi, M., Ceci, M., Mauro, N.D.: Machine learning methods for automatically processing historical documents: From paper acquisition to xml transformation. In: 1st International Workshop on Document Image Analysis for Libraries (DIAL 2004), Palo Alto, CA, USA, pp. 328–335 (January 2004)
8. Coüasnon, B.: Dmos, a generic document recognition method: Application to table structure analysis in a general and in a specific way. International Journal of Document Analysis and Recognition (IJDAR) 8(2-3), 111–122 (2006)
9. Martinat, I., Coüasnon, B.: A minimal and sufficient way of introducing external knowledge for table recognition in archival documents. In: Liu, W., Lladós, J. (eds.) GREC 2005. LNCS, vol. 3926, pp. 206–217. Springer, Heidelberg (2006)

Converting ECG and Other Paper Legated Biomedical Maps into Digital Signals

A.R. Gomes e Silva, H.M. de Oliveira, and R.D. Lins

Federal University of Pernambuco - UFPE, Signal Processing Group, C.P. 7.800,
50.711-970, Recife - PE, Brazil
{hmo,rdl}@ufpe.br

Abstract. This paper presents a digital signal processing tool developed using Matlab^{TM}, which provides a very low-cost and effective strategy for analog-to-digital conversion of legated paper biomedical maps without requiring dedicated hardware. This software-based approach is particularly helpful for digitalizing biomedical signals acquired from analogical devices equipped with a plottingter. Albeit signals used in biomedical diagnosis are the primary concern, this imaging processing tool is suitable to modernize facilities in a non-expensive way. Legated paper ECG and EEG charts can be fast and efficiently digitalized in order to be added in existing up-to-date medical data banks, improving the follow-up of patients.

Keywords: analog-to-digital converter, digitalization of medical maps, digital ECG, digital EEG.

1 Background and Set-Up

Digital equipments are nowadays largely preferred to analogical ones especially due to their high-quality and flexibility of working with their output. Medical equipments that use digital technology have emerged as a true revolution in signal acquisition, analysis and diagnosis [1]. Today, electrocardiograms, electroencephalograms, electromyogram and other biomedical signals are all digital. Digital signals allow very high signal processing capabilities, easy storage, transmission and retrieval of information. The well-recognized advantages of digital technology turns it the first-choice. One of the limiting factors of adopting the digital technology is the high cost of some modern digital equipment, overall some medical ones. This is a serious barrier to be crossed by those who already have a working analogical device and/or face budget limitations. An alternative to device replacement is adopting an A/D-converter and a suitable interface to a digital microcomputer or laptop. This would also allow digitizing legated analogical data, something of paramount importance in many areas, overall in medicine as the history of patients would be kept and case studies may be correlated, etc. A number of laboratories, medical institutes and hospitals have only available analogical equipments, particularly those equipped with plottingters. The storage of these signals is rather inefficient and the data processing unfeasible. In this challenging scenario, a substantial advance can be performed by designing acquisition cards with interface to microcomputers, instead of purchasing sophisticated high-cost computerized equipments. This kind of up-grade can be

W. Liu, J. Lladós, and J.-M. Ogier (Eds.): GREC 2007, LNCS 5046, pp. 21–28, 2008.

beneficial to small laboratories with modest resources. Nevertheless, it is not a trivial task to assemble or to design a set-box to convert signals. This study describes the development of a software tool intended to convert a version of signals and/or spectra digitalized by a scanner (files of the extension .jpg .tif .bmp etc.) to a data file, which can be efficiently processed and stored. It deals with an alternative approach to the classical A/D conversion without requiring any specific hardware.

2 An A/D Image-to-Data Converter Algorithm

Many relevant but old data are only available in a chart-format and the appending new data may be suitable. For instance, this is precisely what happens in many long standing time series. How to perform efficiently such a procedure? The following description is strongly based on ECG, but it can easily be adapted to other signals, either biological or not. An implementation of the A/D platform on MATLAB^{TM} is presented [2], exhibiting a few cases to illustrate the lines of the procedure.

S1. Digitalization of the paper strip
S2. Image binarization
S3. Skew correction
S4. Salt-and-pepper filtering
S5. Axis identification
S6. Pixel-to-vector conversion
S7. Removing the header and trailer of the acquired signal (used for device tuning)
S8. Splitting the ECG chart and re-assembling it.

S1. Digitalization of the Paper Strip by Scanner up to 600 dpi

The digitalizing process of the paper containing the chart to acquire the data can be carried out at different resolutions. Higher resolutions turn feasible details identification on the acquired image, but most applications require only a resolution high enough to achieve acceptable digitalization quality, claiming as small storage and scanning time as requested. Tests were performed over ECG charts scanned at 100, 200, and 300 dpi. The paper strip scale is in millimetres, thus a 100 dpi resolution would theoretically be sufficient to record the signal information. However, in a number of cases data from the ECG map was not retrieved at low resolution (100 dpi) mainly due to the existing similarity between the axis and the plotting trace. Besides that paper folding scanned at 100 dpi may give rise to signal discontinuities.

S2. Image Binarization

Image binarization is a process to translate a colour image into a binary image. It is a widespread process in image processing, especially for images that contain neither artistic nor iconographic value. Since binarization reduces the number of colours to a binary level, there are apparent gains in terms of storage, besides simplifying the image analysis as compared to the true colour image processing [3]. In this analysis, we have first discarded the axis composing the image and then applied the binarization process by Otsu's algorithm [4], since it has been shown to provide satisfactory results in many applications [3],[5]. Binarization is an important step in moving from a biomedical map image towards a digital signal, as the target of the process described herein is to obtain a sequence of values that correspond to the amplitude of a uniform time series (see S6, below).

S3. Skew Correction

A distortion frequently found in scanning processes is the skew caused by the position of the paper on the scanner flatbed. This rotation makes hard the analysis of data embedded in the image and increases the complexity of any sort of automatic image recognition. Whenever extracting data from a digitalized chart even 0.5 degrees or less can introduce errors on the extracted data. The algorithm presented in [6] was used here to correct the skew of the image taking as reference the axis or the border of the paper strip.

S4. Salt-and-Pepper Filtering

Salt and pepper noise is characterized by the presence of isolated white and black pixels in a black-and-white image [7] [8]. It may bring technical hitches in analysis of the data. In order to avoid a false identification of those pixels as piece of the analyzed chart, a filter was implemented to extract isolated pixels in the middle of a 3×3 matrix and a 3×2 matrix.

S5. Axis Identification

Several different types of data plotting were analysed for the sake of the generality of the methodology proposed herein. The piece of paper used to register the data may contain a grid, a box, one horizontal line and one vertical line, or no axis, but in each case a specific analysis needs to be performed in order to adequately interpret the value of the signal obtained, compensating offsets, etc.

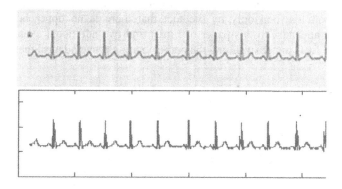

Fig. 1. An ECG with no axis: (a) original scanned ECG, (b) MatlabTM plotting of the ECG from the retrieved data

S6. Pixel-to-Vector Conversion

For a better data retrieving from the image, vectors have twice the number of columns of pixels analyzed, because some peaks are found as successive vertical black lines for low resolution images. Each component of the vector is a complex number. The two components related to a same column have identical real part, which is the column index of the pixel matrix. The two consecutive imaginary parts (same real part) quantify the upper and lower limits of a vertical black line. For instance, the 2D-vector V=[...; 11+25i; 11+26i; ...] has coordinates meaning that the vertical black line spans from line 25 to line 26 at the column 11. Figure 1 deal with an ECG chart with no axis. As an example, a stretch of the vector used to plotting Figure 1(b) is V = [10+26i; 10+26i; 11+25i; 11+26i; 12+25i; 12+26i; 13+26i; 13+27i 14+26i 14+28i 15+28i...].

The algorithm searches the data since the first pixel (bottom-left) until the last one (top-right). When there are no axis present in the map (no clear vertical and horizontal bounds), the value of the bottom vertical and the first column positions are used as a reference for ECG scanning. If step 5 (S5) is able to find axes of the graph, then these axes are assumed as data scanning reference.

In order to convert the 2D-vector into one dimension, the algorithm computes the modal distance (α) between the imaginary part of two consecutive components, which means the amount of vertical black pixels composing the signal at a specific column. If the difference between the imaginary components of the two coordinates is within the limit α, then only the imaginary part of the second element is stored in the 1D version of the vector. This value is assumed as a reference (β) to the peak identification in next column analysis. If this difference is greater than α, the algorithm calculates the module of the difference between the β and each of the two components. The stored 1D-value is the one that gives the greater value.

Whenever leading with plotters, one often finds portions of the chart where the drawing is no longer continuous. In order to provide a one-dimension vector, the components at those places are estimated through linear interpolation. Data retrieving can be performed from a broad range of medical-related plottings. Figure 2 shows an example of a nuclear magnetic resonance (MNR) spectrum. Horizontal and vertical lines differ from the box only by the fact that there is no upper bound in data acquisition. In appendix to this paper one may find two additional examples of data extracted from ECG paper strips and the corresponding plotted data signal.

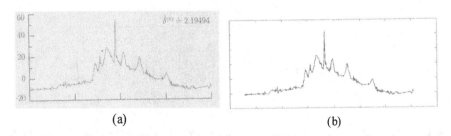

(a)   (b)

Fig. 2. Example of data from a nuclear magnetic resonance (MNR) spectrum (extracted from the public domain software package Wavelab). (a) Original scanned spectrum, (b) MatlabTM plotting from the corresponding data file.

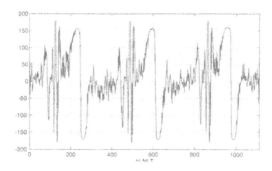

Fig. 3. Image of an ECG in a box, having some horizontal and vertical dotted lines

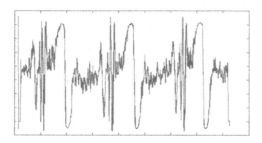

Fig. 4. Plotting of signal vector corresponding to the image from Fig. 3 with header and trailer removed

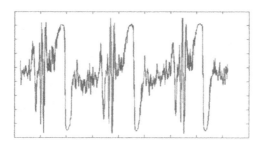

Fig. 5. Plotting of data retrieved from Fig. 3 after throwing away the first and final 16-values of the acquired data vector

S7. Removing the Header and Trailer of the Acquired Signal

Typical ECG and a large number of ordinary graphs are composed either by boxes or horizontal and vertical lines the same colour as the plotting. Whenever those lines encompass parts of the area of the graphic, it becomes rather difficult to remove the interference, since it can happen above and/or below the graphic. As the process of data acquisition is done by searching vertically, only the vertical lines interfere with the process. Besides that, very often at the beginning and ending of the data acquisition process a line is drawn carrying no information. Such headers and trailers carry no real signal information, thus they are removed from the generated vector signal, as illustrated in the example presented in Figs. 3 to 5.

S8. Splitting the ECG Chart and Re-assembling it

A problem that arises whenever scanning an ECG and other legated paper data signals is that usually the scanner flatbed does not cover the whole length of the paper strip, leading the operator to scan it in separated parts. Special care and image processing is needed to avoid loss or redundancy in data. A possible solution to this problem is to insert in the paper strip easily detectable marks. During the tests, handmade marks were inserted on the paper strip, drawn by pen with no other tool or mechanical support. The mark should be of a colour not originally present in the paper data strip. The image processing algorithm scans the image horizontally for one or two of such marks: only one mark on the first and on the last stretch, or two marks on the intermediary ones. Those marks play the role of the horizontal bounds in step 6. The vector is extracted from the beginning of the image until the mark for the first stretch, from the first mark to the second one for the intermediary stretches and from the mark until the end of the image for the last stretch. After obtaining the vectors of each partial ECG, one can handle data by appending segments.

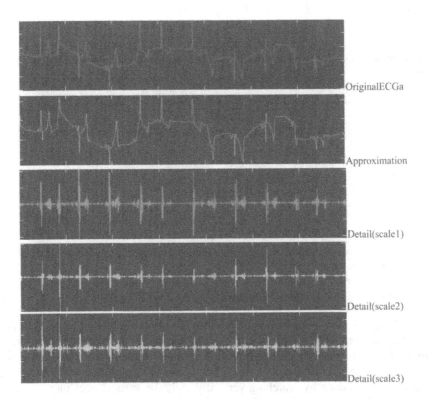

Fig. 6. Example of an ECG processing using the Haar wavelet corresponding to ECG paper strip of the Fig. 2A. Analysis was performed from the data acquired via the proposed algorithm.

3 An Application of Signal Processing Tools to ECG Paper Chart

This section illustrates the application of the wavelet decomposition [9] to the ECG shown in Fig. 2A (Appendix). The aim here is to corroborate that it is feasible to handle a data file derived from a strip of paper through this digitalization approach. No comments about the signal analysis are made: we just check whether or not usual signal processing techniques can be applied to such kind of retrieved data without abnormal or unexpected results. The algorithm proposed in this paper was first used to extract a Matlab data file as to allow the wavelet toolbox to be used straight away [2]. The acquired image was converted into a 1D vector that was loaded using the wavemenu command. As an example, a three level decomposition of this ECG using the Haar wavelet is shown in Figure 6.

4 Closing Remarks

Notwithstanding the fairly amount of scientific results, this paper describes the foundation of an efficient tool to generate digital data signals from legated paper charts. The solution proposed is a low-cost software tool that can be particularly helpful to scientists and engineers. In particular, research institutes, laboratories, clinical centres, hospitals and medical offices can largely have benefit of this up-and-coming technique, particularly due to its user-friendliness, cost-effectiveness, and accuracy. One still can save data as Matlab^{TM} file or as ASCII files and edit or complement the data. One natural follow-up step to the tool presented herein is to generalise the processing capability to multi-plotting paper charts, that also frequently appears in legated data. Besides that at a later stage we hope to generalise the tool to work with signals recorded paper disks.

Acknowledgements. The authors are grateful to Mr. Bruno T. Ávila for making available some C command lines.

References

1. Bruce, E.N.: Biomedical Signal Processing and Signal Modelling. Wiley, Chichester (2001)
2. Kamen, E.W., Heck, B.S.: Fundamentals of Signals and Systems Using Matlab. Prentice Hall, Englewood Cliffs (1997)
3. da Silva, J.M.M., Lins, R.D., da Rocha, V.C.: Binarizing and Filtering Historical Documents with Back-to-Front Interference. In: SAC 2006: Proceedings of the 2006 ACM symposium on Applied computing, pp. 853–858. ACM Press, New York (2007)
4. Otsu, N.: A Threshold Selection Method from Grayscale Document Images. IEEE Trans. Syst. Man Cybern. – SMC 9(1), 62–66 (1979)
5. Trier, O.D., Taxt, T.: Evaluation of Binarization Methods for Document Images. IEEE Trans. Pattern Anal. Mach. Intell. 17(3), 312–315 (1995)
6. Lins, R.D., Ávila, B.T.: A New Algorithm for Skew Detection in Images of Documents. In: Campilho, A.C., Kamel, M. (eds.) ICIAR 2004. LNCS, vol. 3212, pp. 234–240. Springer, Heidelberg (2004)
7. Jähne, B.: Digital Image Processing, 3rd edn. Springer, Heidelberg (1995)
8. Gonzalez, R.C., Woods, R.E.: Digital Image Processing, 2nd edn. Prentice Hall, Englewood Cliffs (2001)

9. Mallat, S.: A theory for multiresolution signal decomposition: The wavelet representation. IEEE Trans. Pattern Anal. Machine Intell. 11, 674–693 (1989)
10. Data bank of Electrocardiograms, http://www.ecglibrary.com

Appendix

In this appendix we present a couple of the data extracting from true ECG strips [10] and the plottings of the retrieved data. (Fig. 1A,2A): we first find out the grid by using the blue colour information, and then switch the pixel by white pixel.

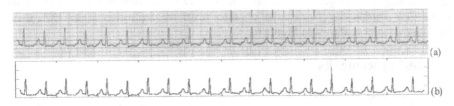

Fig. 1A. Image of an ECG with red grid: (a) original ECG chart, (b) MatlabTM plotting of retrieved ECG with data file available

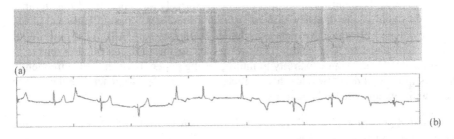

Fig. 2A. A green grid ECG: (a) original ECG chart, (b) MatlabTM plotting of retrieved ECG with available data file

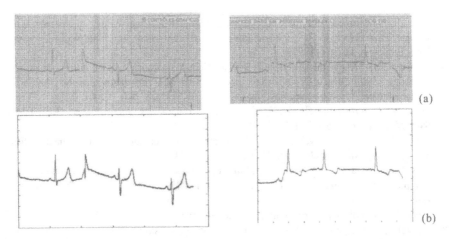

Fig. 3A. Image of an ECG: (a) Two ECG charts with red handmade marks drawn by pen (initial and intermediary portion), (b) Matlab[TM] plotting of retrieved ECG with data file available

Hand Drawn Symbol Recognition by Blurred Shape Model Descriptor and a Multiclass Classifier

Alicia Fornés[1,2], Sergio Escalera[2,3], Josep Lladós[1,2], Gemma Sánchez[1,2], and Joan Mas[1,2]

[1] Computer Vision Center, Universitat Autònoma de Barcelona,
08193, Bellaterra, Spain
{afornes,josep,gemma,jmas}@cvc.uab.es
[2] Department of Computer Science, Universitat Autònoma de Barcelona,
Campus UAB, Edifici Q, 08193, Bellaterra, Spain
[3] Matemàtica Aplicada i Anàlisi, Universitat de Barcelona,
Gran Via 585, Barcelona, Spain
sergio@maia.ub.es

Abstract. In the document analysis field, the recognition of handwriting symbols is a difficult task because of the distortions due to hand drawings and the different writer styles. In this paper, we propose the Blurred Shape Model to describe handwritten symbols, and the use of Adaboost in an Error Correcting Codes framework to deal with multi-class categorization handwriting problems. It is a robust approach tolerant to the distortions and variability typically found in handwritten documents. This approach has been evaluated with the public GREC2005 database and an architectural symbol database extracted from a sketching interface, reaching high recognition rates compared with the state-of-the-art approaches.

1 Introduction

Symbols are a good way to express ideas. A number of graphical languages exist in different domains like engineering, architecture, software modelling, etc. These languages allow users to describe complex models with compact diagrammatic notations. On the other hand, in a technology world, freehand sketching is a very natural and powerful way of communication between humans. Sketch understanding is a research area with long history that brings together the above issues, i.e. a natural way of man-machine interaction in terms of freehand drawings and a pattern recognition ability to interpret sketches according to a diagrammatic notation. The first attempts of sketch recognition in the graphical domain can be found more than two decades ago. The early approaches were off-line systems mainly devoted to diagram beatification [1], or application cases of pattern recognition theory [2]. With the progress of digital pen and paper protocols such as Tablet PC or PDA, on-line sketch recognition systems gained in prominence. New alphabets have been developed for these devices, e.g. the

W. Liu, J. Lladós, and J.-M. Ogier (Eds.): GREC 2007, LNCS 5046, pp. 29–39, 2008.
© Springer-Verlag Berlin Heidelberg 2008

Graffitti alphabet for PDA's. The input of digital pen devices, called digital ink, consists in a sequence of points acquired at regular time intervals and grouped into basic sketch entities called strokes. A stroke is the set of points comprised between a pen down and a pen up movement. In an off-line input mode, strokes are stored in a binary image and each point has its coordinates as attributes. The advantage of on-line modes, in addition to the coordinates, is that each point may be attributed by dynamic information as the time order or the pressure. A number of applications exist that use sketches as input in areas like architecture [3], mechanics [4], logic diagrams [5], proofreading [6], retrieval [7] or iconic search in PDAs [8].

A sketch understanding system can be divided in three stages: primitive extraction, symbol recognition, and interpretation. In this paper we focus on symbol recognition. Recognizing a diagrammatic notation requires the identification of the alphabet symbols, that will subsequently be interpreted in the context of a domain-dependent graphic notation. Symbol recognition is one of the most active Graphics Recognition areas. A symbol recognition architecture requires two components, namely a shape signature able to robustly describe symbol instances, and a classification strategy. When we work with hand drawn inputs, due to the inherent distortions of strokes, the design of the descriptor is of key importance. The main kinds of distortions (see Fig.1) are: inaccuracy on junctions, on the angle between strokes, shape deformation, elastic deformation, ambiguity between line and arc, and errors like over-tracing, overlapping, gaps or missing parts. In addition, the system must cope with the variability produced by the different writer styles and different sizes.

As stated above, a symbol recognition system firstly requires the definition of expressive and compact descriptors. It has to ideally guarantee intra-class compaction and inter-class separability, even when we describe noisy and distorted symbols. A number of well-known shape signatures exist (see a review in [9]) that can be used for describing symbols in Graphics Recognition. It was proved that some descriptors, robust with some affine transformations and occlusions in printed symbols, are not efficient enough for hand drawn symbols.

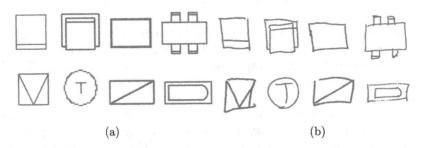

(a)   (b)

Fig. 1. a) Original shapes. b) Distorted shapes, from top to bottom and from left to right: 1) Distortion on junctions. 2) Distorion on angles 3) Overlapping. 4) Missing parts. 5) Distortion on junctions and angles. 6) and 7) Ambiguity arc-segment. 8) Gaps.

Fig. 2. Process scheme

Secondly, the formulation of robust classification methods according to such descriptors is required. Both, the descriptor and the recognition strategy must tolerate the inherent distortions involved in hand drawn inputs. A number of symbol recognition methods have been proposed to modelize such distortions. Examples are spectral models [10], arc-length curvature signatures [11], HMMs [12], deformable models [13], or graph transformation [14]. The reader is referred to [15] for a further review.

In this paper, we present an approach to model and classify handwritten symbols. Symbols are described using the Blurred Shape Model representation. The obtained features show to be high discriminative and tolerant to the transformations produced by the different writing styles. Moreover, we present a multi-class scheme, where Adaboost and Error-Correcting Output Codes are combined to deal with multi-class handwriting recognition problems. One of the most well-known techniques in the Machine Learning domain is the Adaboost algorithm due to its ability for feature selection, detection, and classification problems [17]. The design of a single multi-classifier is a difficult task, and it is common to conceive just binary classifiers and to combine them. One-versus-one voting scheme or one-versus-all strategies are the schemes most frequently applied. In this topic, Error Correcting Output Codes (ECOC) efficiently combines binary classifiers to address the multi-class problem [18]. The results over two multi-class databases show that the present methodology obtains significant performance improvements compared to the state-of-the-art approaches.

The steps of our approach are shown in Figure 2: First, the input hand drawn symbol is obtained as a binary image. Secondly, the Hotelling transform based on principal components [19] is applied to find the main axis of the object so the alignment can be performed. Third, the method defines a blurred region of the shape that makes the technique robust against elastic deformations. Afterwards, Adaboost is applied to each pair of classes to train relevant features that split better object classes. And finally, the set of binary classifiers is embedded in the framework of Error Correcting Output Codes (ECOC) to deal with multi-class categorization problems.

The paper is organized as follows: The proposed distortion tolerant descriptor, called Blurred Shape Model, is described in section 2. The classification stage is presented in section 3. Section 4 shows the experimental results over two multi-class databases. Finally, section 5 concludes the paper.

2   Blurred Shape Model Descriptor

The proposed descriptor describes the distribution of pixels among a predefined set of spatial regions. It is inspired in the well known zoning signature used

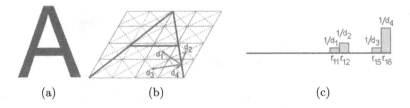

(a)   (b)   (c)

Fig. 3. (a) Input image (b) Shape pixel distances estimation respect to neighbor centroids. (c) Vector actualization of the region 16th, where $\frac{1}{d_1+d_2+d_3+d_4} = 1$.

in a number of OCR systems. Our method has also a foundation in the SIFT descriptor [16] that is one of the preferred strategies to describe image regions. The SIFT descriptor constructs a probability density function of the distribution of orientations within a region. However, in the handwriting recognition topic, orientations suffer from the variations produced by the different writing styles. Our approach consists in describing the symbol by a probability density function of Blurred Shape Model (BSM) that encodes the probability of pixel densities of image regions: The image is divided in a grid of $n \times n$ equal-sized subregions, and each bin receives votes from the shape points in it and also from the shape points in the neighboring bins. Thus, each shape point contributes to a density measure of its bin and its neighboring ones.

In Fig. 3(a), an input symbol is shown. Figure 3(b) shows the distances estimation of a shape point respect to the nearest centroids. To give the same importance to each shape point, all the distances to the neighbors centroids $\{d_1, d_2, d_3, d_4\}$ are normalized to unit. The output descriptor is a vector histogram v of length $n \times n$, where each position corresponds to the amount of shape points in the context of the sub-region. The estimated normalized distances for each affected sub-region r are used to actualize their corresponding vector locations. Fig. 3 (c) shows the vector at this stage for the analyzed point of Fig. 3(b).

The resulting vector histogram, obtained by processing all feature points, is normalized in the range [0..1] to obtain the probability density function (pdf) of $n \times n$ bins. In this way, the output descriptor represents a distribution of probabilities of the object shape considering spatial distortions. In Fig. 4, an input

(a)   (b)   (c)   (d)   (e)

Fig. 4. (a) Input image. (b) 48 regions blurred shape. (c) 32 regions blurred shape. (d) 16 regions blurred shape. (e) 8 regions blurred shape.

shape is processed. Fig. 4b) to e) are the blurred parameterizations considering 48×48, 32×32, 16×16, and 8×8 sub-regions, respectively. In one hand, one can see that the less number of sub-regions, the less clear is the shape, and consequently, more tolerant to distortions. In the other hand, the more blurring effect, the more probability to having confusion between classes. Thus, it is important to find the suitable number of sub-regions in a problem-dependent way, reaching a balance between these two aspects.

Referring the computational complexity, for a region of $n \times n$ pixels, the $k \leq n \times n$ pixel points are considered to obtain the BSM with a cost of $O(k)$ simple operations. The whole algorithm is summarized in Figure 5.

Given a binary image I,
  Obtain the *shape* S contained in I
  Divide I in $n \times n$ equal size sub-regions $R = \{r_i, ..., r_{n \times n}\}$, with c_i the center of coordinates for each region r_i.
  Let $N(r_i)$ be the neighbor regions of region r_i, defined as $N(r_i) = \{r_k | r \in R, ||c_k - c_i||^2 \leq 2 \times g^2\}$, where g is the cell size.

  For each point $\mathbf{x} \in S$,
    For each $r_i \in N(r_{\mathbf{x}})$,
      $d_i = d(\mathbf{x}, r_i) = ||\mathbf{x} - c_i||^2$
  End_For

  Update the probabilities vector v positions as:
  $v(r_i) = v(r_i) + \frac{1/d_i}{D_i}, \quad D_i = \sum_{c_k \in N(r_i)} \frac{1}{||\mathbf{x} - c_k||^2}$
  End_For

  Normalize the vector v as: $v = \frac{v(i)}{\sum_{j=1}^{n^2} v(j)} \forall i \in [1, ..., n^2]$

Fig. 5. Blurred Shape Model algorithm

3   Classification

Concerning the classification step, the Adaboost algorithm is proposed to learn the descriptor features that best split classes, training the classifier from Blurred Shape Model descriptors. The BSM has a probabilistic parametrization on the object shape considering its possible shape distortions. Due to the fact that different types of objects may share local features, Adaboost has been chosen to boost the BSM model in order to define a classifier based on the features that best discriminate one class against the others.

To extend the binary behavior of Adaboost to the multi-class case, we embed the binary classifiers in the framework of Error Correcting Output Codes [18]. The basis of the ECOC framework is to create a codeword for each of the L_c classes. Arranging the codewords as rows of a matrix, a "coding matrix" M is defined, where $M \in \{-1, 0, 1\}^{L_c \times z}$, being z the code length. From the point of view of learning, M is constructed by considering n binary problems, each

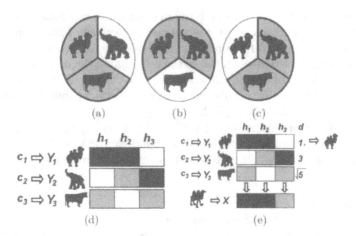

Fig. 6. One-versus-one ECOC design for a 3-class problem. a)b)c) Three bi-partitions of classes. d) ECOC coding. e) ECOC decoding.

corresponding to a matrix column. Joining classes in sets, each classifier defines a partition of classes (coded by +1, -1, according to their class set membership, or 0 if the class is not considered by the classifier).

In Fig. 6 an example of a matrix M is shown. The matrix is coded using 3 classifiers $\{h_1, ..., h_3\}$ for a 3-class problem. The white regions are coded by 1 (considered as positive for its respective dichotomy, h_i), the dark regions by -1 (considered as negative), and the grey regions correspond to the zero symbol (not considered classes for the current classifier). Applying the n trained binary classifiers, a code is obtained for each data point in the test set. This code is compared to the base codewords of each class defined in the matrix M, and the data point is assigned to the class with the 'closest' codeword. In the right side of Fig. 6, an input test sample x is shown (a camel). This input shape is tested using the three classifiers, and assigning the outputs to each codeword position (down of the figure). Finally, the Hamming decoding distance is applied between each class codeword, classifying the test sample by the first class, because it has the minimum distance.

4   Results

Before the experimental results are presented, we discuss the data, methods, and evaluation measurements:

• *Data*: The presented approach has been evaluated using two databases. The first one is a architectural symbol database extracted from a sketching interface. The second one is the GREC2005 database, a public printed symbols database.

• *Methods*: We compare our methodology with the kernel density matching method (KDM) proposed in [22], ART, Zoning, and Zernike descriptors [9],[21].

To test our system, we use the Discrete Adaboost version [17] with 50 iterations of decision stumps, and the one-versus-one ECOC coding with Euclidean distance decoding [18],[19].

• *Evaluation measurements*: The performances are obtained by means of stratified ten-fold cross-validation with a two-tailed t-test at 95% of the confidence interval.

4.1   Architectural Hand-Drawn Categorization

The architectural symbol database is a benchmark database that has been created with the logitech io digital pen [24]. This database, which has been used in a sketch CAD framework [25], is composed of on-line and off-line instances from a set of 50 symbols drawn by a total of 21 users. Each user has drawn a total of 25 symbols and over 11 instances per symbol. The database consists on more than 5000 instances. To capture the data the following protocol has defined: The authors give to each user a set of 25 dot papers, which are paper containing the special pattern from anoto. Each paper is divided into 24 different spaces where the user has to draw in. The first space is filled with the ideal model of the symbol to guide the users on their draw due to they are not experts on the field of Architectural design.

Although the database is composed of 50 symbols, in our experiments we have chosen the 14 architectural symbols most representative from this database. Our experimental set consists in 2762 total samples organized in the 14 classes shown in Fig. 7. Each class consists of an average of 200 samples drawn by 13 different authors. In this experiment, the architectural symbol database has been used to test the performance of different descriptors for different number of classes.

The results obtained from BSM are compared with the ART, Zoning, and Zernike state-of-the-art descriptors [9][21]. The compared descriptors are also introduced in the classification framework to quantify the robustness of each descriptor at the same conditions. The parameters for ART are radial order with value 2 and angular order with value 11; and for the Zernike descriptor, 7 Zernike moments are used. The descriptors for BSM and Zoning techniques are of length 8×8, from the considered sub-regions. This optimum grid size has been estimated applying cross-validation over the training set using a 10% of the samples to validate the different sizes of n, being 8×8 the size with the highest performance in the training set.

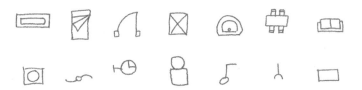

Fig. 7. Architectural handwriting classes

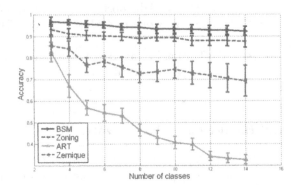

Fig. 8. Descriptors classification accuracy increasing the number of architectural symbol classes

The classification starts using the first 3 classes. Iteratively, one class was added at each step and the classification is repeated. The higher number of classes, the higher confusion degree among them because of the elastic deformations inherent to hand drawn strokes, and the higher number of objects to distinguish. The results of accuracy recognition in terms of an increasing number of classes are shown in Fig. 8. The performance of the ART and Zernike descriptors decreases dramatically when we increase the confusion in terms of the number of classes, while Zoning obtains higher performance. Finally, the accuracy of the BSM outperforms the other descriptors results, and its confidence interval only intersects with Zoning in few cases. This behavior is quite important since the accuracy of the latter descriptors remains stable, and BSM can distinguish the 14 classes with an accuracy upon 90%.

4.2   GREC05 Categorization

The GREC2005 database [23] is not a hand drawn symbol database, but it has been chosen in order to evaluate the performance of our method on a standard, public and big database. It must be said that our initial tests are applied on the first level of distortions (see Fig. 9). We generated 140 artificial images per model (thus, for each of the 25 classes) applying different distortions such as morphological operations, noise addition, and partial occlusions. In this way, the ECOC Adaboost is able to learn a high space of transformations for each class. The BSM descriptor uses a grid of 30×30 bins. In this sense, 900 features are extracted from every image, from which Adaboost selects a maximum of 50. For this experiment, we compare our results with the reported [22] using the kernel density matching method (KDM). The results are shown in Table 1. One can see that the performances obtained with our methodology are very promising, outperforming for some levels of distortions the KDM results.

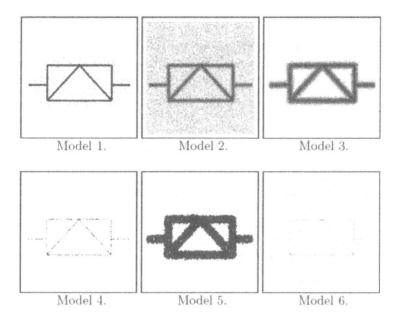

Fig. 9. An example of the distortion levels used in the GREC2005 database

Table 1. Descriptors classification accuracy increasing the distortion level of GREC2005 database using 25 models and 50 test images

Method	Distortion Level 1	Distortion Level 2	Distortion Level 3	Distortion Level 4	Distortion Level 5	Distortion Level 6
KDM	100	100	100	96	88	76
BSM	100	100	100	100	96	92

5   Conclusions and Future Work

In this paper, we have proposed the Blurred Shape Model descriptor and the use of Adaboost in the Error-Correcting Output Codes framework to deal with multi-class handwriting recognition problems. This methodology was evaluated on two multi-class databases, showing promising results in comparison to the state-of-the-art approaches, being robust against noise, scale, and the elastic deformations produced by the different writing styles. Moreover, the complexity of the present methodology shows to be very suitable for real-time multi-class classification problems.

Nowadays, we are extending the experiments on different printed databases, such as the MPEG7 or GREC07, increasing the set of distortions to evaluate the robustness of the present descriptor. Moreover, we are also testing several handwritten symbol databases to show the suitability of the present multi-classification scheme.

Acknowledgements

This work has been partially supported by the spanish projects TIN2006-15694-C02-02, TIN2006-15308-C02-01 and CONSOLIDER-INGENIO 2010 (CSD2007-00018).

References

1. Pavlidis, T., Van Wyk, C.J.: An automatic beautifier for drawings and illustrations. J. SIGGRAPH Comput. Graph. 19(3), 225–234 (1985)
2. Bunke, H.: Attributed programmed graph grammars and their application to schematic diagram interpretation. IEEE Trans. on PAMI 4, 574–582 (1982)
3. Leclercq, P.: Absent Sketch Interface in Architectual Engineering. In: Lladós, J., Kwon, Y.B. (eds.) GREC 2003. LNCS, vol. 3088, pp. 351–362. Springer, Heidelberg (2004)
4. Davis, R.: Understanding in Design: Overview of Work at the MIT Lab. In: AAAI Spring Symposium Sketch Understanding, pp. 24–31. AIII Press, Stanford (2002)
5. Gennari, L., Burak Kara, L., Stahovich, T.F., Shimada, K.: Combining geometry and domain knowledge to interpret hand-drawn diagrams. Computers & Graphics 29(44), 547–562 (2005)
6. Guimbretière, F.: Paper augmented digital documents. In: 16th annual ACM symposium on User interface software and technology, Canada Edition, pp. 51–60. ACM Press, Vancouver(2003)
7. Rigoll, G., Muller, S.: Graphics-Based Retrieval of Color Image Databases Using Hand-Drawn Query Sketches. In: Chhabra, A.K., Dori, D. (eds.) GREC 1999. LNCS, vol. 1941, pp. 256–265. Springer, Heidelberg (2000)
8. Liu, W., Xiangyu, J., Zhengxing, S.: Sketch-Based User Interface for Inputting Graphic Objects on Small Screen Device. In: Blostein, D., Kwon, Y.-B. (eds.) GREC 2001. LNCS, vol. 2390, pp. 67–80. Springer, Heidelberg (2002)
9. Zhang, D., Lu, G.: Review of shape representation and description techniques. Pattern Recognition 37, 1–19 (2004)
10. Liang, S., Sun, Z., Li, B.: Sketch Retrieval Based on Spatial Relations. In: International Conference on Computer Graphics, Imaging and Visualization, Beijing, China, pp. 24–29 (2005)
11. Wilfong, G., Sinden, F., Ruedisueli, L.: On-line recognition of handwritten symbols. IEEE Transactions on Pattern Analysis and Machine Intelligence 18(9), 935–940 (1996)
12. Muller, S., Rigoll, G.: Engineering drawing database retrieval using statistical pattern spotting techniques. In: Chhabra, A.K., Dori, D. (eds.) GREC 1999. LNCS, vol. 1941, pp. 246–255. Springer, Heidelberg (2000)
13. Valveny, E., Martí, E.: Hand-drawn symbol recognition in graphic documents using deformable template matching and a bayesian framework. In: 15th. Int. Conf. on Pattern Recognition, Barcelona, Spain, vol. 2, pp. 239–242 (2000)
14. Lladós, J., Martí, E., Villanueva, J.J.: Symbol Recognition by Error-Tolerant Subgraph Matching between Region Adjacency Graphs. IEEE Transactions on Pattern Analysis and Machine Intelligence 23(10), 1137–1143 (2001)
15. Lladós, J., Valveny, E., Sánchez, G., Martí, E.: Symbol Recognition: Current Advances and Perspectives. In: Blostein, D., Kwon, Y.-B. (eds.) GREC 2001. LNCS, vol. 2390, pp. 104–127. Springer, Heidelberg (2002)

16. Lowe, D.G.: Distinctive Image Features from Scale-Invariant Keypoints. International Journal of Computer Vision 60(2), 91–110 (2004)
17. Friedman, J., Hastie, T., Tibshirani, R.: Additive logistic regression: a statistical view of boosting. The Annals of Statistics 8(2), 337–374 (1998)
18. Dietterich, T., Bakiri, G.: Solving multiclass learning problems via error-correcting output codes. J. Artificial Intelligence Research 2, 263–286 (1995)
19. Dunteman, G.H.: Principal components analysis (Quantitative Applications in the Social Sciences). Sage Publications, Thousand Oaks (1989)
20. Escalera, S., Fornés, A., Pujol, O., Lladós, J., Radeva, P.: Multi-class Binary Object Categorization Using Blurred Shape Models. In: Rueda, L., Mery, D., Kittler, J. (eds.) CIARP 2007. LNCS, vol. 4756, pp. 142–151. Springer, Heidelberg (2007)
21. Kim, W.: A new region-based shape descriptor. Technical report, Hanyang University and Konan Technology (1999)
22. Zhang, W., Wenyin, L., Zhang, K.: Symbol recognition with kernel density matching. Trans. on PAMI 28(12), 2020–2024 (2006)
23. GREC2005 database, http://symbcontestgrec05.loria.fr/formatgd.php
24. Logitech, I(O) digital pen (2004), http://www.logitech.com
25. Sáchez, G., Valveny, E., Lladós, J., Mas, J., Lozano, N.: A platform to extract knowledge from graphic documents. Application to an architectural sketch understanding scenario. In: Marinai, S., Dengel, A.R. (eds.) DAS 2004. LNCS, vol. 3163, pp. 389–400. Springer, Heidelberg (2004)

On the Combination of Ridgelets Descriptors for Symbol Recognition

O. Ramos Terrades[1,*], E. Valveny[1,2,**], and S. Tabbone[3]

[1] Computer Vision Center, Spain
[2] Dept. Computer Science, Universitat Autònoma de Barcelona, Spain
{oriolrt,ernest}@cvc.uab.es
[3] LORIA-Université Nancy 2, France
tabbone@loria.fr

Abstract. In this paper we propose an original solution to combine the scales of multi-resolution shape descriptors. More precisely, a classifier fusion scheme is applied to a set of shape descriptors obtained from the ridgelets transform. The Ridgelets coefficients are grouped into different descriptors according to their resolution. Then a classifier is trained for each descriptor and a final classification is obtained using the classifier fusion scheme. We have applied this approach to symbol recognition using the GREC 2003 database. In this perspective, we increase the recognition rates of previous works on ridgelets-based descriptors.

Keywords: graphics recognition, ridgelets descriptors, classifier fusion.

1   Introduction

One of the challenges of graphics recognition is that shape is nearly the only kind of information that can be used to recognize symbols. Roughly speaking we can assert that graphic documents are composed of shapes which are mainly characterized by features such as the number of connected components, the type geometric shapes (lines, squares, circles, circumferences, etc.), the structural relations between them (position, adjacency, alignment...), etc. Thus, the kind of descriptors that have been proposed in the literature vary according to the properties we want to capture [1,2,3].

In addition to the large number of symbols to be recognized, a main challenge is the fact that a majority of the symbols have common properties. To distinguish them, some kind of local matching or hierarchical representation is required in order to group symbols into more general classes. In addition, the main differences between symbols come from the relative position of lines and arcs. Thus, structural descriptors have been extensively used based on the vector representation. However, despite their good performance with high resolution images, these algorithms present some drawbacks when the resolution is low, in the presence of noise or when graphics are composed of dashed

---

* Supported by TIN2006-15694-C02-02.
** Partially supported by TIN2006-15694-C02-02, Spain and by the Spanish research programme Consolider Ingenio 2010: MIPRCV (CSD2007-00018).

W. Liu, J. Lladós, and J.-M. Ogier (Eds.): GREC 2007, LNCS 5046, pp. 40–50, 2008.

and doted lines. On the other hand, and from a hierarchical viewpoint, multiresolution (MR) descriptors based on the ridgelets transform have been proposed in the last years for pattern recognition [4,5]. The ridgelets transform is defined to detect –and localize– linear singularities, so it becomes a suitable transform to represent graphics. In these cases, MR shape descriptors based on the ridgelets transform can overcome the problems of structural descriptors. However, another problem arises for MR descriptors: which are the most suitable resolutions in order to maximize the performance.

In this paper, we tackle the problem of choosing the suitable resolutions from the perspective of classifier fusion. Thereby, for the representation of a class we choose those resolutions that yield better individual classification results. In a supervised framework, we train a single classifier for each resolution and we evaluate their performance in terms of recognition rates. Then, we fusion the output of all the classifiers by using an aggregation operator so that we return a single result as a candidate class. In this work, as we have used the approach proposed in [6] where the output of the classifiers is linearly weighted according to their individual performance. It has been shown that this scheme gives optimal combination rules when some assumptions are verified on the distribution of classifiers and that, even when these assumption do not hold, it yields better results than other standard combination rules. The main contribution of this work is a pyramidal version of these linear combination rules that can be applied to symbol recognition using MR shape descriptors. Experimental results show a significant improvement compared to previous and similar related works.

This paper is structured as follows. In next sections we briefly review previous works related to ridgelets descriptors (Section 2) and classifier fusion methods in the general framework of pattern recognition (Section 3). Then, in Section 4 we introduce the two pyramidal methods that represent the main contribution of the present paper. Next, in Section 5 we evaluate the method in the context of graphics recognition and we finalize with a discussion of the results and the main conclusions of this work, in Section 6.

2   Ridgelets Descriptors

Ridgelets descriptors are based on the ridgelets transform, which was introduced by Cands [7] in the context of neuronal nets and functional fitting. In his work, the activation function, ψ, is a wavelet which is used to define a ridgelets function, $\rho_{a,t,\theta}$ as follows: for each positive a, any $t \in \mathbb{R}$ and $\theta \in [0, 2\pi)$, we define $\rho_{a,t,\theta} : \mathbb{R}^2 \to \mathbb{R}^2$ as:

$$\rho_{a,t,\theta}(x,y) = a^{-1/2}\psi((x\cos\theta + y\sin\theta - t)/a).$$

This function is constant along the lines defined by $x\cos\theta + y\sin\theta = t$ and transverse to the "ridges" –lines–, it is a wavelet. Cands concludes that ridge functions have better properties than traditional activating functions and that we can use them to approximate some kind of functions, including images. Intuitively, we can think of images as composed of a superposition of ridges. In the field of graphics recognition, as graphics are essentially composed of lines and arcs, the ridgelets transform offers a multiresolution description that permits to represent the variability of graphics in terms of the position of segments.

The computational cost of ridgelets descriptors is equivalent to other descriptors based on the Radon transform –see for instance [8] for more examples on Radon descriptors–. In particular, ridgelets and Radon descriptors are related by the wavelet transform. More explicitly, ridgelets descriptors of an image f at a given scale (j, k) at position $[n, m]$ are expressed by the formula:

$$WRf_{j,k}[n, m] = \left[\mathcal{I}^{-1}f, W_{j,k}(t - n, \theta - m) \right], \tag{1}$$

where the right side of this expression is the inner product between a wavelet basis at scale (j, k) and position $[n, m]$ and $\mathcal{I}^{-1}f$, the representation of the image f in the Radon space –see [4,9] for further details–. In those works, the authors introduce the concept of *decomposition level*, *DL*, that can be applied, in general, to all MR descriptors. The *decomposition level* permits to group all the scales at a given resolution and to achieve zoom invariance. More precisely, when the size of the shape is changed, new scales are added to the MR representation and a new decomposition level is also added to the representation of the shape. Thereby, shapes are compared at each decomposition. As a decomposition level is composed of different number of scales, some strategy to combine the information contained in every scale has to be defined. In particular, in Section 4 we propose two different strategies to combine the scales from the same decomposition level and to combine, as well, the output obtained at each decomposition level.

3   Strategies for the Combination of Classifiers

Classifier fusion methods have been extensively used in pattern recognition community in order to improve overall performance of classifiers in complex recognition tasks. The principle guiding these techniques is based on the complementary nature of classifiers. We mean that not all classifiers will fail with the same samples and therefore, combining the output of the classifiers will reduce the classification error of the set of classifiers.

Depending on the type of classifier we can use different types of aggregation operators. Thus, for classifiers at the *abstract* level, where the output is simply the label of the class, a typical combination rule is the majority vote. Classifiers at the *mank* level return a ranked list of classes. Combination rules based on the Borda Count method [10] or Logistic Regression approaches [11] may be used to combine the output of these types of classifier. Classifiers at the *measurement* level provide a measure of the degree of confidence for each class. They can be combined using a wide range of aggregation operators and rules: *max, mean, product, median, voting*. Indeed, Stejic et al. [12] review a wide set of mathematical aggregation operators and propose a genetic algorithm that select the most appropriate one depending on the query.

In this paper, we propose pyramidal versions of two linear combination rules, namely IN and DN, previously introduced in [6,9] for classifiers at the *measurement* level. These rules are defined using a theoretical probabilistic framework where there is no need to impose conditional independence of classifiers as it is usual in approaches based on a Bayesian framework [13,14]. The error distribution is assumed to be normal for both methods. Besides, the IN method assumes conditional independence whereas for the DN method this hypothesis is not required. In [9] it has been shown that under these assumptions the combination rules give the optimal weights in terms of minimization of the classification error. In addition, experimental results show that if these condition do not hold, classification rates are better than using other standard combination rules. For the sake of completeness, we will briefly summarize these combination rules in the forthcoming subsections (see [9] for further details). Then, in section 4 we will explain the pyramidal versions that have been used to combine all the scales of the multiresolution descriptor described in the previous section.

3.1 IN Method

For the IN method, under the assumption of classifiers with independent and Normal distributions, the optimal weights are obtained after considering three different configurations:

1. Under the assumption of Normal distribution of the classification error, it can be proved that the optimal weights are given by:

$$\alpha_l^{\mathcal{N}} = \frac{\mu_l}{\sigma_l^2}$$

where $\alpha_l, l = 1 \ldots L$ is the weight of each classifier and μ_l and σ_l are the mean and the standard deviation, respectively, of the *validation* of the classifier. The *validation* is obtained by multiplying the output of the classifier by a label $Y = \{-1, +1\}$ depending on whether an element belongs (+1) or not (-1) to the class. In this way, as we have binary classifiers, if the *validation* is positive the classification is correct. Otherwise, the classification is wrong.
2. Using the previous formula, if we have high performance classifiers with very small variance we can have numerical problems. Then, in order to avoid numerical instability, we change the assumption of Normal distribution of the classification error by a Dirac distribution. Under this assumption, it can be proved that the optimal weights are given by:

$$\alpha_l^D = \mu_l$$

3. Finally, in real situations, we can find classifiers with good recognition rates, that can be approximated by Dirac distributions, but not good enough to be used alone. Then, they must be combined with other classifiers that are better approximated by Normal distributions. In these cases the optimal weights are given by a linear combination of the optimal weights computed from Normal and Dirac distributions: $\lambda_{\mathcal{N}} \alpha^{\mathcal{N}} + \lambda_D \alpha^D$. Where the optimal λ are:

$$\begin{cases} \lambda = (\frac{A-B}{A^2}, \frac{1}{A}) & \text{if } A > B \\ \lambda = (0, \frac{1}{B}) & \text{if } A \leq B \end{cases}$$

where $A = \sum \alpha_l^{\mathcal{N}}$ and $B = \sum \alpha_l^D$

3.2 DN Method

The DN method does not assume the conditional independence of classification error and consequently, the estimation of the optimal weights α becomes more difficult. It is not possible to find a formula to compute the optimal weights. Therefore, they must be obtained numerically by solving a constrained optimization problem, where the function to be maximized is the following:

$$\phi(\alpha) = \frac{\mu_\alpha}{\sigma_\alpha} = \left\langle \frac{\alpha}{\sqrt{\alpha^t \Sigma \alpha}}, \mu \right\rangle \qquad (2)$$

with the constraints:

$$\begin{cases} \alpha_l \geq 0 & \forall l \\ \sum_l \alpha_l = 1 \end{cases} \qquad (3)$$

where μ_α and σ_α are the mean and the standard deviation of the validation of the final combination of classifiers, $\alpha = (\alpha_1, \ldots, \alpha_L)$ is the vector of all the weights, $\mu = (\mu_1, \ldots, \mu_L)$ is the vector of the mean of the validation of all the classifiers and Σ is the covariance matrix. The vector α maximizing $\phi(\alpha)$ is obtained using an optimization function based on quadratic programming.

4 Pyramidal IN and DN Method for MR Descriptors

We want to apply the previous classifier fusion strategies to the MR ridgelets descriptors described in section 2. Besides, the recognition system has to be compatible with the resolution of the query image, i.e. if the query is at low resolution and the models of the symbols are represented at high resolution, we must be able to compare both, the query and the models, at the low resolution level. For that, we have already explained that the concept of *decomposition level* (DL) is introduced in order to be able to compare symbols at different resolutions. Every time we change the resolution of a symbol, new scales are introduced in the MR representation. These new scales define a new decomposition level.

Therefore, we have designed a hierarchical combination rule that starts by comparing the MR descriptors at the lowest resolution (1st DL). Then, we increase the level of details and we compare the MR descriptors at the following decomposition level and we repeat the process up to the highest resolution. Thus, we can obtain a different classifier for each decomposition level, namely $C_1, \ldots, C_{DL}$. The combination of all the scales at a given decomposition level is performed using the two optimal combination rules described in the previous section. Thus, we obtain two different hierarchical combination rules named Pyramidal IN (PIN) and Pyramidal DN (PDN).

In the PIN method we can define the combination rule in a straightforward way. Since the classifiers are assumed to be independent, we can compute all the weights together at the beginning of the algorithm. The difficulty lies in computing the factors

---

Algorithm: *PIN*

Input: *Oracle*: $\{(X_n, Y_n)|X_n$: vector of L descriptors, Y_n: label $(+1, -1)\}$
DL, maximum decomposition level
Output: $C_1, C_2, \ldots, C_{DL}$, Pyramidal classifier
begin:
    Obtain α from the IN method;
    for $s = 1, \ldots, DL$,
        $S = (s+1)(s+2)$;   // S: number of scales at level s
        Set: $A = \sum_{l \leq S, t=\mathcal{N}} \alpha_l^t$ and $B = \sum_{l \leq S, t=D} \alpha_l^t$;
        if $A > B$,
            $\lambda_{\mathcal{N}} = \frac{A-B}{A^2}$ and $\lambda_D = \frac{1}{A}$;
        else
            $\lambda_{\mathcal{N}} = 0$ and $\lambda_D = \frac{1}{B}$;
        end;
        update:
            $\alpha_l^{\mathcal{N}} = \lambda_{\mathcal{N}} \alpha_l^{\mathcal{N}}$;
            $\alpha_l^D = \lambda_D \alpha_l^D$;
        $C_s = \sum_{l < S} \alpha_l h_l$;   //$h_l$: output of the classifier for scale l
    endfor
end:

Algorithm 1.1: Pyramidal IN method

$\lambda_{\mathcal{N}}$ and λ_D that, as explained in section 3.1 permit to determine the influence of Normal and Dirac classifiers. These factors are computed independently for each decomposition level (see Alg. 1.1) as the number and the weights of both types of classifiers can vary as we increase the decomposition level. More precisely, the classifier C_1 takes into account the ridgelets descriptors at decomposition levels 0 and 1 (normalized so that the sum of weights is 1). The classifier C_2 is defined as the classifier C_1 but considering descriptors up to the decomposition level 2. Finally, the classifier C_{DL} takes into account all the descriptors up to the decomposition level DL.

On the contrary, in the PDN we have to find the weights at every decomposition level as the classifiers are considered to be dependent. Then we have defined an hybrid fusion rule combining the result of the classifier obtained at the level $s - 1$ with the classifiers for the new scales introduced at level s –see Alg. 1.2–. The minimal ridgelets representation consists of the two first decomposition levels (0 and 1). Hence, we have applied the DN method to obtain the optimal weights for the classifier C_1. Then, we have considered the classifiers trained for the resolutions at the decomposition level 2, in order to define the classifier C_2, and we have combined with the classifier C_1 obtained in the previous iteration. The classifier C_3 is defined way as C_2 and we have repeat the process for all the decomposition levels. At each level the weights are obtained using the classifier obtained for the previous level and all the scales of the current level. Thus, the final classifier for every level can be expressed as:

$$C_{s+1} = \alpha_0 C_s + \sum_{l=S_0}^{S} \alpha_l h_l$$

where C_s is the final classifier for the previous decomposition level and h_l are the classifiers for each of the scales of the current level. α_0 and α_l are the weights obtained with the DN method for all these classifiers.

5 Experiments

We have used the GREC 2003 database in order to validate this approach. We have compared the recognition results to the results obtained by Ramos et al. in [4], who also propose an heuristic algorithm for combining ridgelets descriptors (denoted as CR method). Thus, in order to make the comparison easier, we have repeated the tests of the GREC'03 symbol recognition contest using the ridgelets descriptors but combining the different scales using the PIN and PDN methods. Tables summarizing these results and comparing them to the related methods are given at the end of this section.

Algorithm: *PDN*

      Input: *Oracle*: $\{(X_n, Y_n) | X_n$: vector of L descriptors, Y_n: label $(+1, -1)\}$
                     DL, maximum decomposition level
      Output: $C_1, C_2, \ldots, C_{DL}$, Pyramidal classifier
begin:
      Train L classifiers, h_l, one for each of L descriptors;
      for $l = 1 \ldots L,$
            Get the hypothesis of the classifier for every sample: $z_{l,n} = h_l(X_{l,n})$;
            Obtain the *validation* value of the classifier for every smaple: $u_{l,n} = y_n z_{l,n}$;
      endfor;
      Set $C_0 = 0$;
      for $s = 1, \ldots, DL,$
            $S_0 = s(s+1)$;   // S_0: number of scales at level $s - 1$
            $S = (s+1)(s+2)$;   // S: number of scales at level s
            Obtain the *validation* value of the final classifier at level $s - 1$: $\tilde{u}_n = y_n C_{s-1}$;
            Set $U_s = (U_{S_0}, \ldots, U_S)$, the validation values of all classifiers at level s;
            Set $\tilde{U} = (\tilde{u}_n, U_s)$, the combined validation values of level $s - 1$ and s;
            Obtain weights: $\alpha = DependentWeight(\tilde{U})$;   //Apply the DN method
            $C_s = \alpha_0 C_{s-1} + \sum_{l=S_0}^{S} \alpha_l h_l$;
      endfor
end:

Algorithm 1.2: Pyramidal DN method

Furthermore, two different classifiers have been considered based on the distance distribution between the samples and the model of each class: an Adaboost classifier (DAB) and a *Normal* classifier (CNormal). The name of the *CNormal* classifier is justified by the fact that the distribution of the classifier follows a mixture of two normal variables.

Finally, we have only four decomposition levels in the ridgelets representation because we have worked with images of size 128×128. Thus, the PIN and PDN methods will return three classifiers: C_1, C_2 and C_3.

5.1 Datasets

We have taken the GREC'03 dataset to evaluate the pyramidal methods applied to MR descriptors like ridgelets descriptors. The GREC'03 dataset is considered a benchmark database that has been widely used by the graphics recognition community to test shape descriptors. This dataset is composed of 50 different symbols coming from the architectural, electronic and engineering domain, that have been degraded and distorted according to some specific probabilistic models (cf. [15] for further details on the GREC'03 dataset).

5.2 Classifiers

The pyramidal versions of IN and DN methods need the training of suitable classifiers that accurately capture the properties of shape descriptors. Depending on the choice of classifier, recognition rates may dramatically vary and the estimated weights can change. Thus, we have to take into account the effect of classifiers when we evaluate the classifier fusion scheme applied to shape descriptors. In order to make our discussion independent of the classifiers we have taken two different kinds of classifiers, namely DAB and $CNormal$. The DAB classifier is the typical Adaboost classifier introduced by Schapire et al. –see [16] for further details– trained on the distance distribution between queries and shape models. Conversely, the $CNormal$ classifier is essentially a classifier based on a distance threshold, i.e., using a training set, we have selected the distance threshold that minimizes the classification error. Then, to evaluate the Normal condition required by the IN and DN methods, we have modified the distribution of distances between queries and shape models to make it Normal. This modification is done by linear interpolation of empiric cumulative distribution. More details on $CNormal$ classifier construction are given in [9].

5.3 Invariance to Similarity Transforms

For this experiment we have used three tests containing images with rotation, scaling and combined rotation and scaling, respectively. Each test is composed of 250 images of 50 different symbols. We can see the results in Table 1, where the values in bold are the results obtained using the CR method:

A fast analysis of these results permits us to verify that the ridgelets descriptors are invariant to the change of scale for all the classifiers and the two pyramidal combination

Table 1. Recognition rates for tests with rotation and scale. The numbers in bold correspond to the recognition rates of the method proposed in [4].

	C_1	C_2	C_3	C_1	C_2	C_3
	Scale: **100,00%**			Rotation: **100,00%**		
PIN-DAB	100,00%	100,00%	100,00%	100,00%	100,00%	100,00%
PDN-DAB	99,00%	100,00%	100,00%	94,00%	94,00%	99,00%
PIN-CNormal	99,00%	100,00%	100,00%	92,00%	92,00%	99,00%
PDN-CNormal	100,00%	100,00%	100,00%	96,00%	94,00%	96,00%
	C_1	C_2	C_3			
	Rotation & Scale: **98,80%**					
PIN-DAB	100,00%	100,00%	100,00%			
PDN-DAB	96,00%	96,00%	94,00%			
PIN-CNormal	92,00%	95,00%	94,00%			
PDN-CNormal	94,00%	93,00%	90,00%			

rules. Only the C_1 classifier (working at the lowest resolution) does not reach 100% of recognition rate. However, when we introduce rotation the performance of the pyramidal methods decreases. Only the PIN method applied to the DAB classifier reaches 100% of recognition rate (the same as the CR method) for all the decomposition levels. The other methods have a recognition rate lower than 96% for classifiers C_1 and C_2. Finally, the test with rotated and scaled images confirms that the combination of the PIN method with the DAB classifier obtains the best result (again, 100% of recognition rate). Thus, we can conclude that this combination achieves invariance to rotation and scaling. On the contrary results with all the other combinations achieve worse results than the CR method.

5.4 Robustness to Degradation and Vectorial Distortion

Then, we have tested the robustness to degradation and vectorial distortions of the PIN method using the DAB and $CNormal$ classifiers. We have not used the PDN method in this test because of the low recognition rates achieved in the previous experiment with rotated and scaled symbols.

For this experiment we have used two kinds of test of the GREC'03 database. In the first one, images are simply degraded using a probabilistic model of binary noise. There are nine different models of binary noise and for each model 250 images of the 50 different symbols have been generated. In the second test, in addition to the binary degradation a model of vectorial distortion modifying the original shape of the symbol has also been applied. We have used the test with the highest level of distortion (*level 3*). In table 2 we can see the results of applying the PIN method using DAB and $CNormal$ classifiers to both kinds of tests, comparing them to the results using the CR method. We can observe that we have achieved a similar performance for both tests. We can remark that the performance of the CR method is degraded in most of the tests while the PIN method achieves 100% of recognition rate in most of the cases, even for low resolution levels. At the highest resolution level (classifier $C3$) and with the DAB classifier we obtain 100% recognition rate for all models of degradation.

Table 2. Test with degradation and vectorial distortions using the PIN method with DAB and $CNormal$ classifiers

CR	C_1		C_2		C_3	
	DAB	CNormal	DAB	CNormal	DAB	CNormal
(a) degraded						
model 1 100,00	100,00%	100,00%	100,00%	100,00%	100,00%	100,00%
model 2 100,00	100,00%	100,00%	100,00%	100,00%	100,00%	100,00%
model 3 100,00	100,00%	100,00%	100,00%	100,00%	100,00%	100,00%
model 4 99,60	100,00%	100,00%	100,00%	100,00%	100,00%	100,00%
model 5 100,00	100,00%	100,00%	100,00%	100,00%	100,00%	100,00%
model 6 100,00	100,00%	100,00%	100,00%	100,00%	100,00%	100,00%
model 7 100,00	100,00%	99,00%	100,00%	98,00 %	100,00%	97,00%
model 8 98,40	98,00%	98,00 %	100,00%	100,00%	100,00%	100,00%
model 9 89,20	95,00%	93,00 %	99,00%	97,00 %	100,00%	100,00%

CR	C_1		C_2		C_3	
	DAB	CNormal	DAB	CNormal	DAB	CNormal
(b) deform-degrad-level3						
model 1 98,67	99,00%	100,00%	100,00%	100,00%	100,00%	100,00%
model 2 97,33	100,00%	100,00%	100,00%	100,00%	100,00%	100,00%
model 3 98,67	100,00%	100,00%	100,00%	100,00%	100,00%	100,00%
model 4 98,67	99,00%	100,00%	99,00%	100,00%	100,00%	100,00%
model 5 97,33	99,00%	100,00%	100,00%	100,00%	100,00%	100,00%
model 6 97,33	100,00%	100,00%	100,00%	100,00%	100,00%	100,00%
model 7 100,00	100,00%	100,00%	100,00%	100,00%	100,00%	100,00%
model 8 100,00	100,00%	100,00%	100,00%	100,00%	100,00%	100,00%
model 9 100,00	100,00%	100,00%	100,00%	100,00%	100,00%	100,00%

6   Discussion and Open Issues

The main contribution of this paper is the proposal of two hierarchical combination rules inspired in the classifier fusion methods for the problem of selecting suitable resolutions in MR descriptors for graphics recognition. With this approach, this problem is solved finding suitable weights for the linear combination of a set of classifiers, one for each resolution.

Experiments on the GREC'03 database show the goodness of this approach and motivate further studies in this direction. Based on this experiments, we can state the following conclusions: first, not any combination of the PIN and PDN methods with any kind of classifier performs better that the heuristic CR method. Second, the DAB classifier combined using the PIN method is actually invariant to similarity transforms even using the two first decomposition levels (classifier C_1). Third, the PIN method outperforms the PDN when all classifiers are considered. Finally, the PIN method improves the CR method for images with degradation and vectorial distortion.

Therefore, more experiments in different directions have to be driven in order to achieve more concluding results. On the one hand, we can apply this approach to other MR descriptors like those based on wavelets transform or the Scale Space. On the other hand, other experiments taking into account huge databases can be considered in order to evaluate the effect of hierarchical approaches in the recognition process.

References

1. Pavlidis, T.: Survey: A review of algorithms for shape analysis. Computer Graphics and Image Processing 7(7), 243–258 (1978)
2. Trier, I.D., Jain, A.K., Taxt, T.: Feature extraction methods for character recognition - a survey. Pattern Recognition 29(4), 641–662 (1996)
3. Zhang, D., Lu, G.: Review of shape representation and description techniques. Pattern Recognition 37, 1–19 (2004)
4. Ramos Terrades, O., Valveny, E.: A new use of the ridgelets transform for describing linear singularities in images. Pattern Recognition Letters 27(6), 587–596 (2006)
5. Chen, G.Y., Bui, T., Krzyzak, A.: Rotation invariant pattern recognition using ridgelets, wavelet cycle-spinning and fourier features. Pattern Recognition 38, 2314–2322 (2005)
6. Ramos Terrades, O., Tabbone, S., Valveny, E.: Optimal linear combination for two-class classifiers. In: International Conference in Advances in Pattern Recognition, Kolkata, India (January 2007)
7. Candès, E.J.: Ridgelets: Theory and Applications. Phd, Standford University (September 1998)
8. Tabbone, S., Wendling, L., Salmon, J.P.: A new shape descriptor defined on the radon transform. Computer Vision and Image Understanding 102(1), 42–51 (2006)
9. Ramos Terrades, O.: Linear Combination of MultiResolution Descriptors: Application to Graphics Recognition. PhD thesis, Universitat Autonoma de Barcelona and Universite de Nancy 2 (October 2006)
10. Xu, L., Krzyzak, A., Suen, C.Y.: Methods of combining multiple classifiers and their applications to handwriting recognition. IEEE Transaction on Systems, Man and Cybernetics 22(3) (1992)
11. Ho, T.K., Hull, J.J., Srihari, S.N.: Decision combination in multiple classifiers systems. IEEE Transactions on PAMI 16(1), 66–75 (1994)
12. Stejic, Z., Takama, Y., Hirota, K.: Mathematical aggregation operators in image retrieval: effect on retrieval performance and role in relevance feedback. Signal Processing 85(2), 297–324 (2005)
13. Kittler, J., Hatef, M., Duin, R.P.W., Matas, J.: On combining classifiers. IEEE Transactions on PAMI 20(3), 226–239 (1998)
14. Alkoot, F.M., Kittler, J.: Experimental evaluation of expert fusion strategies. Pattern Recognition Letters 20, 1361–1369 (1999)
15. Valveny, E., Dosch, P., Winstanley, A., Zhou, Y., Yang, S., Yan, L., Wenyin, L., Elliman, D., Delalandre, M., Trupin, E., Adam, S., Ogier, J.M.: A general framework for the evaluation of symbol recognition methods. International Jqurnal on Document Analysis and Recognition (2006)
16. Schapire, R.E., Singer, Y.: Improved boosting algorithms using confidence-rated predictions. Machine Learning 37(3), 297–336 (1999)

Old Handwritten Musical Symbol Classification by a Dynamic Time Warping Based Method

Alicia Fornés, Josep Lladós, and Gemma Sánchez

Computer Vision Center, Dept. of Computer Science, Universitat Autònoma de Barcelona, 08193, Bellaterra, Spain
{afornes,josep,gemma}@cvc.uab.es

Abstract. A growing interest in the document analysis field is the recognition of old handwritten documents, towards the conversion into a readable format. The difficulties when we work with old documents are increased, and other techniques are required for recognizing handwritten graphical symbols that are drawn in such these documents. In this paper we present a Dynamic Time Warping based method that outperforms the classical descriptors, being also invariant to scale, rotation, and elastic deformations typical found in handwriting musical notation.

1 Introduction

In the Graphics Recognition field, Optical Music Recognition (OMR) is a classical application area of interest, whose aim is the identification of music information from images of scores and their conversion into a machine readable format. It is a mature area of study, and lots of works have been done in the recognition of printed scores (see the survey written by Blostein in [1]).

Recently, the analysis of ancient documents has had an intensive activity, and the recognition of ancient musical scores is slowly being taken into account. In fact, the recognition of ancient manuscripts and their conversion to digital libraries can help in the diffusion and preservation of artistic and cultural heritage. It must be said that contrary to printed scores, few works can be found about the recognition of old handwritten ones (see [2], [3]). Working with old handwritten scores makes the recognition task more difficult: Firstly, and due to handwritten documents, one must cope with elastic deformations, the variability in the writer style, with variations in sizes, shapes and intensities, and increasing the number of touching and broken symbols. Secondly, working with old documents obviously increases the difficulties due to paper degradation, the frequent lack of a standard notation and the fact that staff lines are often handwritten. For those reasons, the preprocessing, segmentation and classification phases must be adapted to this kind of scores.

The specific processes required here belong to the field of Graphics Recognition, more than the field of Cursive Script Recognition(symbols are bidimensional). Symbol recognition is one of the central topics of Graphics Recognition. A lot of effort has been made in the last decade to develop good symbol and shape recognition methods inspired in either structural or statistic

W. Liu, J. Lladós, and J.-M. Ogier (Eds.): GREC 2007, LNCS 5046, pp. 51–60, 2008.

pattern recognition approaches. In [4], the state-of-the art of symbol recognition is reviewed. It must be said that the definition of expressive and compact shape description signatures is very important in symbol recognition, and has been an important area of study. In [5] the main techniques used in this field are reviewed. They are mainly classified in contour-based descriptors (such as polygonal approximations, chain code, shape signature, and curvature scale space) and region-based descriptors (such as Zernike moments, ART, and Legendre moments). A good shape descriptor should guarantee inter-class compacity and intra-class separability, even when describing noisy and distorted shapes. It has been proved that some descriptors which are robust with some affine transformations and occlusions in printed symbols, are not efficient enough for handwritten symbols. Thus, the research of other descriptors for elastic and non-uniform distortions are required, coping with variations in writing style.

In this paper we present our work in the recognition of old handwritten musical scores, which are from the 17th-19th centuries. The goal is not only the preservation of these old documents (see Fig.1 for an example), but also the edition and the diffusion of these unknown composers' compositions, which have not been published yet.

As it has been said above, handwritten recognition means dealing with elastic deformation. In Cursive Script Recognition it has been observed that the alignment (warping) of profiles of the shapes can cope with elastic deformations. For that reason, Dynamic Time Warping is a good solution to this problem. In fact, it has been successfully applied to handwritten text recognition in [12]. For

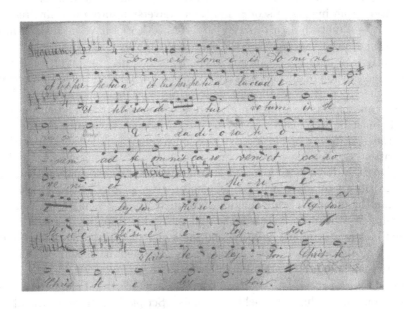

Fig. 1. Example of an old score (XIX century)

the classification of musical symbols we are applying the same concept, extending the method to two-dimensional graphical symbols. In addition, due to isolated symbols present in graphical documents, we must take into account the variations in rotation. For those reasons, the Dynamic Time Warping algorithm and the feature vectors have been adapted to 2D graphical symbols.

This paper is organized as follows: in section 2 the extraction of staff lines and the recognition of graphical primitives are presented. In section 3, the classification of handwritten musical symbols is fully described. It is performed using a Dynamic Time Warping based method, being invariant to rotation, scale and variations in writing style. In section 4 preliminary results over a database of musical symbols are shown. Finally, concluding remarks are presented.

2   Preprocessing, Staff Removal and Recognition of Graphical Primitives

For the sake of better understanding, we first briefly review our previous work for segmenting elements in the score (for further details, see [16]): First of all, the input gray-level scanned image is binarized with an adaptive binarization technique and morphological operations are used to filter the image and reduce noise. Afterwards, the image is deskewed using the Hough Transform method for detecting lines. Then, recognition and extraction of the staff lines (using median filters and a contour tracking process) and graphical primitives (using morphological operations) are performed.

The extraction of staff lines is difficult due to distortions in staff (lines present often gaps in between), and because of the fact that staff lines are rarely perfectly horizontal. This is caused by the degradation of old paper, the warping effect and the inherent distortion of handwritten strokes (staff lines are often drawn by hand). For those reasons, a more sophisticated process is required: After analyzing the histogram with horizontal projections of the image, detecting staff lines, a rough approximation of every staff line is performed using skeletons and median filters. Afterwards, a contour tracking algorithm is performed to follow every staff line and remove segments that do not belong to a musical symbol. Once we have the score without staff lines, vertical lines are recognized using median filters with a vertical structuring element, and filled headnotes are detected performing a morphological opening with elliptical structuring element (see Fig.2).

3   Classification of Handwritten Musical Symbols

Concerning the classification of old handwritten musical symbols, such as clefs, accidentals and time signature, we state two main problems: the enormous variations in handwritten musical symbols and the lack of an standard notation in such these old scores. Thus, the classification process must cope with deformations and variations in writing style. Some of the classical descriptors (such as Zernike moments, Zoning, ART) do not reach good performance for

Fig. 2. Results from a section of 'Salve Regina' of the composer Aichinger: Filled headnotes and beams in black color. Bar lines are the thickest lines.

old handwritten musical symbols, because there is no clear separability between classes (the variability can be seen easily when we compare printed clefs with handwritten ones, see Fig. 3 and Fig. 4).

For those reasons, we are working in the research of other descriptors able to cope with the high variation in handwritting styles. The Dynamic Time Warping

Fig. 3. Printed clefs: (a) Treble clef. (b) Bass clef. (c) Alto clef.

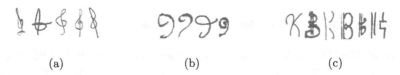

Fig. 4. High variability of handwritten clefs appearance: (a) Treble clefs. (b) Bass clefs. (c) Alto clefs.

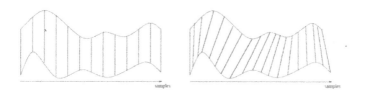

Fig. 5. Normal alignment and DTW alignment

(DTW) algorithm was first introduced by Kruskal and Liberman [6] for putting samples into correspondence in the context of speech recognition. DTW can warp the time axis, compressing it at some places and expanding it at others, avoiding the resample. Thus, it optimizes the best alignment (matching) between two samples, because it minimizes a cumulative distance measure consisting of local distances between aligned samples (see Figure 5).

Beside speech recognition, this technique has been widely used in many other applications, such as bioinformatics [7],[8], gesture recognition [9], data mining [10] and music audio recordings [11]. Rath and Manmatha have applied DTW to the handwritten recognition field [12], [13], coping also with the indexation of repositories of handwritten historical document. Also, Manmatha [14] has proposed an algorithm based on DTW for a word by word alignment of handwritten documents with their (ASCII) transcripts. Concerning online handwriting recognition, some work has also been done. Vuori [15] has also used a DTW based method for recognizing handwritten characters of several different writing styles. Concretely, the system retrieves a set of best matching allographic prototypes based on a query input character from an online handwriting system.

The main contribution of our work is to use a DTW approach for 2D shapes instead of 1D (in handwritten text it is used to align 1-dimensional sequences of pixels from the upper and lower contours). Concretely, we are using this idea for the classification of old handwritten musical symbols, using some features of every symbol as rough descriptors, and the Dynamic Time Warping algorithm (DTW) as the classifier technique used for clef matching: First, every image I is normalized, and for every column c (between 1 and w) of the image (where the width of the image is w pixels) we extract a set of features $X(I)=x_1..x_w$, where $x_c=(f_1,f_2,f_3..f_k)$ (see Fig.6), defined as:

- f_1 is the upper profile.
- f_2 is the lower profile.
- $f_3..f_k$ are the sum of pixels (zoning) of every column region (k-3+1 regions).

In handwritten text recognition (see [12]), features used are: f_1= sum of foreground pixels per column, f_2 = upper profile, f_3= lower profile, f_4= number of transitions from background to foreground. In our case of study, the sum of foreground pixels per column is not accurate enough (too many combinations of the same number of pixels per column can be found), so the column is divided

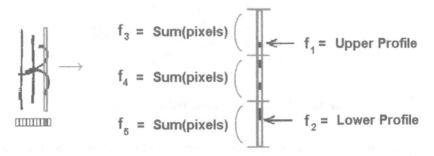

Fig. 6. Features extracted from every column of the image: f_1 = upper profile, f_2 = lower profile, $f_3..f_5$ = sum of pixels of the image of the three regions defined

in several regions and the sum of foreground pixels per region is computed (similar to a zoning). Concerning the number of transitions, when we work with old handwritten graphical symbols, the number of transitions and gaps can be very different from one symbol to another. For that reason, we have not included this feature.

After normalizing these vectors ($0 \leq f_s \leq 1$, s=1..k), the DTW distance between $X(I)=x_1..x_M$ and $Y(J)=y_1..y_N$ is $D(M,N)$, calculated using a dynamic programming approach:

$$D(i,j) = min \begin{cases} D(i,j-1), \\ D(i-1,j), \\ D(i-1,j-1), \end{cases} + d2(x_i, j_j) \qquad (1)$$

$$d2(x_i, j_j) = \sum_{s=1}^{k} (f_s(I,i) - f_s(J,j))^2 \qquad (2)$$

The length Z of the warping path between X and Y (which can be obtained performing backtracking starting from (M, N)) biases the determined distance:

$$D(X,Y) = \sum_{k=1}^{Z} d2(x_{i_k}, y_{j_k}) \qquad (3)$$

Finally, the matching cost is normalized by the length Z of the warping path, otherwise longest symbols should have a bigger matching cost than the shorter ones:

$$MatchingCost(X,Y) = D(X,Y)/Z \qquad (4)$$

The warping path is typically subject to several constraints, and once these conditions are satisfied, then the path that minimizes the warping cost is chosen:

– Boundary conditions: The warping path must start and finish in diagonally opposite corner cells of the matrix.

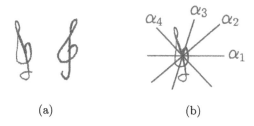

Fig. 7. Clefs: (a)Two treble clefs with different orientations. b)Some of the orientations used for extracting the features of every clef.

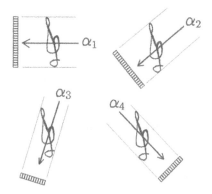

Fig. 8. Features extracted from some orientations ($\alpha_1..\alpha_4$)

- Continuity: The warping path must follow steps in adjacent cells (vertically, horizontally and diagonally adjacent cells).
- Monotonicity (monotonically increasing): This condition is for avoiding the matching from "going back in time".

The basic DTW algorithm will not work for comparing handwritten graphical symbols because the slant and the orientation of every symbol are usually different and it can not be easily computed (see Fig.7a). For that reason, given two symbols to be compared, profiles for the DTW distance are extracted from different orientations (see Fig.7b). Notice that the length of these profiles depends on the number of columns of the image, and varies from an orientation to another (see Fig.8).

Once we have the profiles for all the considered orientations, the DTW algorithm computes the matching cost between every orientation of the two symbols, and decides in which orientation these two symbols match in a better way.

In the classification stage, every input symbol is compared to the representatives of every class using this algorithm. The minimum distance will define the class where the input symbol belongs to.

Fig. 9. Selected representative clefs: (a) Treble representative clef. (b) Bass representative clef. (c)(d) Two Alto representative clefs.

4   Results

The DTW-based method for the classification of handwritten musical symbols has been evaluated using a database of clefs, which has been extracted from a collection of modern and old musical scores (19th century), containing a total of 2128 clefs between 24 different authors. For every class, the representative chosen corresponds to the data sample with the minimum mean distance to the rest of samples from the same class. In Figure 9 the representatives chosen for each class are shown: one representative for treble and bass clefs and two representative alto clefs (because of its huge variability). Thus, only 4 comparisons are computed for classifying every input symbol, where the 1-NN distance will decide which symbols belongs to each class.

The results from the DTW-based descriptor are compared with the classical Zernike moments and ART descriptors, because they are robust and invariant to scale and rotation. Table 1 shows the rates achieved using Zernike moments (number of moments = 7), ART (radial order = 2, angular order = 11) and our DTW-based proposed descriptor, where a 95% rate is achieved.

Table 1. Classification of clefs: Recognition rates of these 3 classes using 4 models

Method	Zernike moments	ART	DTW
Accuracy	65.07	72.74	95.81

An extension of these experiments has been performed including accidentals (sharps, naturals, flats and double sharps) in the musical symbol database. They are a total of 1970 accidentals drawn by 8 different authors. In Figure 10, one can see that some of them (such as sharps and naturals) can be easily misclassified due to their similarity. Contrary to double sharp, a double flat is just two flats drawn together, and for that reason double flats are not included in the accidentals' database.

Similarly to the experiments previously showed, we have chosen one representative for each class (see Fig. 11). The system will have now 8 models (4 clefs and 4 accidentals), and for every input symbol, 8 comparisons will be made.

Fig. 10. Accidentals (printed) in music notation

(a)   (b)   (c)   (d)

Fig. 11. Selected representative accidentals: (a) Sharp model. (b) Natural model. (c) Flat model. (d) Double sharp model.

Table 2. Classification of clefs and accidentals: Recognition rates of these 7 classes using 8 models

Method	Zernike moments	ART	DTW
Accuracy	43.97	52.26	89.55

Results are shown in Table 2, where the DTW-based proposed descriptor reaches a 89.55% classification rate, outperforming the results obtained by the other descriptors.

5 Conclusion and Future Work

In this paper we have presented an approach to classify musical symbols extracted from modern and old handwritten musical scores. This method is based in the Dynamic Time Warping method that has been extensively used in many applications. It is an extension of the approach used for handwriting text recognition, adapted to the 2D graphical symbols that are present in musical scores. One can see the outperform of our method in front of Zernike and ART descriptors. In addition, the method is invariant to scale, rotation and elastical deformations typically found in handwriting musical notation.

Further work will be focused on the extension of the experiments over several handwritten symbols databases, in order to show the robustness and scalability of the method. Finally, it must be said that the method could be improved using an expert system to learn every way of writing. Thus, the identification of the author in a musical score could be used for extracting knowledge information from a database, and helping in the classification stage.

Acknowledgements

This work has been partially supported by the spanish projects TIN2006-15694-C02-02 and CONSOLIDER-INGENIO 2010 (CSD2007-00018).

References

1. Blostein, D., Baird, H.: A Critical Survey of Music Image Analysis. In: Baird, H., Bunke, H., Yamamoto, K. (eds.) Structured Document Image Analysis, pp. 405–434. Springer, Heidelberg (1992)
2. Pinto, J.C., Vieira, P., Sosa, J.M.: A new graph-like classification method applied to ancient handwritten musical symbols. International Journal of Document Analysis and Recognition 6(1), 10–22 (2003)
3. Carter, N.P.: Segmentation and preliminary recognition of madrigals notated in white mensural notation. Machine Vision and Applications 5(3), 223–230 (1995)
4. Lladós, J., Valveny, E., Sánchez, G., Martí, E.: Symbol Recognition: Current Advances and Perspectives. In: Blostein, D., Kwon, Y.B. (eds.) GREC 2001. LNCS, vol. 2390, pp. 104–127. Springer, Heidelberg (2002)
5. Zhang, D., Lu, G.: Review of shape representation and description techniques. Pattern Recognition 37, 1–19 (2004)
6. Kruskal, J., Liberman, M.: The symmetric time-warping problem: from continuous to discrete. In: Sankoff, D., Kruskal, J. (eds.) Time Warps, String Edits, and Macromolecules: The Theory and Practice of Sequence Comparison, pp. 125–161. Addison-Wesley Publishing Co., Reading (1983)
7. Aach, J., Church, G.: Aligning gene expression time series with time warping algorithms. Bioinformatics 17(6), 495–508 (2001)
8. Clote, P., Straubhaar, J.: Symmetric time warping, Boltzmann pair probabilities and functional genomics. Journal of Mathematical Biology 53(1), 135–161 (2006)
9. Gavrila, D.M., Davis, L.S.: Towards 3-D Model-based Tracking and Recognition of Human Movement. In: Bichsel, M. (ed.) Int. Workshop on Face and Gesture Recognition, pp. 272–277 (1995)
10. Keogh, E., Pazzani, M.: Scaling up dynamic time warping to massive datasets. In: Żytkow, J.M., Rauch, J. (eds.) PKDD 1999. LNCS (LNAI), vol. 1704, pp. 1–11. Springer, Heidelberg (1999)
11. Orio, N., Schwarz, D.: Alignment of monophonic and polyphonic music to a score. In: 2001 International Computer Music Conference, Havana, Cuba, pp. 155–158. International Computer Music Association, San Francisco (2001)
12. Rath, T., Manmatha, R.: Word image matching using dynamic time warping. In: IEEE Computer Society Conference on Computer Vision and Pattern Recognition, Madison, WI, vol. 2, pp. 521–527 (2003)
13. Rath, T.M., Manmatha, R.: Lower-Bounding of Dynamic Time Warping Distances for Multivariate Time Series. Technical Report MM-40, Center for Intelligent Information Retrieval, University of Massachusetts Amherst (2003)
14. Kornfield, E.M., Manmatha, R., Allan, J.: Text Alignment with Handwritten Documents. In: First International Workshop on Document Image Analysis for Libraries, pp. 195–209. IEEE Computer Society, Washington (2004)
15. Vuori, V.: Adaptive Methods for On-Line Recognition of Isolated Handwritten Characters. PhD thesis, Helsinki University of Technology (2002)
16. Fornés, A., Lladós, J., Sánchez, G.: Primitive segmentation in old handwritten music scores, In: Liu, W., Lladós, J. (eds.) Graphics Recognition: Ten Years Review and Future Perspectives. LNCS, vol. 3926, pp. 279–290. Springer-Verlag (2006)

On the Joint Use of a Structural Signature and a Galois Lattice Classifier for Symbol Recognition

Mickaël Coustaty, Stéphanie Guillas, Muriel Visani, Karell Bertet,
and Jean-Marc Ogier

L3I, University of La Rochelle, av M. Crépeau, 17042 La Rochelle Cedex 1, France
{mcoustat,sguillas,mvisani,kbertet,jmogier}@univ-lr.fr

Abstract. In this paper, we propose a new approach for symbol recognition using structural signatures and a Galois Lattice as classifier. The structural signatures are based on topological graphs computed from segments which are extracted from the symbol images by using an adapted Hough transform. These structural signatures, which can be seen as dynamic paths which carry high level information, are robust towards various transformations. They are classified by using a Galois Lattice as a classifier. The performances of the proposed approach are evaluated on the GREC03 symbol database and the experimental results we obtain are encouraging.

Keywords: Symbol recognition, Concept lattice, Structural signature, Hough transform, Topological relation.

1 Introduction

This paper deals with the symbol recognition problem. The literature is very abundant in this domain [1,2,3,4]. Symbol recognition can be basically defined as a two-step process: signature extraction and classification. Signature extraction can be achieved by using statistical-based methods or syntactic/structural approaches while most of the statistical-based methods use the pixels distribution. Syntactic and structural approaches are generally based on a characterization of elementary primitives. These primitives (basic description, relations, spatial organization, ...) are extracted from the symbols. They are generally coupled with probabilistic or connexionist classifiers. In this paper, a new approach for symbol recognition is introduced. It is based on the use of a Galois lattice (also called concept lattice) [5] as a classifier. The combined use of statistical-based signatures and a Galois lattice has already been introduced by Guillas *et al.* in [6]. Our proposed approach is based on the joint use of structural signatures inspired by the work of Geibel *et al.* [7] and a Galois lattice classifier. The paper is organized as follows. Section 2 describes the proposed technique. Section 3 gives experimental results. Section 4 provides a conclusion and presents our future work.

2 Description of the Approach

The technique that is introduced in this paper is based on the combined use of structural signatures and of a Galois lattice classifier. The elementary primitives on which

W. Liu, J. Lladós, and J.-M. Ogier (Eds.): GREC 2007, LNCS 5046, pp. 61–70, 2008.

are based the structural signature are segments which are extracted by using the Hough transform. For each symbol, we compute a topological graph by describing the spatial organisation of the segments. Then, signatures are constructed from the topological graphs. Finally, these signatures are classified using a Galois Lattice classifier. Our method is inspired of the work of Geibel *et al.* [7] but differs from that work on many points. Firstly, we use a Galois lattice instead of a decision tree. Secondly, we do not use the same set of topological relations. Finally, our method is based on a Hough-based segments extraction method from images of symbols while [7] works on chemical compounds and do not use any primitive extractor.

2.1 Segments Extraction

The structural primitives we use for symbol description are segments. The segments extraction method we have implemented is an adaptation of the Hough transform (HT), initially defined in the sixties [8] for line extraction by Hough. Indeed, among the existing methods, the HT is known for its robustness property [9], especially in the context of noisy symbols images. The HT has been widely used for different purposes in image processing and analysis ([10]). The HT key idea is to project pixels of a given image onto a parametric space where the shapes can be represented in a compact way. This space is used to find curves that can be parameterized like straight lines, polynomials, circles, ... Each line in the image corresponds to a peak in the associated Hough space. Therefore, the line extraction problem is solved by processing peak detection.

For our purpose we are especially interested in the detection of straight lines. The Figure 1 shows how pixels of an image, represented with their (x, y) coordinates, can be mapped in the Hough space where any straight line of the image is represented by the couple (ρ_i, θ_i) of its polar coordinates.

The practical use of the Straight Line Hough Transform (SLHT) raises different problems [10]. First of all the HT is of quadratic complexity, it is therefore necessary to use a pre-processing step in order to decrease the number of pixels to map during the transform. Next, on real-life images, the mapped points produce heterogeneous sine curves in the Hough space and multiple crossing points can appear. So, a peak detection algorithm is needed in order to group these crossing points and to detect their corresponding mean line.

In this paper, we introduce an adapted version of the HT that does not suffer the preceding drawbacks and that is designed to extract segments instead of lines. The end points of detected lines cannot be known from the analysis of the Hough space. So, it

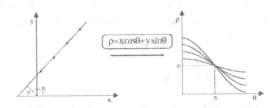

Fig. 1. Straight Line Hough Transform (SLHT)

is necessary to map the lines detected in the Hough space on their corresponding document image in order to achieve the detection process. Based on these considerations an HT-based segments detection system can be divided into four main steps:

1. **Reduction of the search space:** Characteristic points are to be selected before performing the HT, in order to reduce the number of pixels to map and as a consequence the processing time. In our method, we just use a mean filtering in combination with a skeletonization processing [8].
2. **Projection onto the Hough Space:** Each of the previously selected point is mapped onto the Hough space. This step corresponds to the process shown in Figure 1. An accumulator array is commonly used during this step in order to record the number of sine curve for a given point in the Hough space. We use the initial HT implementation of [8].
3. **Peak detection:** It consists in identifying the points in the accumulator associated to a large number of sine curves. Our peak detection algorithm is based on the analysis of the gravity centres of the line sets.
4. **Segments extraction:** The lines detected in the Hough space are mapped on their corresponding document image in order to extract segments (begin and end points). It consists in detecting sequence of strictly adjacent pixels along the detected line. This is realized by using the Euclidean distances $d(p_i, L)$ between the line L and the crossing points P of the image.

Evaluation of the robustness. Our algorithm performs robust extraction of maximal segments. An example of the obtained results is shown in Figure 2. The maximal length of the segments implies a reduction of the possible junctions between adjacent segments. Indeed, an "X" will be described by 2 segments instead of 4.

The Fig.3 shows the robustness of the SLHT. This table shows the recognition rate obtained with different symbols of GREC'03 corpus ([11]). What we call $RecognitionRate$ here corresponds to percentage of good associations between symbols tested and models they refer. Those associations were realized from matching distances between segments. The model we attribute to the treated symbol corresponds

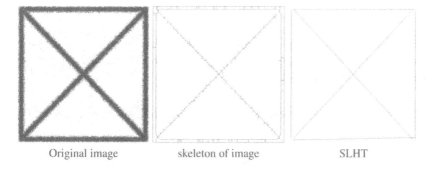

| Original image | skeleton of image | SLHT |

Fig. 2. Examples of différents segments extraction

Symbole	degrad1	degrad2	degrad3	degrad4	degrad5	degrad6	degrad7	degrad8	degrad9	Total
A1	100	100	100	100	100	100	100	100	100	100
A2	100	100	100	100	80	100	100	100	100	97,77778
A3	100	80	100	100	100	100	100	100	100	97,77778
A4	100	100	100	100	100	100	100	100	100	100
A5	100	100	100	100	80	100	100	100	100	97,77778
A6	100	100	80	60	20	100	100	100	80	82,22222
A7	100	100	100	100	100	100	100	60	100	95,55556
E1	100	100	100	100	100	100	100	100	80	97,77778
E2	100	100	100	100	100	100	100	100	100	100
E3	100	100	100	100	100	100	100	100	100	100
E4	100	100	100	100	100	100	100	100	100	100
E5	100	100	100	100	100	100	100	100	100	100
E6	100	100	100	100	100	100	100	100	100	100
E7	100	100	100	100	100	100	100	100	100	100
E8	100	100	100	100	100	100	100	100	100	100
Total	100	98,66667	98,66667	97,33333	92	100	100	97,33333	97,33333	97,92593

Fig. 3. Evaluation of the robustness of the SLHT

to the minimal distance. In all the degradation levels we can see that the proposed approach perform a robust segments-based symbols extraction.

2.2 Topological Graph Computation

Description. Once the segments are extracted, each topological relation between two segments s and s' is described by the following triplet of information:

$$< relation\ type, relation\ value, length\ ratio > \tag{1}$$

- **relation type:** We use the finite set of relations types X, Y, V, P, O as in [12,3,1,13] to fully describe the possible relations between pairs of segments (see Table 1).
- **relation value:** To be more exhaustive and to discriminate more precisely the relations, we add a value to the relation. This value aims at precising topological relations between segments, such as angle between intersecting segments (available for X, Y, V and O), or distance for parallel segments (relation P).
- **length ratio:** The last value of each triplet is a ratio between the lengths of the longest and shortest segments of each pair.

We build a topological graph per symbol where nodes are segments and edges are relations (see Figure 5). The topological graph we obtain is a complete graph where each pair of segments is uniquely described.

Table 1. The different types of relations we consider (from left to right: X, Y, V, P, O)

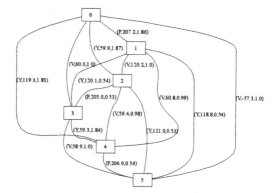

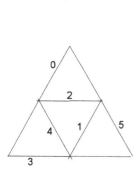

Fig. 4. Example of extracted segments

Fig. 5. Associated topological graph

In order to reduce the cardinality of the possible triplets ensemble (see Eq. 1), we discretize them. After performing a statistical analysis of the symbol shapes, we choose to limit the set of possible values for the angles of junctions X, Y and V to the following set: $\{30°, 45°, 60°, 90°\}$ (possibly, a relation value may be assigned to the closest value in that set). It is also possible to specialize the distances between parallel segments in groups (collinear, near and far for example). The length ratios can be separated into three groups (equal, globally near or very different). We could also consider only the type of relation (or any of the pairs <relation type, relation value> or <relation type, length ratio>), or reduce the set of types of relations we consider.

Discussion. For each symbol, we obtain a set of triplets which fully describes the structural organization of the segments (*eg.*, the relation type differentiates a cross from a rhombus, the relation value a rhombus from a rectangle and the length ratio a rectangle from a square). Moreover, the use of this triplet-based representation has three main advantages:

- each pair of segments is described by one unique triplet;
- each symbol is characterized by one unique and complete graph;
- this description is invariant towards rotation, scale and vectorial distortion.

But, this representation also has some drawbacks:

- It does not consider circle arcs
- n^2 triplets are needed to characterize one symbol (at most n^2, where n is the number of segments). This number of triplets can be reduced when using a restriction of the types of relations we consider.

2.3 Computation of the Structural Signatures

Description. The triplets which are extracted from each pair of segments characterize the paths of length 1. These paths are equivalently described by the topological graph (see Figure 5) or its associated adjacency matrix (see Table 2), as in [1,13]. However,

Table 2. Adjacency matrix (M) associated to the graph of Figure 4 where triplets are only given by the relation type

	0	1	2	3	4	5
0		P	Y	V	Y	V
1	P		V	Y	V	O
2	Y	V		P	V	Y
3	V	Y	P		Y	V
4	Y	V	V	Y		P
5	V	O	Y	V	P	

paths of length 1 are insufficient for discriminating different types of structures, such as regular shapes (square, rectangle, triangle, ...).

That is why, as in [7], we compute the paths of different lengths by using the adjacency matrix and its powers (see Tables 2 and 3). Let us denote M the adjacency matrix. As M conveys information about paths of length 1, M^3 corresponds to 3-length paths (useful to describe triangles), M^4 to 4-length paths (squares and rectangles),...

The adjacency matrices we work with are not boolean or integer, so we generalize the usual product of boolean or integer matrices (see Eq. 2) :

$$\forall (i, j) \in [0, L]^2, \ (A \times B)_{ij} = \sum_{k=1}^{L} (a_{ik} \times b_{kj}) \tag{2}$$

to the union of string concatenation (see Eq. 3) :

$$\forall (i, j) \in [0, L]^2 ; (A \times B)_{ij} = (\bigcup_{k=1}^{L} (a_{ik} + b_{kj})) \tag{3}$$

where L is the size of the matrix and $+$ is the string concatenation operator. Once this product has been computed, we keep only the elementary paths and group the equivalent or symmetric paths. For instance, two equivalent paths XV are grouped as $2\times$XV and the symmetric paths POV and VOP are grouped as $2\times$POV. The matrix M^2 corresponding to the square of the matrix M (given in Table 2) is provided in Table 3.

Table 3. Matrix M^2 (where M is given in Table 2)

	0	1	2	3	4	5
0		4YV	2PV 2YV	2PY 1YY 1VV	2PV 2YV	2PY 1YY 1VV
1	4YV		2PY 1VV 1YY	2PV 2VY	2PY 1VV 1YY	2PV 2VY
2	2PV 2VY	2PY 1VV 1YY		4VY	1YY 1VV 2PY	2VY 2PV
3	2PY 1YY 1VV	2PV 2VY	4VY		2VY 2PV	1VV 1YY 2PY
4	1VY 1YV 2PV	1VV 1YY 2PY	1YY 1VV 2PY	2VY 2PV		4VY
5	1YY 1VV 2PY	2VY 2PV	2VY 2PV	1VV 1YY 2PY	4VY	

Table 4. Structural signature of Fig. 4 with paths of length 1 & 2 (only relation type)

Signature	P	PV	PX	PY	V	VV	X	XV	XX	Y	YV	YX	YY
Value	2	8	2	10	6	5	3	8	2	8	28	12	11

Once all the power matrices have been computed, a set of paths (features) of different lengths and their number of occurrences are available. We organize these features in a hierarchical way, as in [3], in order to compute the signature. Indeed, the presence of a 4-length path is more discriminative than the presence of a 1-length path, but the longest paths are the most affected by distortions. For each symbol image, we compute its structural signature by concatenating the type of path and its number of occurrences in the topological graph associated to that symbol.

A lot of paths might be needed to describe a symbol and therefore the signatures may be huge and contain much redundant information. That is why we only consider paths of length inferior or equal to 4.

Discussion. The structural signatures we obtain are not based on the search for predefined shape templates. Instead, we dynamically compute the shapes observed from our sample images, which confers genericity to our approach.

2.4   Classification

We developed a recognition system named NAVIGALA (NAVIgation into GAlois LAttice), dedicated to noisy symbol recognition [14]. As denoted by its name, this system is based on the use of a Galois lattice as classifier. A Galois lattice is a graph which represents, in a structural way, the correspondences between a set of symbols and a set of attributes. These correspondences are given by a binary table (see Figure 6 where each attribute corresponds to an interval of occurrences for a given path) where crosses are membership relations. In the Galois lattice, nodes are denoted as concepts and contain a subset of symbols and a corresponding subset of attributes and edges represent an inclusion relation between the nodes (see Figure 7). The principle of classification is to navigate through the lattice from the top of the graph to its bottom by validating attributes and thus to reduce the candidates symbols to match. This navigation is similar to the one used for classification with a decision tree. However, in the Galois lattice, several ways are proposed to reach the same node of the graph. We noticed that

	X[0]	X[1]	PP[0]	PP[1]	V[0]	V[3-12]	VV[0]	VV[3]	VV[4-12]
✕		X	X		X		X		
☐	X			X		X			X
⊠		X		X		X			X
△	X		X			X		X	

Fig. 6. Example of binary table used for lattice construction

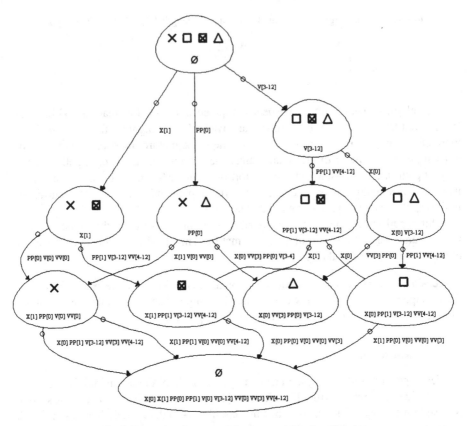

Fig. 7. Example of a concept lattice used for classification

this property is interesting for noisy symbols because, experimentally, concept lattice is more efficient than decision tree in the presence of noise.

3   Experimental Results

We perform our experiments on the GREC03 database of symbol images [11]. We evaluate the effectiveness of the proposed approach on symbols extracted from 8 classes (see Figure 8) and 9 levels of deterioration (see Figure 9). We use the original symbol, more one symbol per level of deterioration (*ie.* 10 symbols per class) for training. The recognition results are computed from 72 deteriorated query symbol images per class. Tables 5 and 6 provide the recognition rates we obtain by using a) only the relation types and not the full triplet given in (1) (Table 5) and b) the full triplet (Table 6).

For comparison, we perform tests on the same sets of symbols (for learning and recognition) with a method based on the use of statistical signatures (Radon Transform) and a Galois lattice as classifier [6]. The recognition rate we obtain is 98.9%. 14 attributes and 96 concepts were created in the lattice for recognition. We can see that the use of statistical signatures gives a better global recognition rate. But the two

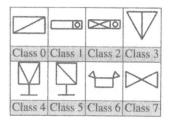

Fig. 8. 8 classes of symbol used for tests

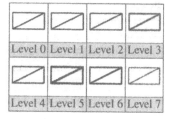

Fig. 9. Different levels of noise for class 0

Table 5. Experimental results using partial triplets

Lengths of paths	1 and 2	2 and 3	3 and 4	1, 2 and 3
Recognition rate	85,3%	87,3%	86,1%	87,7%
Number of paths	4	50	161	54
Number of attributs	20	25	27	24
Number of concepts	410	533	658	511

Table 6. Experimental results using full triplets

Lengths of paths	1	2	3	4	1 and 2	2 and 3	1, 2 and 3	1, 2, 3 and 4
Recognition rate	96%	86,1%	86,7%	82,6%	95,1%	80,4%	94,4%	95,7%
Number of paths	38	175	959	3270	202	1134	1161	4427
Number of attributs	20	22	22	28	20	20	18	16
Number of concepts	452	577	1475	9077	410	689	214	140

approaches can be complementary in some way. For example, for the symbols from class 6, statistical signature leads to confusions with classes 0 or 3. Using the structural signatures, we recognize symbols from class 6 without any ambiguity with classes 0 and 3 (with a structural signature, for class 6 two symbols among 81 are misclassified). We can infer from these results that these two signatures may be combined in order to improve the performances. We are actually working on an iterative combination of statistical and structural signatures to enhance the performances of the proposed approach.

4   Conclusion and Future Work

In this paper, we propose a new structural signature dedicated to symbol recognition using a Galois lattice as classifier. This structural signature relies on segments extracted by using an adapted Hough transform. The structural signature extraction is in 2 main steps. First, for each symbol, we compute a topological graph to describe the spatial organization of the segments. Then, from these topological graphs, we can extract the structural signature by counting the number of occurrences of each path of the graphs. The signatures are further classified by using a Galois Lattice classifier. The experiments we perform on the GREC03 database show the robustness of the proposed approach towards various sources of noise. The structural signatures we obtain are not based on the search for predefined shape templates. Instead, we dynamically compute the shapes observed from our sample images, which confers genericity to our approach.

In order to ameliorate this structural signature, we are further working on the extraction of circle/ellipse arcs and on their integration into our structural signature. Next, we aim at evaluating the performances of the proposed approach not only on single symbols, but in real-life applications. Finally, a procedure based on an iterative combination of statistical and structural signatures may enhance the performances of the proposed approach.

References

1. Lladós, J., Dosch, P.: Vectorial Signatures for Symbol Discrimination. In: Lladós, J., Kwon, Y.-B. (eds.) GREC 2003. LNCS, vol. 3088, pp. 154–165. Springer, Heidelberg (2004)
2. Geibel, P., Wysotzki, F.: Learning relationnal concepts with decision trees. ECAI 1996. In: 12th European Conference on Artificial Intelligence (1996)
3. Rusiñol, M., Lladós, J.: Symbol Spotting in Technical Drawings Using Vectorial Signatures. In: Wenyin, L., Lladós, J. (eds.) GREC 2005. LNCS, vol. 3926, pp. 35–46. Springer, Heidelberg (2006)
4. Tombre, K., Lamiroy, B.: Graphics Recognition - from Re-engineering to Retrieval. In: Proceedings of ICDAR (2003)
5. Ganter, B., Wille, R.: Formal Concept Analysis. Springer, Heidelberg (1999)
6. Guillas, S., Bertet, K., Ogier, J.M.: A Generic Description of the Concept Lattices'Classifier: Application to Symbol Recognition. In: Graphics Recognition: Ten Years Review and Future Perspectives - Selected papers from GREC 2005 (2006)
7. Geibel, P., Wysotzki, F.: Learning Relational Concepts with Decision Trees. In: Saitta, L. (ed.) Machine Learning: Proceedings of the Thirteenth International Conference, pp. 166–174. Morgan Kaufmann Publishers, San Francisco (1996)
8. Hough, P.: Machine Analysis of Bubble Chamber Pictures. In: International Conference on High Energy Accelerators and Instrumentation, pp. 554–556 (1959)
9. Wenyin, L., Dori, D.: From Raster to Vectors: Extracting Visual Information from Line Drawings. Pattern Analysis and Applications (PAA) 2(2), 10–21 (1999)
10. Leavers, V.F.: Survey: which hough transform. Computer Vision and Image Understanding (CVIU) 58(2), 250–264 (1993)
11. GREC03 Database (Graphics RECognition), www.cvc.uab.es/grec2003/symreccontest/index.htm
12. Etemadi, A., Schmidt, J.P., Matas, G., Illingworth, J., Kittler, J.: Low-Level Grouping of Straight Line Segments. In: British Machine Vision Conference (1991)
13. Iqbal, Q., Aggarwal, J.: Retrieval by Classification of Images Containing Large Manmade Objects Using Perceptual Grouping. Pattern recognition 35, 1463–1479 (2002)
14. Guillas, S., Bertet, K., Ogier, J.M.: Reconnaissance de Symboles Bruités à l'Aide d'un Treillis de Galois. Colloque International Francophone sur l'Ecrit et le Document, 85–90 (2006)

A Discriminative Representation for Symbolic Image Similarity Evaluation

Guanglin Huang, Wan Zhang, and Liu Wenyin

Dept of Computer Science, City Univ. of Hong Kong, KLN, Hong Kong, China
csliuwy@cityu.edu.hk

Abstract. Visual similarity evaluation plays an important role in intelligent graphics system. A basic problem of it is how to extract the content information of an image and how to describe the information with an intermediate representation, namely, image representation, because the image representation has great influence on the efficiency and performance of the similarity evaluation. In this paper, we focus on the domain of symbolic image recognition and introduce the Directional Division Tree representation, which is the image representation used in our algorithm. The conducted experiment shows that similarity evaluation algorithm based on this representation can yield satisfactory efficiency and performance.

Keywords: Image representation, computer vision, similarity evaluation.

1 Introduction

Symbolic image recognition is an important part of many pervasive intelligent systems, such as notation recognition system, shape recognition system, sketch recognition system and trademark recognition system. Furthermore, more and more systems adopt a sketch-based GUI to achieve better user-friendliness. In these systems, symbolic image recognition is also needed to recognize the user's drawing. Symbolic image, such as logo, trademark, icon, sketch stroke, etc., are special kinds of images whose structures are usually simple and regular. Unlike text document, whose content can be used as indices directly, recognition on image content can not use the raw data directly, because an image's raw data have no semantic meaning.

The core operation of symbolic image recognition system is to compute the similarity between two symbolic images. To benefit from the similarity evaluation procedure, recognition algorithm needs to extract the content information of an image from its raw data, and to describe the information with an intermediate representation, i.e. an image representation [1]. The performance of similarity evaluation procedure mainly depends on the type and correctness of the image representation used in it. An ideal image representation should be able to describe all vision content of an image, and can be manipulated easily and efficiently. However, it is very difficult to find such representation, because the way how human vision works is still unknown; and furthermore, human perception is usually very casual and fuzzy, while algorithm requires an accurate representation in order to work out a deterministic solution.

Semantically, a symbolic image is usually composed of many simple entities, which have specific geometrical or logical meaning. These objects' individual visual

W. Liu, J. Lladós, and J.-M. Ogier (Eds.): GREC 2007, LNCS 5046, pp. 71–79, 2008.

features and their spatial relations are important factors that contribute to the content of this image. Spatial relations among entities can be divided into two categories: directional relations such as "left-of", "above", "north-of", etc; and topological relations such as "disjoint", "touch", "overlap", etc. [2]. Construction of the symbolic image representation should consider these factors carefully.

In this paper, we propose the Directional Division Tree (DDT) representation for symbolic image. An image is first segmented into several entities. Based on the entity segmentation, we select a visually critical entity, and then organize the rest entities according to their locations in 9 regions around the critical entity. The basic idea of DDT is similar to Spatial Division Tree [3]. The main difference is that DDT performs space division in a two dimensional manner while SDT does it only in one dimensional manner thus it is more complex. The key of our DDT representation is to give privilege to the entity which can attract more visual attention, so that the representation can snatch the facts of the image visual effect in an economical way.

The rest of this paper is organized as follows: section 2 presents the background and related works in this topic; section 3 introduces the Directional Division Tree (DDT), which is the image representation in our approach, in detail; section 4 explains the similarity evaluation method based on this representation; section 5 shows the result of the experiment for validating the effectiveness of our representation and the performance of our recognition algorithm; section 6 presents the conclusion and future work.

2   Related Work

A traditional representation of image is feature vector. The methodology of this representation is to extract some key visual factors from an image, and compose them into a fixed length "feature vector" [4]. The content of the image is then represented by this feature vector.

Feature vector is universally used in CBIR. Generally speaking, these features used in existing approaches can be classified into two categories: global features, such as image's color, area, aspect ratio and curvature histogram, and local features, such as individual edge and corner. For example, the QBIC system [5] a pioneer system developed by IBM, employs 22 algebraic properties, like area, perimeter, and axis orientation, to form the feature vector. Cheng et al. [6] use irreducible covers of maximal rectangles. E. Petrakis and E. Milios [7] partition a symbolic image into tokens in correspondence with their protrusion, and each of them forms a feature vector according to a set of perceptually salient attributes. However, appropriate features are usually hard to choose, furthermore, lots of experiment have shown that there is a gap between vector representation and vision, in which the images with similar vectors sometimes are very different in vision.

2D string [8] is another frequently used structure to depict the semantic properties of a symbolic image. This technique provides a simple and compact representation of spatial symbolic image properties in a form of two strings. It also depends on a prior segmentation to the image, after which, a set of disjoint entities is extracted. By taking the names or types of these entities from left to right and from below to above, two one-dimensional strings are produced respectively. The original image is finally represented by the two strings, namely, 2D string.

2D C string [9] is an improvement form of 2D string, which relaxes the constraint and allows segmented entities overlapping with each other. Overlapping entities are segmented into smaller subparts so that their relations can be classified uniquely as "disjoint", "same position" or "edge to edge". In this manner, 13 different spatial relations in 1 dimension and 169 relations in 2 dimensions can be identified and encoded in a 2D C string representation of the original image.

Graph representation is also a widely used image representation because it owns some common characteristics with human vision, such as transformation insensitive and scale insensitive. In graph representation, an image is usually segmented into several entities. Each entity in the image is represented by a vertex in the graph, and the relation between two entities is regarded as an edge on the graph, which connects two vertexes corresponding to the two entities.

Some graph representations, such as Attribute Relation Graph (ARG), Region Adjacent Graph (RAG) and Spatial Relation Graph (SRG), have been proposed in [10], [11] and [12]. Graph representation is a powerful image representation; by assigning suitable semantic meaning to the vertices and edges, it can form a complete representation of the symbolic image.   However, the main drawback of graph representation is its great complexity, which makes it difficult to manipulate very efficiently.

To explore the spatial relations within the symbolic image more precisely, tree structure is also used.   In fact, tree representation is a simple version of graph representation. It discards some unimportant features to make the final representation more elegant, so that it can yield higher efficiency.

Y. Liu and W. Liu [3] use a Spatial Division Tree (SDT) representation for recognizing incomplete graphic object in the domain of sketch symbol recognition. In their approach, at each node in the tree, a stroke of the drawing symbol is first selected, and then the remaining strokes are divided into 2 groups: strokes on its left side and strokes on its right side. The selection and division procedures are recursively repeated until there is only one stroke left in a node, i.e., the left node. They claim that, on term of this representation, similarity evaluation algorithm can heuristically prune the searching branch, and yields a satisfactory efficiency.

3 Construction Algorithm

The main idea of our proposed representation is that the contributions of different entities to the overall vision are different. For example, although Fig.1 (a) and Fig.1 (b) are constituted by the same entities and their spatial relations are also the same, most people will think Fig.1(c) is more similar to Fig.1 (a) than Fig.1 (b). However, graph representation and 2D string don not refer to this point, and they treat all the entities equally. Therefore, in their expression, (a) and (b) are the same, and (b) is more similar to (a) than (c). Our proposed representation tackles this problem by identifying the "critical entity" of the symbolic image. The construction of our representation can be divided into 3 phases: in the first phase, preparation work is done; after that, the "critical entity" is selected in phase 2, and 2D space of the symbolic image is divided into 9 regions; phase 3 recursively repeats the work of phase 2 to each of these 9 regions, and finally construct the tree representation.

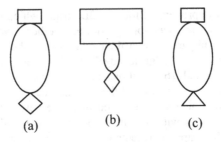

Fig. 1. Different vision contributions

3.1 Preparation Work

Because the raw data of the symbolic image can not be used directly to describe the meaningless of it content, it is necessary to decompose the symbolic image and identify its component entities. Therefore, the major work in preparation phase is to segment the image into several entities which are spatially or logically independent from each other. A lot of techniques for image segmentation [1], [13] are currently available. However, it is an independent research area and beyond the scope of our work in this paper. Hence we would not discuss it here. In practice, we adopt a conventional approach [14] resulting in polygonal and circular approximation of the entity contours. It should be pointed out that the segmentation to the image in the database is semi-automatic, which means we can edit the segmentation result to correct the mistake if exists. However, the segmentation to the query image is automatic. Result of the segmentation is a list of polygonal and circular entities. Fig 2 shows an example.

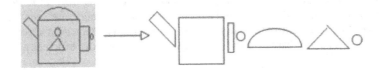

Fig. 2. Example of a symbolic image and its segmentation result

3.2 Critical Entity Selection and Space Division

One visual attention theory named "object-based" theory [15], [16] suggests that human attention is allocated to perceptual entities rather than undifferentiated region of the visual field. Another Gestalt "spotlight" model [17] also argues that the attentive resources of a human are usually concentrated in a particular region in space, i.e. "visual center". These evidences from psychology strongly suggest that human's vision is discriminative and there is certain entity with a higher visual privilege so that it makes critical contribution to the overall vision. Our methodology is to find out the critical entity and place it on a privileged position in the representation.

Critical entity selection is mainly based on the covering area of the entities, because usually large entities can attract human's attention easier than small entities.

Other criteria can be the entity's color, contrast, brightness, location and contour. In our practice, a simple model, which considers covering area only, is employed. The entity with maximum covering area is selected as the critical entity.

According to the location of critical entity, the whole 2D space of the symbolic image is then divided into 9 regions. The bounding-box of the critical entity is regarded as "center" region, and on its eight directions, eight regions named "east", "northeast", "north", "northwest", "west", "southwest", "south", "southeast" are established correspondingly, as illustrated in Fig. 3.

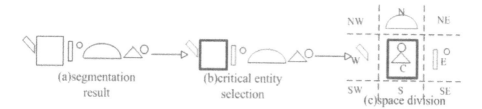

(a)segmentation result   (b)critical entity selection   (c)space division

Fig. 3. Example of critical entity selection and space division

3.3 Tree Construction

Except critical entity, every other entity can be allocated exclusively to one of the nine regions according to the position of its geometrical center. Each region that isn't empty could be regarded as an individual symbolic image so that the previous phase can be recursively repeated on this sub-image. The following pseudo-code illustrates this process:

```
procedure DDTSetup(EntityList el)
        criEntity=select-cri-entity(el);
        for each direction in DirectionSet do
                EntityList  DirRegion =space-div (el, criEntity, direction);
                if not isempty(DirRegion) then
                        DDTSetup(DirRegion);
                end if
        end for
end procedure
```

We can view the critical entity as a node of a tree, the region as a branch rooted by the node. Therefore, a hierarchical tree is established when the procedure ends. Because the space division is based on the directional relations between the critical entity and the rest entities, we name it Directional Division Tree (DDT).

After a DDT is setup, the topological relation between every two nodes which are connected to each other is examined. In practice, we define 34 topological relations; the relation between two connected nodes is one of these 34 relations, it's treated as the attribute of the edge that links the two nodes. Therefore, we finally obtain an attributed DDT.

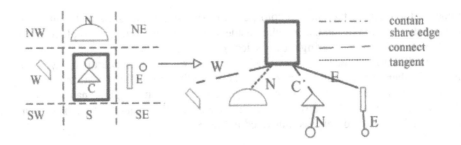

Fig. 4. Example of an attributed DDT setup (empty branch is not drawn)

Fig. 4 shows such an example. The edges are colored differently to indicate their different relation attributes.

4 Similarity Evaluation

Similarity evaluation is based on a matching scheme of individual entities of the two compared symbolic images. In term of traditional matching scheme, there is a big gap between human and computer's vision in some situations. Take the symbols in Fig. 1 for example; (a) and (b) have identical structure, so in traditional matching scheme, they're matched perfectly, their similarity is calculated very high; but in human' eyes, they are quite different.

It's because the positions of critical entities of the two symbolic images are greatly different, so their "visual center" are different. Human always compare two images based on their "visual center", that is, he first matches the critical entities of the two

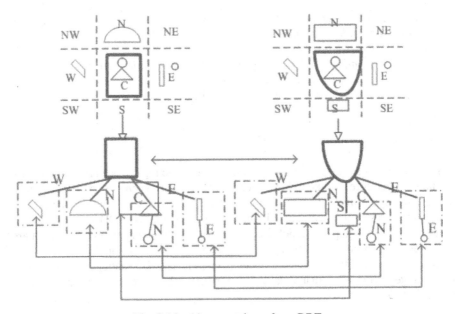

Fig. 5. Matching procedure of two DDTs

images, and then matches other entities. We design our matching algorithm to make it consistent with human vision.

Given two DDT forms of symbolic images I_q and I_t, similarity evaluation is conducted by matching their root nodes, their edges connecting to the root nodes and corresponding branches from top to down. As illustrated in Fig.5.

Every matched pair contributes a vote value to the final similarity score. Similarity computation is performed from top to down, such that, entities with higher positions have more influence on the final similarity score, as specified in the following pseudo-code:

> *double procedure DDTSimi(DDT I_a, DDT I_t)*
>   *If isempty(I_a) and isempty(I_t) return 1;*
>   *If isempty(I_a) or isempty(I_t) return CONST_EMPTY;*
>   *double branchsimi=0;*
>   *for each direction in DirectionSet do*
>   *branchsimi=branchsimi+DDTSimi(I_a.direction,I_t.direction);*
>   *end for*
>   *double relationsimi=0;*
>   *for each edge in EdgeSet do*
>   *relationsimi=relationsimi+RelationSimi(Ia.edge, It.edge);*
>   *end for*
>   *double nodesimi=NodeSimi(I_a.root, I_t.root);*
>   *return W_{node}*nodesimi+$W_{relation}$*relationsimi+W_{branch}*branchsimi;*
> *end procedure*

For node similarity evaluation, a lot of method [7], [18], [19], [20], [21] are proposed and can be used here. For relation similarity evaluation, similarity score is given by a predefined 34×34 score matrix. As the space limited, we don't discuss them in detail.

5 Experiment

We conduct an experiment to validate the correctness and effectiveness of the proposed representation. First we setup a database which contains 345 symbolic images. Semi-automatic segmentation is performed to every image in the database, and the results are recorded and corrected to generate the representations of all the database images.

A noise generator is used to add noise to the segmentation results of database images. The noise is some short segments generated randomly. We choose 30 symbolic images randomly from database, and add noise to them. The produced noisy images form the set of query images.

Ground truth data is generated manually. For a query image Q, a result set $R(Q)$ is produced. However, we find that there are some result images in $R(Q)$ which can be regarded as either similar to query image, or dissimilar to query image, it depends on person. It is necessary to distinguish them from other result images. Therefore we further divide $R(Q)$ into two sets: $R_{exact}(Q)$, which contains the exactly similar images; and $R_{marginal}(Q)$, which contains the marginally similar images, whether they are similar to query image is different for different persons.

For the same query image Q, retrieval system also returns a result set $As(Q)$. We define:

$$RA_{exact}(Q) = R_{exact}(Q) \cap A_s(Q)$$

$$RA_{m\arg inal}(Q) = R_{m\arg inal}(Q) \cap A_s(Q)$$

Retrieval recall and precision are defined as:

$$recall(Q) = \frac{|RA_{exact}(Q)| + |0.5 \times RA_{m\arg inal}(Q)|}{|R_{exact}(Q)| + |0.5 \times R_{m\arg inal}(Q)|}$$

$$precision(Q) = \frac{|RA_{exact}(Q)| + |0.5 \times RA_{m\arg inal}(Q)|}{|A_s(Q)|}$$

In our experiment, average "recall" is 83.2%, and average "precision" is 77.1%.

We measure the efficiency of our approach by the response time. Each symbolic image in database is selected to serve as the query image. Fig. 6 shows the distribution of the response time:

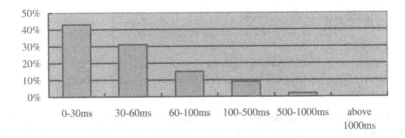

Fig. 6. Distribution of response time

The average response time is 31.5 ms, which is within the range of real-time response time. Therefore, retrieval system based on our proposed representation is efficient enough to give real-time response for a several-hundred-sized database.

6   Conclusion and Future Work

In this paper, we present a novel representation of symbolic image. The key point of it is to discriminate the vision contribution of the entities, and according to their different contribution, place them different positions. Based on this representation, retrieval system shows satisfactory efficiency and performance, as validated in experiment.

There are several aspects of this representation which can be improved. The model for critical entity selection used here is a little bit simple. In some cases, such as the symbol of "dumbbell", there may be several entities which are nearly symmetrical to each other, thus choosing anyone of them as critical entity may be improper. Another point to improve could be the algorithm of deciding which region an entity belongs to; here we judge that by only its position of geometrical center, while in some cases, this method can't work satisfactory. We will investigate these situations to give corresponding solutions to enhance our representation.

References

[1] Patrakis, E.: Image Representation, Indexing and Retrieval Based on Spatial Relationships and Properties of Objects. PhD dissertation, Univ. of Crete, Greece (1993)

[2] Nabil, M., Shepherd, J., Ngu, A.: 2D Projection Interval Relationships: A Symbolic Representation of Spatial Relationships. In: Symposium on Large Spatial Databases, pp. 292–309 (1995)

[3] Liu, Y., Liu, W., Jiang, C.J.: A Structural Approach to Recognizing Incomplete Graphic Objects. In: International Conf. on Pattern Recognition, pp. 371–375 (2004)

[4] Ricardo, B., Berthier, B.: Modern Information Retrieval. Addison-Wesley, Reading (1999)

[5] Flickner, M., Sawhney, H., Niblack, W., Ashley, J., Huang, Q., Dom, B., et al.: Query by Image and Video Content: The QBIC System. IEEE Computer 28(9), 23–32 (1995)

[6] Cheng, Y., Iyengar, S., Kashyap, R.: A New Method of Image Compression Using Irreducible Covers of Maximal Rectangles. IEEE Trans. Software Engineering 14(5), 651–658 (1988)

[7] Petrakis, E., Milios, E.: Efficient Retrieval by Shape Content. In: IEEE International Conf. Multimedia and Computing Systems, pp. 616–621 (1999)

[8] Chang, S., Shi, Q., Yan, C.: Iconic Indexing by 2-D Strings. IEEE Trans. on PAMI 9(3), 413–428 (1987)

[9] Lee, S., Hsu, F.: 2D C-String: a New Spatial Knowledge Representation for Image Database Systems. Pattern Recognition 23(10), 1077–1088 (1990)

[10] Li, C., Yang, B., Xie, W.: On-line Hand-sketched Graphics Recognition Based on Attributed Relation Graph Matching. In: Proc. of 3rd World Congress on Intelligent Control and Automation, Hefei, China, pp. 2549–2553 (2000)

[11] Llados, J., Marti, E., Jose, J.: Symbol Recognition by Error-Tolerant Subgraph Matching Between Region Adjacency Graphs. IEEE Trans. on PAMI 23(10), 1137–1143 (2001)

[12] Jin, X.: Sketch-based User Interface Study. M.S. thesis, Nanjing Univ., China (2002)

[13] Chazelle, B., et al.: Application Challenges to Computational Geometry, CG Impact Task Force Report. Advances in Discrete and Computational Geometry, Contemporary Mathematics 223, 407–463 (1999)

[14] Bezek, S., Ehrlich, R., Full, W.: FCM: The Fuzzy C-means Clustering Algorithm. Computational Geo. Science 10(2), 191–203 (1984)

[15] Lavie, N., Driver, J.: On the spatial extent of attention in object-based visual selection. Perception & Psychophysics 58, 1238–1251 (1996)

[16] Jarmasz, J.: Towards the Integration of Perceptual Organization and Visual Attention: the Inferential Attentional Allocation Model. PhD dissertation, Carleton Univ. Canada (2001)

[17] Downing, C., Pinker, S.: The spatial structure of visual attention: the spotlight metaphor brakes down. Journal of Experimental Psychology: Human Perception and Performance 15, 448–456 (1985)

[18] Milios, E., Petrakis, E.: Shape Retrieval Based on Dynamic Programming. IEEE Trans. Image Processing 9(1), 146–147 (2000)

[19] Berretti, S., Bimbo, A., Pala, P.: Retrieval by Shape Similarity with Perceptual Distance and Effective Indexing. IEEE Trans. Multimedia 2(4), 225–239 (2000)

[20] Ravela, S., Manmatha, R., Croft, W.: Retrieval of Trademark and Gray-Scale Images Using Global Similarity. Tech. Rep. MM-25, Univ. of Massachusetts, Amherst, Mass (1998)

[21] Jain, A., Vailaya, A.: Shape-Based Retrieval: A Case Study with Trademark Image Databases. Pattern Recognition 31(9), 1369–1390 (1998)

ARG Based on Arcs and Segments to Improve the Symbol Recognition by Genetic Algorithm

J.-P. Salmon and L. Wendling

LORIA-UMR 7503, INPL,
BP 239, 54506 Vandoeuvre-lès-Nancy, France
{salmon,wendling}@loria.fr

Abstract. A genetic matching algorithm is extended to take into account primitive arcs in a pattern recognition process. Usually approaches based on segments are sensitive to over segmentation effects. Handling with more accurate description allows to improve the recognition by limiting the number of vertices to be matched and so to decrease processing time. Experimental studies using real data attest the robustness of the proposed method.

Keywords: Graphic recognition, vector representation, inexact graph matching, technical drawings.

1 Introduction

Symbol recognition is especially well suited to structural pattern recognition techniques, unlike usual statistical classification techniques. When dealing with specific families of symbols, techniques similar to OCR could be used; this is the case for symbols having all a loop [1] or for music recognition. However these techniques have their own limitations, in terms of computational complexity and of discrimination power. Currently, graph matching techniques are especially adapted to the specificities of symbol recognition[2]. Generally, methods match symbols to be recognized with model graphs using graph distance computations. This kind of approach is sensitive to errors and noise; as we usually cannot assume that segmentation is perfect nor reliable, this means that the graphs to be processed can also have a number of extra or missing vertices. To deal with this problem, error-tolerant subgraph isomorphism algorithms have been proposed [3,4]. Another problem with graph matching is the computational complexity of subgraph isomorphism methods. To deal with this, even if a lot of efforts have therefore been devoted to optimizing the matching process through continuous optimization [5] or constraint propagation techniques to perform discrete [6,7] or probabilistic relaxation [8].

Arcs and circles are basic elements contained in technical drawings (engineering, electrical networks, mechanical...) and often relate to parts of symbols (for instance diode or logical nor representations) [9]. Considering raster data, powerful operator dedicated to their recognition should be required for both

W. Liu, J. Lladós, and J.-M. Ogier (Eds.): GREC 2007, LNCS 5046, pp. 80–90, 2008.

analysis and understanding of documents [10,11,12]. In the early years, many approaches have been developed (or extended) to extract such primitives [13,14] as arc-fitting methods [15,16], using Hough transform [17,18] and stepwise arc extension methods [19,20,13]. We can also denote that several methods were embedded in comparative studies using GREC contest [21,22]. For instance the last edition has shown than RANVEC method [23] supersedes most of them in many cases. This approach has been used here to ensure an accurate description from raster image.

The outline of the paper is as follows. In section2, some background on graph description and invariant attributes are introduced. The matching step is described in Section 3. At last a study of the behaviour of the method is proposed in section 4.

2   Graph Description

2.1   Structure

Attributed relational graph vertices correspond to graphic primitives and associated edges characterize topological relations. First, a basic description [24] of filled shape is used (see Fig. 1).

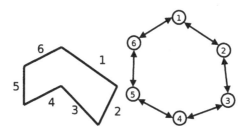

Fig. 1. A shape and its associated graph

In Fig. 1 the planar graph is entirely connected, that is each vertex is linked to two others. The variability of symbols encountered in graphic documents, requires to extend such process to study strings of points corresponding to symbol parts. In this case, constraints are weaker: a vertex can be connected at least to another (see Fig. 2).

Nevertheless such representation does not allow to really take into account symbols composed of several connected components (see Fig. 3). Remaining noise may remove junctions and so associated edges. A minimal distance has been calculated between each pair of connected components. A particular relation is created to link such component from the nearest associated primitives. Then, all the vertices of the ARG are connected at less with another.

This graph provides a more accurate description of symbols. Nevertheless symbols may have arcs in their composition. Generally they are approximated by a set of segments depending on the scale (see Fig. 4).

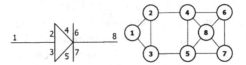

Fig. 2. A Symbol composed of several strings and its associated graph

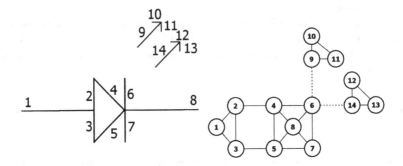

Fig. 3. A symbol composed of 3 connected components and its associated graph

The ability to handle with arcs has been studied to improve the accuracy during the recognition step.

2.2 Sharing Attributes

According to the conclusion of the previous section, the first step to build the ARG is to extract the skeleton of the shape using the algorithm of Sanniti Gabriella di Baja [25]. Secondly, RANVEC algorithm [23] vectorize the skeleton in order to find the circular arcs and segments. Finally, for each graphical primitive, some attributes are computed and brought together into a vertex of ARG. As the same way, some attributes are deduced of relation between two graphical primitives and are brought together into a edge. In this section, the attributes used are presented.

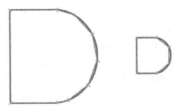

Fig. 4. Problem of vectorization : increase in the number of segments due to scale factor

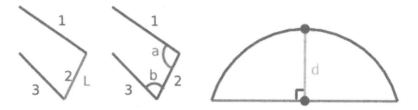

Fig. 5. Vertex attributes: length, sum of internal angles, distance between a primitive and its chord

A set of sharing attributes is required to be able to compare arcs and segments. Unary attributes are usually calculated within the recognition step to characterize graphic primitives, as the length, the sum of internal angles of adjacent primitives and the distance between the middle of a primitive and the middle of the associated chord which is equal to 0 using segments (see Fig. 5).

Binary relations between primitives are also defined such as the angle between two primitives, the distance between the middle points of primitives and the angles between the extremities of primitives (see Fig. 6).

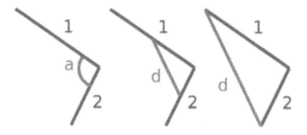

Fig. 6. Edge attributes: the angle, the distance between middle points, the maximum distance between end points

Calculating an angle between a segment and an arc has no sense. So we consider the angle between the current segment and an adjoining segment of the arc (see Fig. 7).

All these attributes are rotation and translation invariants. The acute angle between two primitives is invariant to scale factor as well as the sum of internal angles. Li [26] proposes to integrate a logarithm function to ensure scale factor invariance following both size and distance attributes. Normalized height and normalized distance following maximal values computing on the shape under consideration are also used to ensure invariance scale property. Nevertheless these approaches are sensitive to vectorization errors. The fitness function, proposed by Khoo and Suganthan [24], is adapted here to overcome such problem. This

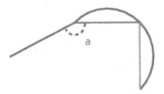

Fig. 7. Angle between a segment and an adjoining segment of an arc

method allows to efficiently differentiate the shapes while keeping an accurate recognition despite unexpected vectorization errors remain (see example Fig. 8).

3   Genetic Algorithm

Considering real data, noise can disturb the vectorization process. Thus, some graphical primitives can be merging or split and the ARG attribute values can be changed. Therefore, the matching method must be error-tolerant and fast to guarantee acceptable response time. That's the reason why, exhaustive research methods(ie: maximal clique based methods [27]) or the algorithms finding isomorphism are excluded. Among heuristics [28], [29], [30], [31], genetic the algorithm is really suitable, efficiently parallelizable and able to provide a good solution in short time.

3.1   Fitness Function

For each attribute u, the matching cost is calculated as follows.

$$C_{i,f_i,u} = |a_{f_i,u} - a_{i,u}| \tag{1}$$

with f_i the vertex matched with the vertex i. Let us consider a vertex i, the global cost A_{i,f_i} is calculated as follows.

$$A_{i,f_i} = \begin{cases} \displaystyle\sum_{u=1}^{U} \tanh(k_u(t_u - C_{i,f_i,u})), & f_i \neq Null \\ 0 & Otherwise \end{cases} \tag{2}$$

with U the number of attributes. This sum relies on the global matching cost obtained from vertex i. The use of function $tanh$ allows to normalize values

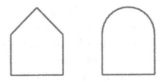

Fig. 8. Two different shapes having same description based on line segmentation

between -1 and 1 and to act on the importance of each attribute. The parameter k relies on the slope of $tanh$ and t is a threshold assuming that the matching cost is too high to consider symbols are similar. We introduce a similar process to handle with edges. Let us i and j be adjacent vertices and let b be an attribute of an edge.

$$C_{i,j,f_i,f_j,b} = |a_{f_i,f_j,b} - a_{i,j,b}| \quad (3)$$

$$D_{i,j,f_i,f_j} = \begin{cases} \sum_{b=1}^{B} \tanh(k_b(t_b - C_{i,j,f_i,f_j,b})), & f_i \neq Null, f_j \neq Null, \\ 0 & Otherwise \end{cases} \quad (4)$$

The fitness function is given by:

$$F = \sum_{i=1}^{S} A_{i,f_i} + \lambda \sum_{i=1}^{S} \sum_{j=i+1}^{S} D_{i,j,f_i,f_j} \quad (5)$$

with S the number of the vertices. A value of fitness greater than 0 implies an interesting match. A value equal to zero relies on unknown information (NULL) and a value lower than 0 implies a bad match. Parameter λ weights the importance of vertex attributes with respect to edge attributes during the initialization step.

3.2 Matching

A gene is associated here to the matching between a vertex of a scene and one of a model. Strings of genes are randomly set in the initialization step. The size of the strings relies on the complexity of the scene following the number of connected components and the number of primitives. For each generation, crossovers are performed on strings and new genes are randomly added. The crossover step consists in randomly swapping genes between two matched strings. When the number of generations is completed, only genes having the best score according to the fitness function are kept. Optimal values of main algorithm parameters are discussed in experimental section § 4. Two approaches of matching have been studied. First, a scene is compared to each model using the genetic algorithm. The algorithm returns the model having the greatest similarity value. In a second way, a scene is compared to all the models of the database to extract the most similar parts. Such a method is interesting to detect sub-symbols included in the composition of other symbols (see Fig. 9) but this approach relies on the number of model contained in the database. The proposed method integrates these two kinds of matching.

4   Experimental Study

4.1   Parameters k and t

Let us consider a set of learning samples. The aim is to find optimal values for parameters k and t in order to improve the recognition rate. In order to reach

Fig. 9. a) Scene composed of two models b) first model (5 vertices) c) second model (6 vertices).

this goal, different values of k are tested. For each value of k, n queries are preformed. Using of stochastic optimization algorithm not guarantees repeatability of results, that the reason why, each learning image is used n times and the recognition rate for one image corresponds to the ratio between good answers of GA and n, this means a representative result. The value of k maximizing the percentage of right matchings is kept. Same process is carried out to determine the optimal value of t.

For example the Fig. 10 shows the recognition rates obtained for k and t using $n = 1000$ queries. Optimal values are set to 0.1 for k and 5 for t.

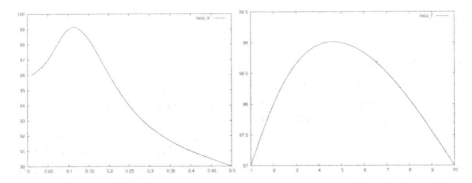

Fig. 10. Recognition rates following t and k values

4.2   Variability of the Results

The accuracy of the method relies on the number of generations. The influence of this parameter has been studied to improve the robustness of the matching by refining the stopping criterion of the genetic algorithm. Let us consider $n = 1000$ queries (see Fig. 11.a), the maximal recognition rates is used after 300 queries.

A threshold has been defined to determine the number of queries to be performed in order to keep a stable recognition rate. Its expression is given by:

$$rate(n) = \frac{((n-1) * rate(n-1) + \delta(n))}{n}$$

with: $rate(n)$ the mean recognition rate after n queries and $\delta(n)$ is equal to 1 if the query n is right and 0 otherwise. The Fig. 11.b shows that recognition rates are relatively stable about 100 queries.

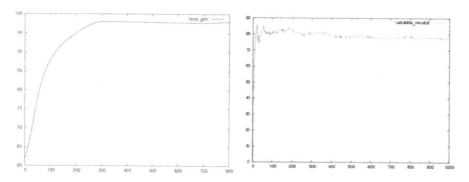

Fig. 11. Recognition rates following the number of generations (left) and cumulated recognition rates (right)

4.3 Experimental Study

First, GREC database (distortion 2) has been used to test the ability of the method to handle with several strings of points. This database consists in 74 symbols split into 15 classes. The Fig. 12 presents few symbols and the recognition rates achieved for associated class. A mean rate around 93.2% is reached which underlines the interest of the decomposition scheme.

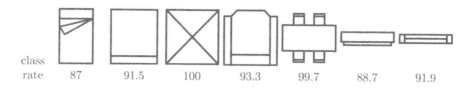

class							
rate	87	91.5	100	93.3	99.7	88.7	91.9

Fig. 12. Symbols and associated recognition rates

A database composed of 9 classes of 11 symbols including arcs in their composition has been defined to show the contribution of this primitive. A random model of distortions has been applied to the extremities of the primitives (see Fig. 13) in order to test the robustness of the method.

The basic genetic algorithm approach [24] gives rise to a mean recognition rate of 52.1% considering a set of segments obtained from a vectorization step.

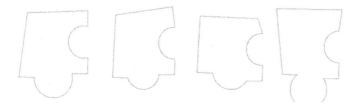

Fig. 13. Distorted symbols

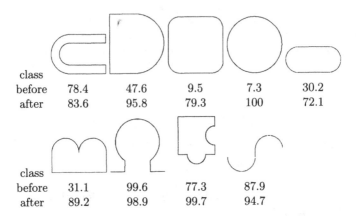

class					
before	78.4	47.6	9.5	7.3	30.2
after	83.6	95.8	79.3	100	72.1

class				
before	31.1	99.6	77.3	87.9
after	89.2	98.9	99.7	94.7

Fig. 14. Symbols and associated recognition rates

Integrating arcs in the method allows to reach a mean recognition rate of 90.4% (see Fig. 14).

5   Conclusion

An attributed relational graph based on arcs and segments has been proposed in this paper. Such ARG has been embedded in a genetic algorithm to improve the recognition of symbols which can be composed of several strings of points and several connected components. Furthermore invariant attributes which are invariant to geometric transformations have been proposed to efficiently take into account both primitives in the ARG. Handling with more complex graphic primitives is under consideration.

References

1. Okazaki, A., Kondo, T., Mori, K., Tsunekawa, S., Kawamoto, E.: An Automatic Circuit Diagram Reader with Loop-Structure-Based Symbol Recognition. IEEE Transactions on PAMI 10(3), 331–341 (1988)
2. Lladós, J., Valveny, E., Sánchez, G., Martí, E.: Symbol Recognition: Current Advances and Perspectives. In: Blostein, D., Kwon, Y.-B. (eds.) GREC 2001. Lladós, J., Valveny, E., Sánchez, G., Martí, vol. 2390, pp. 104–127. Springer, Heidelberg (2002)
3. Bunke, H.: Error Correcting Graph Matching: On the Influence of the Underlying Cost Function. IEEE Transactions on PAMI 21(9), 917–922 (1999)
4. Messmer, B.T., Bunke, H.: A New Algorithm for Error-Tolerant Subgraph Isomorphism Detection. IEEE Transactions on PAMI 20(5), 493–504 (1998)
5. Kuner, P.: Efficient Techniques to Solve the Subgraph Isomorphism Problem for Pattern Recognition in Line Images. In: Proceedings of 4th Scandinavian Conference on Image Analysis, Trondheim, Norway, pp. 333–340 (1985)

6. Habacha, A.H.: Structural Recognition of Disturbed Symbols Using Discrete Relaxation. In: Proceedings of 1st International Conference on Document Analysis and Recognition, Saint-Malo, France, vol. 1, pp. 170–178 (1991)
7. Wilson, R.C., Hancock, E.R.: Structural Matching by Discrete Relaxation. PAMI 19(6), 634–648 (1997)
8. Christmas, W.J., Kittler, J., Petrou, M.: Structural Matching in Computer Vision Using Probabilistic Relaxation. IEEE Transactions on PAMI 17(8), 749–764 (1995)
9. Kasturi, R., Bow, S., ELMasri, W., Shah, J., Gattiker, J., Mokate, U.: A system for interpretation of line drawings. IEEE Trans. on Pattern Analysis and Machine Intelligence 12(10) (1990)
10. Cordella, L.P., Vento, M.: Symbol recognition in documents: a collection of techniques? International Journal on Document Analysis and Recognition 3(2), 73–88 (2000)
11. Messmer, B.T., Bunke, H.: Automatic learning and recognition of graphical symbols in engineering drawings. Graphics Recognition: Methods and Applications, 123–134 (1996)
12. Tombre, K., Dori, D.: Interpretation of engineering drawings. Handbook of Character Recognition and Document Image Analysis, 457–484 (1997)
13. Song, S., Lyu, M.R., Cai, S.: Effective multiresolution arc segmentation: Algorithms and performance evaluation. IEEE Trans. on Pattern Analysis and Machine Intelligence 16(11), 1491–1506 (2004)
14. Wenyin, L., Zhai, J., Dori, D.: Extended summary of the arc segmentation contest. In: Blostein, D., Kwon, Y.-B. (eds.) GREC 2001. LNCS, vol. 2390, pp. 343–349. Springer, Heidelberg (2002)
15. Rosin, P., West, G.: Segmentation of edges into lines and arcs. Image and Vision Computing 7(2), 109–114 (1989)
16. Elliman, D.: An algorithm for arc segmentation in engineering drawings. In: Blostein, D., Kwon, Y.B. (eds.) GREC 2001. LNCS, vol. 2390, pp. 350–358. Springer, Heidelberg (2002)
17. Conker, R.: A dual plane variation of the hough transform for detecting nonconcentric circles of different radius. Computer Vision and Image Processing 43, 115–132 (1988)
18. Leavers, V.: The dynamic generalized hough transform: Its relationship to the probabilistic hough transforms and an application to the concurrent detection of circles and ellipses. Computer Vision, Graphics, Image Understanding 56(3), 381–398 (1992)
19. Dori, D.: Vector-based arc segmentation in the machine drawing understanding system environment. IEEE Trans. Pattern Analysis and Machine Intelligence 17(11), 1057–1068 (1995)
20. Liu, W., Dori, D.: Incremental arc segmentation algorithm and its evaluation. IEEE Trans. Pattern Analysis and Machine Intelligence 20(4), 424–431 (1998)
21. Blostein, D., Kwon, Y.-B. (eds.): GREC 2001. LNCS, vol. 2390. Springer, Heidelberg (2002), http://www.cs.cityu.edu.hk/liuwy/ArcContest/ArcSegContest.htm
22. Lladós, J., Kwon, Y.-B. (eds.): GREC 2003. LNCS, vol. 3088. Springer, Heidelberg (2004), http://www.cvc.uab.hk.es/grec03/contest.htm
23. Hilaire, X.: Ranvec and the arc segmentation contest: Second presentation. In: Liu, W., Llados, J. (eds.) Postproceedings of GREC 2005. LNCS. Springer, Heidelberg (to appear, 2006)

24. Khoo, K., Suganthan, P.: Evaluation of genetic operators and solution representations for shape recognition by genetic algorithm. Pattern Recognition Letters 23(13), 1589–1597 (2002)
25. Sanniti di Baja, G.: Well-Shaped, Stable, and Reversible Skeletons from the (3,4)-Distance Transform. Journal of Visual Communication and Image Representation 5(1), 107–115 (1994)
26. Li, S.Z.: Matching: Invariant to Translations, Rotations and Scale Changes. Pattern Recognition 25(6), 583–594 (1992)
27. Ambler, A.P., Barrow, H.G., Brown, C.M., Burstall, R.M., Popplestone, R.J.: A versatile computer-controlled assembly system. In: International Joint Conference on Artificial Intelligence, vol. 1, pp. 298–307 (1973)
28. Cross, A.D.J., Hancock, E.R.: Inexact Graph Matching with Genetic Search. In: Perner, P., Wang, P., Rosenfeld, A. (eds.) SSPR 1996. LNCS, vol. 1121, pp. 150–159. Springer, Heidelberg (1996)
29. Wang, Y.K., Fan, K.C., Horng, J.T.: Genetic-based search for error-correcting graph isomorphism. IEEE Transactions on Systems, Man, and Cybernetics – Part B: Cybernetics 27(4) (1997)
30. Sammoud, O., Sorlin, S., Solnon, C., Ghédira, K.: A comparative study of ant colony optimization and reactive search for graph matching problems. In: European Conference on Evolutionary Computation in Combinatorial Optimization, pp. 317–326 (2006)
31. Qureshi, R.J., Ramel, J.Y., Cardot, H.: Graph based shapes representation and recognition. In: Escolano, F., Vento, M. (eds.) GbRPR. LNCS, vol. 4538, pp. 49–60. Springer, Heidelberg (2007)

Spotting Symbols in Line Drawing Images Using Graph Representations

Rashid Jalal Qureshi, Jean-Yves Ramel, Didier Barret, and Hubert Cardot

Université François-Rabelais de Tours
Laboratoire d'Informatique (EA 2101)
64, Avenue Jean Portalis, 37200 Tours – France
{rashid.qureshi,jean-yves.ramel,
didier.barret,hubert.cardot}@univ-tours.fr

Abstract. Many methods of graphics recognition have been developed throughout the years for the recognition of pre-segmented graphics symbols but very few techniques achieved the objective of symbol spotting and recognition together in a generic case. To go one step forward through this objective, this paper presents an original solution for symbol spotting using a graph representation of graphical documents. The proposed strategy has two main step. In the first step, a graph based representation of a document image is generated that includes selection of description primitives (nodes of the graph) and organisation of these features (edges). In the second step the graph is used to spot interesting parts of the image that potentially correspond to symbols. The sub-graphs associated to selected zones are then submitted to a graph matching algorithm in order to take the final decision and to recognize the class of the symbol. The experimental results obtained on different types of documents demonstrates that the system can handle different types of images without any modification.

Keywords: Document Image Analysis, Graphics Symbols Recognition, Symbol Spotting, Raster-to-Vector Techniques, Line drawings.

1 Introduction

The classical approaches of symbol recognition use knowledge about the document to drive the analysis of the image [1] [2]. The segmentation algorithms are often looking directly in the bitmap images for precise information specified in the model databases. So, the image analysis is achieved by using mostly a split strategy forced by a priori knowledge. Two classical processing ways can be mentioned:

- Methods looking for some particular configurations of pixels or of low level primitives in the image (horizontal – vertical lines, specific textures, closed loops, connected components …) corresponding to potential symbols [3].
- Methods based on computation of signatures for identification and recognition of symbols. Recently presented in [4] [5], signatures are a collection of features extracted from pre-segmented shapes. Signatures of candidate symbols are computed and matched against the existing signatures of all the models stored in the library of a particular document. In most of the cases, the "localisation" is

W. Liu, J. Lladós, and J.-M. Ogier (Eds.): GREC 2007, LNCS 5046, pp. 91–103, 2008.

realised using the connected components or buckets (small square part of the image). In this last case, the test images are divided into buckets and signature of each bucket is computed and is matched against the symbols signatures to determine what kind of symbols a bucket is likely to contain. With such methods, as pointed out by P. Dosch [4], a lot of false alarms are still present, especially with parts of the images not presented in the pre-definite library. Furthermore, the recognition rate may decrease on much larger databases and the strength of classification may suffer in case of noisy images.

An interactive approach to recognition of graphic objects in engineering drawings has been recently proposed by L. Wenyin [6]. Interactively, the user provides examples of one type of graphic object by selecting them in an engineering drawing and then the system learns the graphical knowledge and uses this learnt knowledge to recognize or search for other similar graphic objects in the same image. However, it is desired that we could segment graphic symbols automatically in large dataset without human involvement and user feedback for each analysed image. A weak aspect which is common in all these algorithms is the lack of information about the handling of regions which do not belong to one of the expected categories.

2   Graph Representation of Image Content

The proposed approach relies on structural methods and is based on capturing the spatial and topological relationships between graphical primitives. The pre-processing includes vectorization of the raster image. The technique which we had proposed in [7] is used to constructs a quadrilaterals based representation of lines in a drawing (figure 1, figure 2b). Each quadrilateral has attributes like length (ℓ) of the median axis, angles of the two vectors (v_1, v_2), width on each side (w_1, w_2) and a zone of influence (as shown by dotted rectangle in figure 2c). The dimensions of the zone of influence depends on the length (ℓ) and the two widths (w_1, w_2) of the quadrilateral i.e., $U_x = \ell / 4$ and $U_y = ((w_1 + w_2) / 2) \times 4$. In fact, in our structural representation, nodes represent quadrilaterals and edges represent the topological relationships between neighbouring quadrilaterals (figure 2d). The length of the quadrilateral is associated to the nodes as an attribute, and relative angle i.e., angle between two neighbouring quadrilaterals is associated to edges as an attribute. All edges are also associated with one label representing the type of the topological relationship (L-junction, S-junction, T-junction, X-intersection or P-parallelism) that exists between the two neighbouring quadrilaterals (figure 2e, figure 2f). A detailed description of this graph representation can be found in [7] [8].

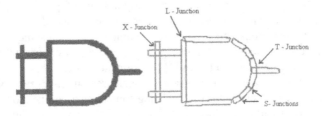

Fig. 1. Initial image (on left), Vectorization results (on right)

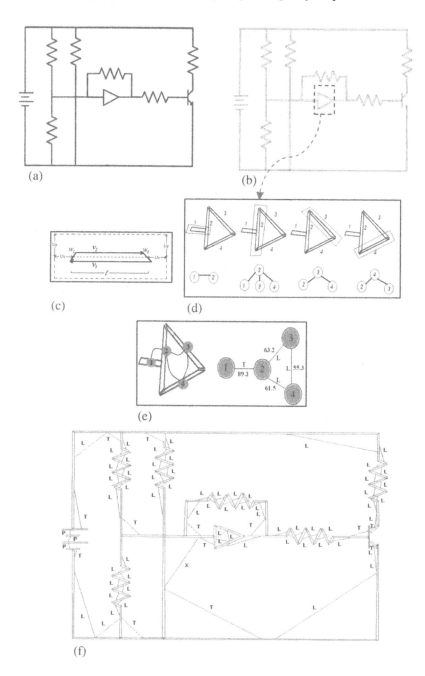

Fig. 2. (a) Initial image, (b) Vectorization results, (c) Zone of influence of a quadrilateral (d) Influence zone of the quadrilaterals and their corresponding sub-graphs respectively, (e) and (f) Graph representation

3 Seeds Detection in the Graph

The key idea of our method is to detect the parts of the graph that may correspond to symbol without a priori knowledge about the type of the document. Such nodes and edges will constitute symbol seeds. Then, the seeds will be analyzed and grouped together to generate sub-graphs that are potentially corresponding to symbols in the document image. We observed from structure analysis of drawings, circuits, maps and diagrams that symbols are composed of quadrilaterals with specific characteristics easily detectable in the graph representation. So, scores (probabilities of being part of a symbol) can sequentially be associated to all the edges and all the nodes of the graph to provide the symbol seeds. The hypothesis used to compute the scores of the nodes and the edges are:

H_1 - Symbols are composed of small segments compared to other elements of the drawings

H_2 - The segments constituting a symbol have comparable lengths

H_3 - Two successive segments with a relative angle far from 90° have a higher probability of being part of a symbol

H_4 - Symbols are often composed of parallel segments

H_5 - A symbol segment is rarely connected to more than 3 other segments

H_6 - Shortest loops are most often corresponding to symbols

This is the basic set of hypothesis that our system uses for the purpose of symbols spotting in documents images. However, the documents containing symbols that do not respect these hypotheses can not be analyzed using the proposed system.

3.1 Studies of the Nodes and the Edges

A study of graph nodes and edges characteristics and attributes is made and a score between 0 and 1 is computed for each node and edge using the above hypothesis. The score of an edge is a function of the two relative lengths and angles of the connected nodes (quadrilaterals) and of the type of topological relationship between them (eq.1). The score of a node (eq. 2) is computed by using the accumulated scores of its connecting edges, the number of its connecting edges and the length of the corresponding quadrilateral.

$$Score(E_i) = \alpha.P_{E1}(E_i \rightarrow Type) + \beta.P_{E2}(E_i \rightarrow Angle) + \gamma.P_{E3}(E_i \rightarrow N_1 \rightarrow Length, E_i \rightarrow N_2 \rightarrow Length) \quad (1)$$

$$Score(N_i) = \lambda \left(\frac{\sum_{j=1}^{\Omega(Ni)} Score(Ej)}{\Omega(Ni)} \right) + \mu . P_{N2}(\Omega(N_i)) + \omega .P_{N3}(N_i \rightarrow Length) \quad (2)$$

In eq.1, E_i is an edge of the graph, α, β and γ are weights and P_{Ei} are functions returning scores between 0 and 1 according to the %age agreement with the above mentioned hypothesis. Here α, β and γ are set to $1/3$ corresponding to the weights

Table 1.

Functions	Scores
$P_{E1}(E_i \rightarrow Type)$	0.50 (T), 0.25 (L), 0.50 (X), 1.0 (P), 0.50 (S)
$P_{E2}(E_i \rightarrow Angle)$	$\|90 - (E_i \rightarrow Angle)\| / 90$
$P_{E3}(E_i \rightarrow N_1 \rightarrow Length, E_i \rightarrow N_2 \rightarrow Length)$	$\dfrac{\min(N_1 \rightarrow Length, N_2 \rightarrow Length)}{\max(N_1 \rightarrow Length, N_2 \rightarrow Length)}$
$P_{N2}(\Omega(N_i))$ (Ω : degree of a node)	0.25 $(\Omega < 2)$ 1.00 $(2 \leq \Omega \leq 4)$ 0.25 $(\Omega > 4)$
$P_{N3}(N_i \rightarrow Length)$ (Length of the primitive in pixels)	0.25 $(Length < 20)$ 1.00 $(20 \leq Length \leq 200)$ $50/Length$ $(Length > 200)$

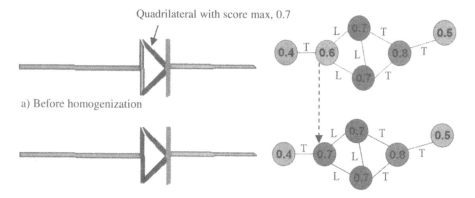

a) Before homogenization

b) After homogenization, 0.7 allocated to all the quadrilaterals connected in loop

Fig. 3. Propagation of the maximum score to all the nodes in the path

associated to each hypothesis. In eq.2, N_i is a node of the graph, $\Omega(Ni)$ is the degree of the node Ni, ω, μ and λ are weights and P_{Ni} are functions returning scores between 0 and 1 according to $H_1 - H_6$, $\lambda = 1/2$ and $\mu = \omega = \frac{1}{4}$ corresponding to the weights associated to each hypothesis.

The values computed by functions P depends on the attributes associated with nodes and edges of the graph, as shown in Table.1. All these weights have been fixed in the light of statistics collected from different images of the databases (electrical diagrams, logic diagrams, architectural maps).

3.2 Score Propagation in the Graph

The different seeds are associated and extended by merging with other seeds depending on their relative scores and relation. This score propagation process seeks and analyses the different loops with shortest length between seeds (nodes in the graph). This step only modifies the scores of the nodes modeling quadrilaterals that form closed loop. The scores of all the nodes belonging to a detected path are homogenized (propagation of the maximum score to all the nodes in the path) until convergence. This is the phase where edges with L and S attributes act as bridges between the different parts of a spotted symbol.

This step only modifies the scores of the symbols composed of closed loops but has no influence on other parts of the graph.

4   From Seeds to Symbol Spotting and Recognition

4.1   Generation of the Bounding Boxes

After the computation of the scores of the nodes, a sub-graph extraction has to be achieved in order to provide the symbol recognizer module with potential symbol present in the document. To obtain these sub-graphs corresponding to symbols and the associated bounding boxes (figure 5), our methods looks for all the sets of directly

---

Algorithm-1

Input : Graph representation of a graphic document
Output : Number of bounding boxes(spotted symbols), Best value of seed threshold Ts
List of Bounding boxes(BB), List of sub-graphs(SG)

for $Ts = 0$ to 1 with step = 0.1 **do**
  begin
      $Max\_BB \leftarrow 0$ // Maximum number of bounding boxes
      $Best\_Ts \leftarrow 0$ // Best threshold value

    for $j = 1$ to $|E|$ **do**
        compute: score(E_j)

    for $i = 1$ to $|N|$ **do**
        compute: score(N_i)

        Seeds $\leftarrow$ $\forall (N_i) \mid$ score$(N_i) \geq Ts$
        Propagation of scores of quads in loops with L and/or S type
connections

        $SG \leftarrow$ Get Sub-graphs
        $BB \leftarrow$ Get Bounding Boxes

    if $(BB > Max\_BB)$
        $Max\_BB \leftarrow BB$
        $Best\_Ts \leftarrow Ts$
    end if
    end for

Fig. 4. Selection of seed threshold (Ts) automatically

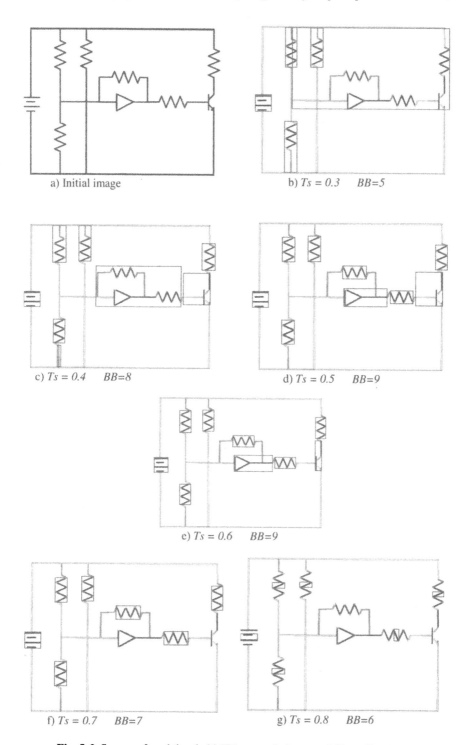

a) Initial image

b) *Ts = 0.3*   *BB=5*

c) *Ts = 0.4*   *BB=8*

d) *Ts = 0.5*   *BB=9*

e) *Ts = 0.6*   *BB=9*

f) *Ts = 0.7*   *BB=7*

g) *Ts = 0.8*   *BB=6*

Fig. 5. Influence of seed threshold (*Ts*) on symbols spotted (Bounding Box)

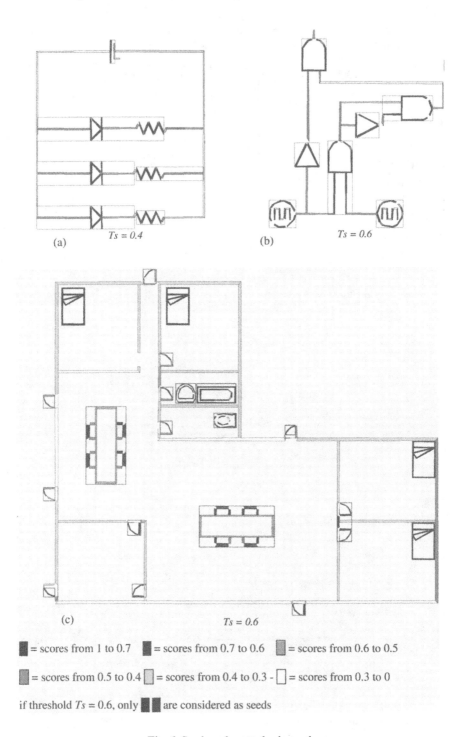

Fig. 6. Seeds and spotted sub-graphs

```
<Graph id="Symbole4">
  <node id="node0">
    <attr name="forme"><string>Quad</string></attr>
    <attr name="x1i"><string>232</string></attr>
    <attr name="y1i"><string>497</string></attr>
    <attr name="x1f"><string>231</string></attr>
    <attr name="y1f"><string>417</string></attr>
    <attr name="x2i"><string>227</string></attr>
    <attr name="y2i"><string>418</string></attr>
    <attr name="x2f"><string>229</string></attr>
    <attr name="y2f"><string>498</string></attr>
    <attr name="angle1"><string>90</string></attr>
    <attr name="angle2"><string>91</string></attr>
    <attr name="thickness1"><string>8</string></attr>
    <attr name="thickness2"><string>7</string></attr>
    <attr name="length"><string>81</string></attr>
    <attr name="score"><string>0.703</string></attr>
  </node>
  ...
  ...
  ...
<edge id="edge1" from="node0" to="node1">
    <attr name="angle"><string>89</string></attr>
    <attr name="type"><string>T</string></attr>
    <attr name="score"><string>0.409</string></attr>
</edge>
<edge id="edge2" from="node1" to="node2">
    <attr name="angle"><string>46</string></attr>
    <attr name="type"><string>L</string></attr>
    <attr name="score"><string>0.388</string></attr>
</edge>
...
...
...
</graph>
</gxl>
```

Fig. 7. Example of sub-graph encoded with GXL and extracted from Fig. 3a

connected seeds (nodes) in the graph using a recursive process. Only, nodes having highest score (probability of being part of symbols) have to be considered as part of the symbols (figure 5).

It is possible to vary the seed threshold (Ts) and to compare the number of symbols generated manually as well as automatically (figure 4). A simple rule for automatic selection of seed threshold (*Ts*) is to consider the value which gives maximum number of bounding box (figure 5). The experiments demonstrated that the better results are obtained by keeping the threshold providing the maximum number of symbols in the image.

However, we can test other complex rules by using statistics about these bounding boxes and ultimately the seeds inside, for example, the dimensions of the bounding boxes.

4.2 Symbol Recognition or Rejection

Hence, we used our graph matching routine [8] to match these sub-graphs with graph representation of models. The detected zones i.e., sub graphs are matched against model graphs using polynomial bound greedy algorithm for the symbol recognition task.

This recognition algorithm outputs a score of similarity and the best mapping of nodes found. The graph matching is error-tolerant and works well in case of under or over segmentation of symbols. The score of similarity produced by the matching algorithm can easily be used (using a threshold) to automatically decide if the extracted sub-graph corresponds to a symbol or not (rejection of the zone).

The approach is parallel, and is capable of spotting all the symbols present in a drawing in one pass. The different steps of proposed strategy are shown in figure 8, however in this paper we have detailed the steps related to symbol spotting only.

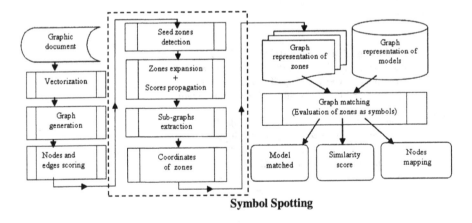

Fig. 8. Proposed system architecture

5 Results and Conclusion

Tests have been conducted on three types of graphic documents, electronic circuits, logic diagrams and architectural maps.

To evaluate the performance of the proposed system we have followed the general framework based on the notion of precision and recall presented in [9]. For a given test, let T be the number of targets belonging to the ground-truth, and R the set of

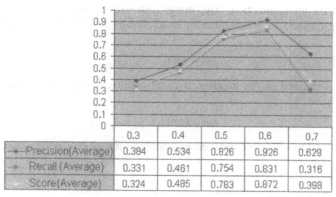

	0.3	0.4	0.5	0.6	0.7
Precision(Average)	0.384	0.534	0.826	0.926	0.629
Recall (Average)	0.331	0.461	0.754	0.831	0.316
Score(Average)	0.324	0.485	0.783	0.872	0.398

a) Performance evaluation graph for electrical diagrams

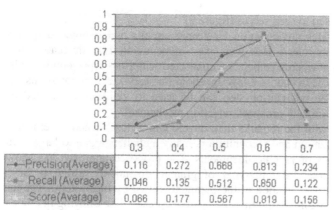

	0.3	0.4	0.5	0.6	0.7
Precision(Average)	0.116	0.272	0.668	0.813	0.234
Recall (Average)	0.046	0.135	0.512	0.850	0.122
Score(Average)	0.066	0.177	0.567	0.819	0.158

b) Performance evaluation graph for logic diagrams

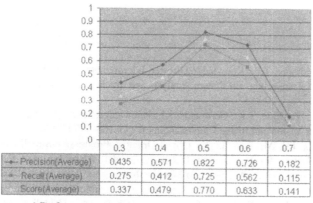

	0.3	0.4	0.5	0.6	0.7
Precision(Average)	0.435	0.571	0.822	0.726	0.182
Recall (Average)	0.275	0.412	0.725	0.562	0.115
Score(Average)	0.337	0.479	0.770	0.633	0.141

c) Performance evaluation graph for architectural maps

Fig. 9. Precision / Recall graphs for few prototypes of electrical diagrams, logical diagrams and architectural maps

results supplied by an application. The number of exact results is called e. The precision is then defined as the number of exact results divided by the number of results:

$$p = e / |R|$$

The recall r is defined as the number of exact results divided by the number of targets:

$$R = e / |T|$$

The precision and recall then combined to determine the global score s, expressing the recognition rate:

$$s = \frac{2}{(1/p) + (1/r)}$$

The localization rate (without considering the recognition step) depends on the threshold associated to seed scores (figure 9).

Localization rates are better in electronic circuits and in logic diagrams where symbols are quite clear. However, the segmentation rate decreases in case of architectural maps where symbols are connected with lines representing walls. In this case, it is preferable to choose a low threshold for seed selection not to miss too many symbol seeds. The recognition step can be used to validate the spotted bounding boxes in connection with proposed rejection mechanism.

We conclude with the remarks that, developing a flexible and general framework for symbol localization and recognition is really a challenge due to large variation in the basic elements of graphic documents. The proposed approach has shown good results on the three different types of graphic documents tested so far (without complex/difficult modification of the parameters used by the system). The method is parallel, and is capable of spotting the symbols present in a drawing in one pass. The graph matching is error-tolerant and works well in case of under or over segmentation of symbols.

References

1. Joseph, S.H., Pridmore, T.P.: Knowledge Directed Interpretation of Mechanical Engineering Drawings. IEEE Transactions on Pattern Analysis and Machine Intelligence 14(9), 928–940 (1992)
2. DenHartog, J.E., TenKate, T.K., Gerbrands, J.J.: Knowledge Based Interpretation of Utility Maps. Computer Vision and Image Understanding 63(1), 105–117 (1996)
3. Song, J., Su, F., Tai, M., Cai, S.: An Object-Oriented Progressive-Simplification-Based Vectorization System for Engineering Drawings: Model, Algorithm, and Performance. IEEE Transaction on Pattern Analysis and Machine Intelligence 24(8), 1048–1060 (2002)
4. Lladós, J., Dosch, P.: Vectorial Signatures for Symbol Discrimination. In: Lladós, J., Kwon, Y.-B. (eds.) GREC 2003. LNCS, vol. 3088, pp. 154–165. Springer, Heidelberg (2004)
5. Wang, Y., Phillips, I.T., Haralick, R.M.: Document Zone Content Classification and its Performance Evaluation. Pattern Recognition 39, 57–73 (2006)

6. Wenyin, L., Zhang, W., Yan, L.: An Interactive Example-Driven Approach to Graphics Recognition in Engineering Drawings. International Journal on Document Analysis and Recognition 9, 13–29 (2007)

7. Ramel, J.Y., Vincent, N., Emptoz, H.: A Structural Representation for Understanding Line Drawing Images. International Journal on Document Analysis and Recognition 3(2), 58–66 (2000)

8. Qureshi, R.J., Ramel, J.Y., Cardot, H.: Graphic Symbol Recognition Using Flexible Matching Of Attributed Relational Graphs. In: Proceeding of 6th IASTED International Conference on Visualization, Imaging, and Image Processing (VIIP), Palma de Mallorca-Spain, pp. 383–388 (2006)

9. Valveny, E., et al.: A general framework for the evaluation of symbol recognition methods. IJDAR 9, 59–74 (2007)

A Region-Based Hashing Approach for Symbol Spotting in Technical Documents

Marçal Rusiñol and Josep Lladós

Centre de Visió per Computador / Computer Science Department
Edifici O, Campus UAB 08193 Bellaterra (Cerdanyola), Barcelona, Spain
{marcal,josep}@cvc.uab.es
http://www.cvc.uab.es

Abstract. In this paper a geometric hash function able to cluster similar shapes and its use for symbol spotting in technical documents is presented. A very compact representation of features describing each primitive composing a symbol are used as key indexes of a hash table. When querying a symbol in this indexing table a voting scheme is used to validate the hypothesis of where this symbol is likely to be found. This hashing technique aims to perform a fast spotting process to find candidate locations needing neither a previous segmentation step nor a priori knowledge or learning step involving multiple instances of the object to recognize.

1 Introduction

Architects and engineers use to store their designs in large databases of documents. They use to re-use data from previous designs for new projects. Graphical documents such as architectural, electronic or mechanical drawings use to have a big size with hundreds of graphical parts. In this framework, fast methods for locating document zones where a given symbol appears are very useful for browsing, navigation or even categorization purposes. Nowadays, the format of these documents does not allow an intelligent way to browse them. The manual inspection of technical documents becomes a tedious task, and when talking about large collections, it becomes impossible to perform an exhaustive search. In this paper we present a method to retrieve the locations of document images in which a certain queried graphical symbol is likely to appear by a fast indexing technique.

Although well-known shape description and recognition approaches perform quite good recognition rates in difficult conditions as in presence of distortions, occlusions and geometric transformations, they can not be applied to locate graphical symbols in large collections of documents. Most of these techniques are time consuming and only work for recognizing isolated shapes. The interested reader is referred to Zhang's [11] review on shape description techniques.

This is the reason why in the last years, Symbol Spotting has become an emerging topic of interest among the Graphics Recognition community. It aims to detect graphical symbols in large and complex document images by a fast

W. Liu, J. Lladós, and J.-M. Ogier (Eds.): GREC 2007, LNCS 5046, pp. 104–113, 2008.

technique and needing neither a segmentation step nor a priori knowledge. Symbol Spotting discriminates symbols by means of a very compact representation of expressive features which are organized in a lookup table (usually a hash table with its associated hash function) allowing graphic indexation of symbols appearing in documents. Usually a validation of the hypothetic locations is performed using a voting scheme as a post-processing step. Two different Symbol Spotting approaches can be used depending on the chosen primitives. The methods working at pixel level, as for instance [6,9,12], and the ones focusing in geometric primitives as segments, polylines, arcs, etc. as in [2,5,8]. Pixel-based methods are usually more robust to noise, but they are more complex and need a multi-level structure (as the dendrogram used in [12]) to segment and recognize at the same time. On the other hand, more compact, fast and elementary descriptions can be formulated when working with vectorial primitives even if they usually lead to more false alarms and are more sensible to noise.

Given a feature vector x defining a shape in a N-dimensional space, a hash function $H(x) = y$ is a deterministic function projecting the feature vectors x to an even distribution of one-dimensional key indexes y. They are useful for finding an entry in a database by a key value. In our case, the index keys are computed from the different closed regions composing a symbol which are the primitives describing a given symbol. The ideal behavior of our hash function is to provoke collisions (associate the same indexing key to different entry values) for similar shaped primitives, and to associate different key indexes to different shape inputs. The resulting hash table clusters similar shapes under a single key index.

In the presented method, polygonally approximated regions composing a symbol are used as primitives which describe it and a key index is associated to each region. Geometric Hashing [3] is a well-known method to extract an index of a given shape. However this technique only works well if the set of segments of the model and the set of segments of the test shape have the same number of segments. The actual raster-to-vector techniques are yet very sensitive to little distortions and can result in a different number of segments for similar shapes, as stated by Tombre *et al.* in [10]. Our presented method applies a hash function to each region to extract an index identifying similar regions, being independent of the number of segments forming the region since it copes with global shape information.

The remainder of this paper is organized as follows: we introduce in the next section how the graphical symbols are represented in a two-center bipolar coordinate system. In section 3, the hash function and the whole indexing scheme is presented. We provide the preliminar experimental results in section 4. A brief discussion on the presented approach is provided in section 5. Finally conclusions and future work are presented in section 6.

2   Symbol Representation in a Two-Center Bipolar Coordinate System

Graphical symbols found in technical documents are usually composed of a number of neighboring regions. Regions result from solid areas or closed contours in line drawing images. We describe a symbol in terms of the structural combination of regions, described by sequences of adjacent segments that approximate their boundary. Let us further detail how the symbols are represented and encoded in a two-center bipolar coordinate system which is an appropriate representation for the computation of the presented indexing keys.

2.1 Primitive Extraction

A graphical symbol appearing in a technical document (such as a furniture symbol in an architectural drawing) is decomposed and described in terms of its n composing closed contours. The presence, in a given zone of the drawing of several similar regions that the ones forming the queried symbol, turns this location into a candidate for containing the queried symbol.

To represent symbols, a connected component analysis of the technical document images is computed to first extract the closed regions. Their contours are subsequently polygonally approximated to take the chains of adjacent segments –polylines– as representing primitives. We illustrate in Figure 1 an overview of all the representation process.

Although the presented method do not need the polylines to be closed, we find that working with closed contours as primitives is much more stable to vectorization errors. We focus on an application dealing with architectural floor plans which can be easily represented by regions. If the method has to be used with another type of technical documents (for example electronic diagrams) which contains symbols less represented by regions, another kind of primitives much more representative should be considered.

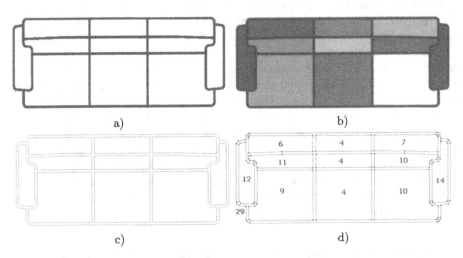

Fig. 1. Symbol representation steps. (a) Original symbol. (b) Connected components. (c) Contour of each closed region. (d) Polygonally approximated regions with the number of segments composing the polylines.

2.2 Two-Center Bipolar Coordinates

A two-center bipolar coordinate system gives the distances (r_1, r_2) of a point (x, y) to two reference centers. We choose the two reference centers among the set of points of a given polyline p and we center them at $(\pm a, 0)$. Afterwards, each other point composing the polyline is represented in two-center bipolar coordinates (r_1, r_2), defined in terms of Cartesian coordinates by

$$
\begin{aligned}
r_1 &= \sqrt{(x + a)^2 + y^2} \\
r_2 &= \sqrt{(x - a)^2 + y^2}
\end{aligned}
\tag{1}
$$

2.3 Cassinian Ovals

The Cassinian ovals were first studied by Giovanni D. Cassini as a model for the orbit of the sun around the earth. We use the parameters of these curves as features of the set of points forming the primitives. Let us detail how Cassinian ovals are described.

In two-center bipolar coordinate system, each point (r_1, r_2) belongs to a characteristic plane curve (known as Cassinian ovals [4]) described by all the points such that the product of its distances from the two centers is a constant $b^2 = r_1 r_2$. In Figure 2 we illustrate a series of Cassinian ovals with $a = 1$ and with values of $b \in [1.001, 2]$. In our case, Cassinian ovals are the basis of the hashing keys described in the next section.

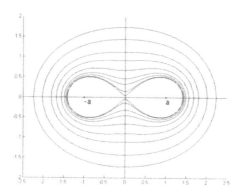

Fig. 2. A series of Cassinian ovals with $a = 1$ and for values of $b \in [1.001, 2]$

3 Hash Function and Indexing Symbols with a Hash Table

In this section we describe how the hash function is formulated to be able to describe shapes in terms of the Cassinian ovals parameters, and how these parameters are used as indexing keys. The whole indexation process is presented afterwards. Let us begin with the primitive description.

3.1   Hash Function

Given a polyline p, it is encoded in terms of a tuple (a, b) defined as follows. Let us first denote as n_i a point over the polyline, $n_i \in p$, and $n_i' \in p$, as the furthest point from n_i in the direction of the line passing through the gravity center gc of p. Let $2a$ be defined as the maximum distance among all the possible pairs of points (n_i, n_i'). A transform matrix M is computed to rotate and scale the whole polyline to transform the points n_i to $(-a, 0)$ and n_i' to $(a, 0)$. Then using $(\pm a, 0)$ points as centers, The value b is defined as the minimum constant to build a Cassinian oval surrounding all the i points of p. An example is shown in Figure 3. We illustrate in Figure 4 the possible surrounding ovals depending on the reference centers.

The hash function $H(p) = (\bar{a}, \bar{b})$ projects p to a tuple by means of minimizing the value of b for the maximum value of a. The value a is normalized to the total length of p then having $a \in [0, 1]$ and $b \in [0, \sqrt{2}]$. Finally, applying the equation 2 we discretize the values of a and b. Experimentally, we found that a

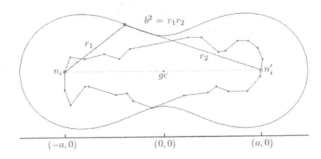

Fig. 3. A polygonally approximated shape p with its gravity center gc. Given a point $n_i \in p$, $n_i' \in p$ is the furthest point from n_i passing through gc. The value b determine the smaller Cassinian oval surrounding the whole shape when n_i and n_i' are transformed to $(-a, 0)$ and $(a, 0)$.

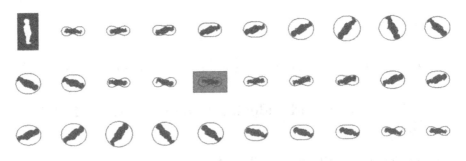

Fig. 4. A shape and the minimum surrounding Cassinian ovals depending on the reference centers. The gray surrounded shape is the one with minimum value b and maximum distance a.

value $m = 30$ provokes collisions when computing keys of similar shapes while minimizing the collisions of dissimilar shapes.

$$\overline{x} = round(x * m) \qquad (2)$$

The tuple $(\overline{a}, \overline{b})$ gives information about the ellipticity and circularity of the set of points composing the analyzed region. As $\overline{a}$ is chosen as the maximum value of a, it gives information of the major axis of the primitive p. If $\overline{b}$ is equal to $\sqrt{2} \times \overline{a}$ that means that the primitive is a circular shape, the more $\overline{b}$ decreases for the same $\overline{a}$ values, the more the primitive shape would be elongated. The use of these parameters is an attempt to combine the information bringed by a very simple well-known global shape descriptors: eccentricity and non-circularity.

3.2   Indexation Process

In an off-line step, the locations of the gravity centers of all the polygonally approximated regions p appearing in the documents of the collection are stored in a hash table in their corresponding entry with index $H(p) = (\overline{a}, \overline{b})$. Then, when we query a symbol formed by a set of polylines $S = \{p_1, ..., p_n\}$, the querying process only needs to compute the entries of the hash table to activate, applying the hash function $H(p_x)$ to all the regions $p_x \in S$. Ideally, in a given table entry i with indexing key $(\overline{a_i}, \overline{b_i})$ we will find all and only the similar primitive regions. A query symbol formed by several primitives will activate as many table entries as dissimilar regions primitives compose it. All the locations pointed by the activated table entries are likely to contain a similar primitive than one of the query, the locations accumulating a high number of these primitives are then hypothetic locations to contain the whole symbol. To deal with a database containing several documents, the locations are expressed with three dimensions including the identification number of the document where the gravity center appears.

Following the idea of the Generalized Hough Transform [1], a voting process is performed to reinforce the locations where diverse regions appear. At each zone in which we find a given indexed primitive, a vote is casted. As the symbols are formed by several primitives, high voting values in a given zone of the technical document points out the presence of several regions composing the query, thus a high probability to find the queried symbol in this zone. The accumulation of queried primitives in a given zone always forms clusters of votes while the indexed primitives belonging to other symbols will just scatter votes without any coherence in the whole space loosing significance.

4   Experimental Results

We have tested our method using a large real architectural floor plan in PDF format. The document image is 7150×6290 b/w pixels. After the connected component analysis and the polygonal approximation using Rosin and West's method [7], the vectorial DXF file is composed of more than 3000 polylines representing regions. In this floor plan several furniture symbols appear, an instance

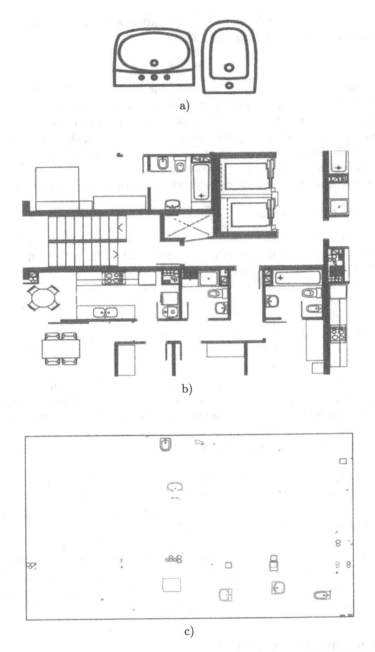

Fig. 5. Symbol spotting. (a) Models, a toilet sink and a bidet. (b) Original vectorial image. (c) Results of the spotting process. Regions labeled as bidet are drawn in red and regions labeled as toilet sink are drawn in green. Notice that most of false alarms are circular or squared simple regions.

Table 1. Number of regions and segments composing each model. Number of regions not belonging to the queried symbol labeled as positives. computation time (in seconds).

	Bath	Bidet	Toilet	Kitchen sink	Toilet sink	Table
Num. of regions	7	8	6	10	12	26
Num. of segments	101	110	112	144	157	351
Num. of false alarms	27	59	40	82	93	589
Time	0.288	0.267	0.347	0.344	0.568	1.331

of each symbol (arising from a different floor plan, but designed by the same architect so the symbol design is the same) has been taken as the model symbol to perform the queries. The whole vectorial file has been labeled to be used as ground truth. Although the results are preliminar, they are encouraging. We can appreciate in Figure 5 the result of spotting the toilet sink and the bidet symbols. Several false positive regions are also spotted since they are very simple circular or squared regions, similar to some of the details of the query symbols.

In Table 1 the numeric details concerning the performance of the proposed spotting method are reported. As we can appreciate, the more complex is the symbol, the more false alarms appear. This result is due to the fact that more entries in the hash table are concerned and consequently, more collisions are considered to form part of the queried symbol. In addition, the more complex is the symbol to search, the more computationally expensive is the method.

5   Conclusions, Discussion and Future Work

The presented method aims to index large documents contained in large databases by the graphical symbols appearing in it by means of a hash table. When indexing shapes by a hash table, neither a segmentation of the shapes is needed nor previous knowledge is required. Learning processes are not needed, in the sense that there is no stage needing several instances of the object to locate to learn its characteristics.

When using hash functions, obviously, if similar shapes have to fit the same table entry, we will have some collisions with shapes which are not similar but are projected to the same tuple. As we can appreciate in the results, these collision make that spotting approaches based on hash functions, usually result in a set of false alarms which are not a really significant problem since elevated recognition rates are not needed in these kind of applications.

Symbol Recognition techniques are a quite mature research topic in the Graphics Recognition field. The results of symbol recognition methods are usually close to the 100% recognition rate. In addition of the symbol descriptor, which usually is a long feature vector, symbol recognition methods combine these descriptors with learning and classification algorithms to cluster these descriptor feature vectors in symbol classes easy to classify.

These approaches involve having an a-priori knowledge about the symbols to recognize since they need enough examples to learn the descriptors distribution.

In addition, usually they are formulated only to recognize isolated symbols. In most of applications it is obvious that this is a correct approach, for instance, an OCR system has to recognize a finite set of shapes which can be "easily" segmented and which can have a lot of variability inside a given class. To reach acceptable results a learning step is necessary. However, if we think in an indexing framework able to spot symbols appearing in a large collection of documents and usually queried by example, we can see that the problem is not the same. Symbol Spotting techniques do not require a-priori knowledge nor learning step. As they have to lead with un-segmented data, they have to tackle with "recognition" and "segmentation" at the same time.

This paper is focused on the use of a very simple primitive description hashed to build an indexing table. The results show that when designing a spotting framework there is no need to use complex shape descriptors, a feature vector formed by few digits (we use only two integers to describe a given primitive), a hash function to cluster the primitives, and an adequate voting scheme is enough to locate the regions of interest of a document where a given symbol is likely to be found.

It is obvious that the use of more complex feature vectors can lead to less false alarms and better identification of correct symbols since they are usually more representative and have more discriminant power. But the use of more and more long feature vectors, also makes harder the definition of a proper hash function able to cluster similar shapes in the same table entries. We strongly believe that symbol spotting algorithms do not need complex shape descriptors but can perform quite well using simple features combined with an appropriate hash function and a posterior voting scheme.

This paper has presented a work in process, and some improvements are planned. Let us enumerate and briefly explain the most important research directions.

First, the performance of some other simple primitive description techniques have to be tested. In this paper we shown that indexing symbols does not need for complex or big feature vectors, but with only a couple of numbers promising results can be reached. Well-known shape descriptors as Hu's geometric moment invariants, Zernike moments, generic fourier descriptors, etc. could be used as well to generate simple key indexes to hash symbols.

Secondly, at this moment, the voting process to validate the hypothetic regions of interest, just accumulates votes in the locations where there is presence of the queried regions. At this moment we do not look if the structural organization of these primitives is consistent with the queried symbol. It is common to have some parts of technical drawings formed by a repetition of a simple region forming a given pattern, as for instance dashed lines are usually used to represent walls in floor plans. These cases are critical since if the queried symbol contains a single instance of these primitives, the zones formed by the repetition of the primitive will accumulate more votes than the zones containing the symbol itself. A voting scheme looking at the structural organization of indexed primitives is planned.

Finally, a performance evaluation framework for symbol spotting techniques has to be developed. Metrics to evaluate the performance of the recognition on the one hand and the localization on the other, are not easy to define. Many different cases has to be taken into account: false alarms, over and under segmentation of the regions of interest, missing symbols, etc.

Acknowledgments

This work has been partially supported by the Spanish projects TIN2006-15694-C02-02 and CONSOLIDER-INGENIO 2010 (CSD2007-00018).

References

1. Ballard, D.H.: Generalizing the Hough Transform to Detect Arbitrary Shapes. Pattern Recognition 13(2), 111–122 (1981)
2. Dosch, P., Lladós, J.: Vectorial Signatures for Symbol Discrimination. In: Lladós, J., Kwon, Y.B. (eds.) GREC 2003. LNCS, vol. 3088, pp. 154–165. Springer, Heidelberg (2004)
3. Lamdan, Y., Wolfson, H.J.: Geometric Hashing: A General and Efficient Model-Based Recognition Scheme. In: 2nd International Conference on Computer Vision, pp. 238–249 (1988)
4. Lockwood, E.H.: A Book of Curves. Cambridge University Press, Cambridge (1967)
5. Locteau, H., Adam, S., Trupin, E., Labiche, J., Héroux, P.: Symbol Spotting Using Full Visibility Graph Representation. In: 7th International Workshop on Graphics Recognition, pp. 49–50 (2007)
6. Müller, S., Rigoll, G.: Searching an Engineering Drawing Database for User-Specified Shapes. In: 5th International Conference on Document Analysis and Recognition, pp. 697–700 (1999)
7. Rosin, P.L., West, G.A.: Segmentation of Edges into Lines and Arcs. Image and Vision Computing 7(2), 109–114 (1989)
8. Rusiñol, M., Lladós, J.: Symbol Spotting in Technical Drawings Using Vectorial Signatures. In: Wenyin, L., Lladós, J. (eds.) GREC 2005. LNCS, vol. 3926, pp. 35–46. Springer, Heidelberg (2006)
9. Tabbone, S., Wendling, L., Zuwala, D.: A Hybrid Approach to Detect Graphical Symbols in Documents. In: Marinai, S., Dengel, A.R. (eds.) DAS 2004. LNCS, vol. 3163, pp. 342–353. Springer, Heidelberg (2004)
10. Tombre, K., Ah-Soon, C., Dosch, P., Masini, G., Tabbone, S.: Stable and Robust Vectorization: How to Make the Right Choices. In: Chhabra, A.K., Dori, D. (eds.) GREC 1999. LNCS, vol. 1941, pp. 3–18. Springer, Heidelberg (2000)
11. Zhang, D., Lu, G.: Review of Shape Representation and Description Techniques. Pattern recognition 37(1), 1–19 (2004)
12. Zuwala, D., Tabbone, S.: A Method for Symbol Spotting in Graphical Documents. In: Bunke, H., Spitz, A.L. (eds.) DAS 2006. LNCS, vol. 3872, pp. 518–528. Springer, Heidelberg (2006)

A System for Historic Document Image Indexing and Retrieval Based on XML Database Conforming to MPEG7 Standard

Wafa Maghrebi[1], Anis Borchani[1], Mohamed A. Khabou[2], and Adel M. Alimi[1]

[1] REsearch Group on Intelligent Machines (REGIM), University of Sfax
ENIS, DGE, BP. W-3038, Sfax, Tunisia
{wafa.maghrebi@fsegs.rnu.tn, anisborchani@yahoo.fr,
adel.alimi@ieee.org}
[2] Electrical and Computer Engineering Dept, University of West Florida
11000 University Parkway, Pensacola, FL 32514, USA
mkhabou@uwf.edu

Abstract. We present a novel image indexing and retrieval system based on object contour description. Extended curvature scale space (CSS) descriptors composed of both local and global features are used to represent and index concave and convex object shapes. These features are size, rotation, and translation invariant. The index is saved into an XML database conforming to the MPEG7 standard. Our system contains a graphical user interface that allows a user to search a database using either sample or user-drawn shapes. The system was tested using two image databases: the Tunisian National Library (TNL) database containing 430 color and gray-scale images of historic documents, mosaics, and artifacts; and the Squid dataset containing 1100 contour images of fish. Recall and precision rates of 94% and 87%, respectively, were achieved on the TNL database and 71% and 86% on the Squid database. Average response time to a query is about 2.55 sec on a 2.66 GHz Pentium-based computer with 256 Mbyte of RAM.

Keywords: Image indexing, image retrieval, eccentricity, circularity, curvature space descriptors, MPEG7 standard, XML database.

1 Introduction

Many image content retrieval systems were lately developed, tested, and some even made available online (e.g. Beretti [1, 2], QBIC [3], FourEyes [4], and Vindx [5, 6]). These systems use image content-based indexing methods to represent images. Image content can be represented using global features, local features, and/or by segmenting the images into "coherent" regions based on some similarity measure(s). For example, the QBIC system [3] uses global features such as texture and color to index images. Global features have some limitations in modeling perceptual aspects of shapes and usually perform poorly in the computation of similarity with partially occluded shapes. The FourEyes system [4] uses regional features to index an input image: An

W. Liu, J. Lladós, and J.-M. Ogier (Eds.): GREC 2007, LNCS 5046, pp. 114–125, 2008.
© Springer-Verlag Berlin Heidelberg 2008

image is first divided into small and equal square parts then, shape, texture and other local features are extracted from these squares. These local features are then used to index the whole image. The system developed by Berretti et al [1, 2] indexes objects in an input image based on their shape and offers the user the possibility of drawing a query to retrieve all images in the database that are similar to the drawn query. The Vindx system [5, 6] uses a database of 17th century paintings. The images are manually indexed based on the shapes they contain. This method is accurate but very time consuming especially when dealing with a huge data base of images. Similar to the system developed by Berretti et al, the Vindx system also offers the user the possibility of drawing a query to retrieve all images in the database that match the query to a certain degree.

Among all image indexing methods described in literature, only two methods conform to the MPEG7 standards of image contour indexing: the Zernike moment (ZM) descriptors [7] and the curvature scale space (CSS) descriptors. A good descriptor should be invariant to scale, translation, rotation, and affine transformation and should also be robust and tolerant of noise. The ZM descriptors are scale, translation and rotation invariant. However, they have the disadvantage of losing the important perceptual meaning of an image. They are generally used with binary images because they are very computationally expensive when applied to gray-scale or color images. The CSS descriptors were introduced by Mokhtarian et al [8, 9]. They are invariant to scale, translation, and rotation and have been shown to be very robust and tolerant of noise. The main disadvantage of the CSS descriptors is that they only represent the concave sections of a contour. Kopf et al [10] proposed an enhanced version of the CSS descriptors that remedied this problem and allowed them to represent both concave and convex shapes. They successfully used these enhanced descriptors to index videos segments [10]. The indexing and retrieval system we are proposing in this paper accepts a drawing query from the user through a graphical user interface, computes a set of global and CSS-based local features of the query, and retrieves all images from the database that contain similar shapes to the query.

This paper is organized as follows: Section 2 describes the CSS descriptors while section 3 describes the extended CSS descriptors used by our system. In section 4 we describe in details the architecture of our system. The performance evaluation of our system is presented in section 5 followed by the conclusion in section 6.

2 Curvature Scale Space Descriptors

Introduced by Mokhtarian et al [8, 9], the CSS descriptors register the concavities of a curve as it goes through successive filtering. The role of filtering is to smooth out the curve and gradually eliminate concavities of increasing size. More precisely, given a form described by its normalized planar contour curve

$$\Gamma(u) = \left\{ (x(u), y(u)) \big| u \in [0,1] \right\}, \tag{1}$$

the curvature at any point u is defined as the tangent angle to the curve and is computed as

$$k(u) = \frac{x_u(u)\,y_{uu}(u) - x_{uu}(u)\,y_u(u)}{\left(x_u(u)^2 + y_u(u)^2\right)^{\frac{3}{2}}}. \tag{2}$$

To compute its CSS descriptors, a curve is repeatedly smoothed out using a Gaussian kernel $g(u,\sigma)$. The contour of the filtered curve can be represented as

$$\Gamma(u) = \left\{(x(u,\sigma), y(u,\sigma)) \,\middle|\, u \in [0,1]\right\} \tag{3}$$

where, $x(u, \sigma)$ and $y(u, \sigma)$ represent the result of convolving $x(u)$ and $y(u)$ with $g(u, \sigma)$, respectively. The curvature $k(u, \sigma)$ of the smoothed out curve is represented as

$$k(u,\sigma) = \frac{x_u(u,\sigma)\,y_{uu}(u,\sigma) - x_{uu}(u,\sigma)\,y_u(u,\sigma)}{\left(x_u(u,\sigma)^2 + y_u(u,\sigma)^2\right)^{\frac{3}{2}}}. \tag{4}$$

The main idea behind CSS descriptors is to extract inflection points of a curve at different values of σ. As σ increases, the evolving shape of the curve becomes smoother and we notice a progressive disappearance of the concave parts of the shape until we end up with a completely convex form (Fig. 1). Using a curve's multi-scale representation, we can locate the points of inflection at each scale (i.e. points where $k(u, \sigma) = 0$). A graph, called CSS image, specifying the location u of these inflection points vs. the value of σ can be created (Fig. 1):

$$I(u, \sigma) = \{(u,\sigma) \mid k(u,\sigma) = 0\}. \tag{5}$$

Different peaks present in the CSS image correspond to the major concave segments of the shape. The maxima of the peaks are extracted and used to index the input shape.

Even though the CSS descriptors have the advantage of being invariant to scale, translation, and rotation, and are shown to be robust and tolerant of noise, they are inadequate to represent the convex segments of a shape. In addition, the CSS descriptors can be considered as local features and hence do not capture the global

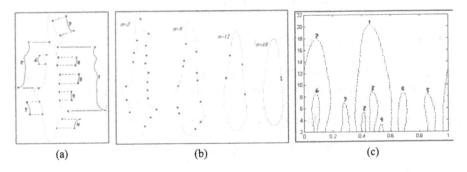

(a)   (b)   (c)

Fig. 1. Creation of CSS image (c) of a contour (a) as it goes through successive filtering (b)

shape of an image contour. The following section presents a remedy (the extended CSS descriptors) for these drawbacks.

3 Extended CSS Descriptors

Kopf et al [10] presented a solution to remedy the inability of the CSS descriptors to represent convex segments of a shape. The idea they proposed is to create a dual shape of the input shape where all convex segments are transformed to concave segments. The dual shape is created by mirroring the input shape with respect to the circle of minimum radius R that encloses the original shape (Fig. 2). More precisely, each point $(x(u),y(u))$ of the original shape is paired with a point $(x'(u),y'(u))$ of the dual shape such that the distance from $(x(u),y(u))$ to the circle is the same as that from $(x'(u),y'(u))$ to the circle. The coordinates of the circle's centre $O(M_x,M_y)$ are calculated as

$$M_x = \frac{1}{N}\sum_{u=1}^{N} x(u)$$
(6)

$$M_y = \frac{1}{N}\sum_{u=1}^{N} y(u).$$
(7)

The projected point $(x'(u),y'(u))$ is located at

$$x'(u) = \frac{2R - D_{x(u),y(y)}}{D_{x(u),y(y)}}(x(u) - M_x) + M_x.$$
(8)

$$y'(u) = \frac{2R - D_{x(u),y(y)}}{D_{x(u),y(y)}}(y(u) - M_y) + M_y$$
(9)

where, $D_{x(u),y(u)}$ is the distance between the circle's centre and the original shape pixel.

Since CSS descriptors as considered local features, we decided to use two extra global features to help in the indexing of shapes: circularity and eccentricity. Circularity is a measure of how close a shape is to a circle, which has the minimum circularity measure of 4π. Circularity is a simple (and hence fast) feature to compute. It is defined as

$$cir = \frac{P^2}{A}.$$
(10)

where, P is the perimeter of the shape and A is its area. Eccentricity is a global feature that measures how the contour points of a shape are scattered around its centroid. It is defined as

$$ecc = \sqrt{\frac{\lambda_{max}}{\lambda_{min}}}$$
(11)

where, λ_{max} and λ_{max} are the eigenvalues of the matrix B

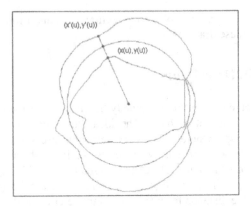

Fig. 2. Creating a dual shape with respect to an enclosing circle

$$B = \begin{bmatrix} \mu_{2,0} & \mu_{1,1} \\ \mu_{1,1} & \mu_{0,2} \end{bmatrix} \tag{12}$$

$\mu_{2,0}$, $\mu_{1,1}$, and $\mu_{0,2}$ are the central moments of the shape defined as

$$\mu_{p,q} = \sum_x \sum_y (x - \bar{x})^p (y - \bar{y})^q \tag{13}$$

with $\bar{x}$ and $\bar{y}$ representing the coordinates of the shape's centroid. Both circularity and eccentricity features are size, rotation and translation invariant.

The descriptors we used for image indexing in our system are thus a combination of four sets of features:

- Circularity feature (one global feature)
- Eccentricity feature (one global feature)
- CSS descriptors of original shape (n local features, where n depends on the object shape)
- CSS descriptors of dual shape (m local features, where m depends on the object shape)

4 System Description

A block diagram of our system showing its major components is shown in Fig. 3. The system takes an input shape, extracts its global and local features as described in the previous section, uses these features/descriptors to index it, and provides the user with a graphical user interface to query the database of indexed shapes. The index of a shape is saved in an XML database conforming to the MPEG7 standard (Fig. 4). The database can be interrogated using a drawing query. The query is formalized in

XQUERY language. The XQUERY processor compiles and executes the query and returns all relevant XML documents and consequently all pertinent images that are similar to the query shape.

Since an image can be composed of more than one simple shape, each image in the database is identified by its name and the list of shapes that it contains. Each shape is indexed using its extended CSS descriptors described in section 3. Fig. 4 below shows an example of an XML description of a sample image containing two objects. The matching of a query to entries in the database is done in two steps:

Step 1: only shapes in the database that have global features "close" (difference in circularity and eccentricity measures less than 12.5% and 25%, respectively) to those of the query are considered potential matches; all other shapes in the database are ignored, thus quickly reducing the pool of potential matches and speeding up the retrieval process.

Step 2: The similarity measures (Euclidian distance) between the CSS descriptors of the query and each of the potential matches from step 1 are computed and the closest entries are returned.

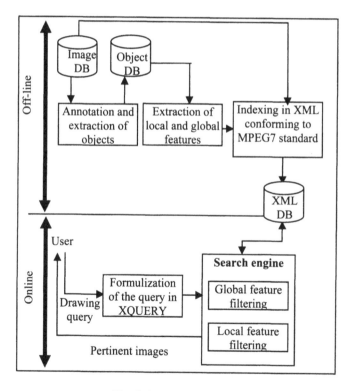

Fig. 3. System architecture

```xml
<?xml version="1.0" encoding="iso-8859-1" ?>
- <Mpeg7 xmlns="urn:mpeg:mpeg7:schema:2001" xmlns:xsi="http://www.w3.org/2001/XMLSchema-instance"
    xsi:schemaLocation="urn:mpeg:mpeg7:schema:2001 schema/Mpeg7-2001.xsd">
  - <DescriptionUnit xsi:type="DescriptorCollectionType">
    - <Descriptor xsi:type="ContourShapeType">
        <imgage_name name="../images/typ11/kk11.jpg" />
      - <objet>
          <classe_name name="animal" />
          <object_descrip name="gazelle" />
          <GlobalCurvature>1.3710531617728288</GlobalCurvature>
          <PrototypeCurvature>125.36538727046803</PrototypeCurvature>
          <HighestPeakY />
          <Peak peakX="0.485" peakY="25.0" />
          <Peak peakX="0.815" peakY="12.8" />
          <Peak peakX="0.65" peakY="12.3" />
          <Peak peakX="0.33" peakY="11.5" />
          <Peak peakX="0.05" peakY="10.1" />
          <Peak peakX="0.815" peakY="5.8" />
          <Peak peakX="0.195" peakY="5.8" />
          <Peak1 peakX="0.48" peakY="26.2" />
          <Peak1 peakX="0.76" peakY="14.4" />
          <Peak1 peakX="0.66" peakY="11.9" />
          <Peak1 peakX="0.24" peakY="9.7" />
          <Peak1 peakX="0.315" peakY="9.3" />
          <Peak1 peakX="0.275" peakY="8.2" />
          <Peak1 peakX="0.355" peakY="8.0" />
          <Peak1 peakX="0.95" peakY="6.4" />
          <Peak1 peakX="0.955" peakY="4.6" />
          <Peak1 peakX="0.995" peakY="4.3" />
        </objet>
      - <objet>
          <classe_name name="personne" />
          <object_descrip name="chasseur" />
          <GlobalCurvature>1.9995490113479923</GlobalCurvature>
          <PrototypeCurvature>99.10659431496987</PrototypeCurvature>
          <HighestPeakY />
          <Peak peakX="0.29" peakY="18.5" />
          <Peak peakX="0.825" peakY="15.5" />
          <Peak peakX="0.125" peakY="9.2" />
          <Peak peakX="0.0" peakY="7.9" />
          <Peak peakX="0.715" peakY="4.9" />
          <Peak peakX="0.385" peakY="4.4" />
          <Peak1 peakX="0.255" peakY="13.2" />
          <Peak1 peakX="0.585" peakY="12.3" />
          <Peak1 peakX="0.69" peakY="11.4" />
          <Peak1 peakX="0.175" peakY="10.5" />
          <Peak1 peakX="0.645" peakY="10.1" />
          <Peak1 peakX="0.975" peakY="9.3" />
          <Peak1 peakX="0.47" peakY="7.4" />
          <Peak1 peakX="0.0" peakY="5.3" />
          <Peak1 peakX="0.995" peakY="5.2" />
        </objet>
      </Descriptor>
    </DescriptionUnit>
  </Mpeg7>
```

Object1

Object2

Fig. 4. Example of XML image description conforming to MPEG7 standard

5 System Evaluation

We tested our indexing and retrieval system using two databases: the Squid database and the Tunisian National Library (TNL) database. The Squid database contains 1100 contour images of fish (Fig. 5). It was used by many researchers [8, 9] to test their indexing and retrieval systems and hence allows us to objectively compare the performance of our system to others. The TNL database [11] consists of 430 color and gray-scale images of ancient documents, mosaics, and artifacts of important historic value (some date back to the second century CE). This database was used by many researchers to test various image processing, indexing, and retrieval techniques [12, 13, 14]. The images are very rich with complex content consisting of many objects of different shape, color, size, and texture (Fig. 6). This makes the automatic extraction of meaningful objects from these images very challenging, if not impossible. Meaningful objects from images in the TNL database are extracted by outlining their contours using an annotation module [15] consists on a dynamic web site which use the user perception (Fig. 7).

The use of meta-data in our system made the size of the databases small (872 Kbyte for the Squid database and 378 Kbyte for the TNL database). The system was

Fig. 5. Sample contour images from the SQUID database

Fig. 6. Sample images from the TNL database

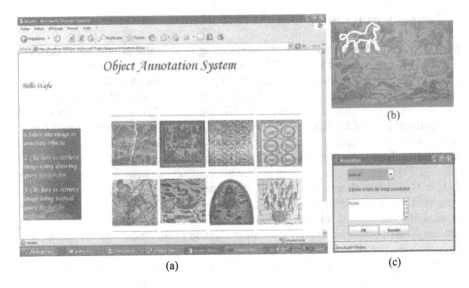

Fig. 7. The TNL database object annotation graphical user interface showing (a) the welcome screen, (b) a sample object contour extraction, and (c) its semantic textual description (not used in our system).

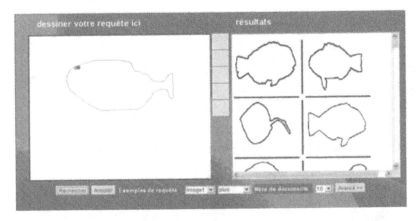

Fig. 8. Example of a drawing query from the Squid database and the top four pertinent images returned by the system

implemented in Java language using client-server architecture and threads. The simplicity in communication between the java language and XML made the system very fast at finding pertinent XML documents and consequently relevant images in the database. The system's query graphical user interface has a lot of flexibility built into it. For example, the user can specify a sample image from the database as a query or he/she can draw the shape of the query using a computer stylus and pad. Fig. 8 and 9 show two examples of drawing queries and the top four pertinent images returned by the system from Squid database and the TNL database, respectively.

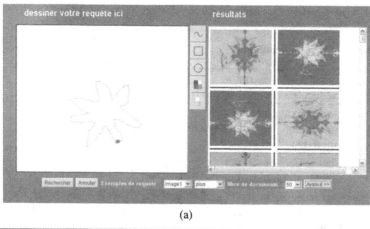

(a)

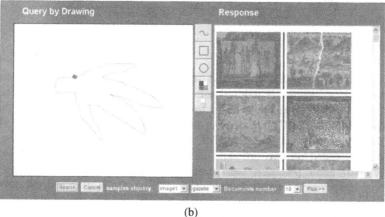

(b)

Fig. 9. Examples of drawing queries from the TNL database and the top four pertinent images returned by the system. (a) the first query resembles the shape of a star and (b) the second query resembles the shape of a person (rotated 90 deg counterclockwise).

The evaluation of our system is based on its recall rate (R), precision rate (P) and average response time to a query. We conducted two sets of experiments: in the first set we used only the CSS descriptors to index the images in the databases and in the second set of experiments we used the extended CSS descriptors. The evaluation results are shown in Table 1. The response times reported in Table 1 are obtained using a 2.66 GHz Pentium-based computer with 256 Mbyte of RAM. As can be seen in Table 1, the extended CSS descriptors helped improve the performance of our system in terms of recall and precision rates and average response time. The use of the global features in the first step of the database interrogation helped eliminate many entries that could not be a match for the query and hence reduced the average response time by about 52%. They also helped increase the recall and precision rates as they added global information about the shapes that could not be captured by the standard CSS descriptors.

Table 1. System performance evaluation using classic CSS and extended CSS descriptors

Database	Features	P (%)	R (%)	Average response time (s)		
				Index	Retrieve	Total
Squid	Classic CSS descriptor (Sequential retrieval method)	75	67	2.677	3.967	6.644
	Our indexing method (Integration of XQUERY language)	86	71	2.136	0.282	2.418
TNL	Classic CSS descriptor (Sequential retrieval method)	66	83	2.187	2.296	4.483
	Our indexing method (Integration of XQUERY language)	87	94	2.396	0.292	2.688

6 Conclusion

We presented an image indexing and retrieval system based on object contour description. Input images are indexed using global (circularity and eccentricity) and local features (CSS descriptors). The images are indexed in an XML database conforming to the MPEG7 standard. Our meta-data permits a standard representation of JPEG images and an indexation of images containing multiple shapes. Our approach was tested on two different databases: the NLT database and the Squid database with good precision and recall rates. The use of the extended CSS descriptors improved the recall and precision rates of the system and cut down the query response time by more than 50%. Future work includes the addition of global and local features (e.g. color and texture) to increase the recall and precision rates.

Acknowledgements

This research was funded by the Tunisian-Egyptian project "Digitization and Valorization of Arabic Cultural Patrimony" and the Tunisian Ministry of Higher Education and Scientific Research. We would like to thank the National Library of Tunisia [11] for giving us access to their large image database of historic Arabic documents.

References

1. Berretti, S., Del Bimbo, A., Pala, P.: Retrieval by Shape Similarity with Perceptual Distance and Effective Indexing. IEEE Transactions on Multimedia 2, 225–239 (2000)
2. Berretti, S., Del Bimbo, A., Pala, P.: Efficient Matching and Indexing of Graph Models in Content-Based Retrieval. IEEE Transactions on Pattern Analysis and Machine Intelligence 23, 1089–1105 (2001)

3. Faloutsos, C., Barber, R., Flickner, M., Flickner, J., Niblack, W., Petkovic, D., Equitz, W.: Efficient and Effective Querying by Image Content. J. Intelligent Information Systems 3, 231–262 (1994)

4. Pentland, A., Picard, R.W., Sclaroff, S.: Photobook: Tools for Content-Based Manipulation of Image Databases. In: SPIE Proc. Storage and Retrieval for Image and Video Databases II, vol. 2185, pp. 34–47 (1994)

5. Schomaker, L., Vuurpijl, L., Deleau, E.: New Use for the Pen: Outline-Based Image Queries. In: Proc. of the 5th International Conference on Document Analysis and Recognition (ICDAR), Piscataway (NJ), pp. 293–296 (1999)

6. Vuupijl, L., Shomaker, L., Broek, E.: Vind(x): Using the User Through Cooperative Annotation. In: Proc. of the 8th International Workshop on Frontiers in Handwriting Recognition (IWFHR.8), pp. 221–225 (2002)

7. Teague, M.R.: Image Analysis Via the General Theory of Moments. Optical Soc. Am. 70, 920–930 (1980)

8. Mokhtarian, F., Abbasi, S., Kittler, J.: Efficient and Robust Retrieval By Shape Through Curvature Scale Space. In: Proc. First International Workshop on Image Databases and Multimedia Search, pp. 35–42 (1996)

9. Mokhtarian, F., Abbasi, S., Kittler, J.: Robust and Efficient Shape Indexing Through Curvature Scale Space. In: Proc. British Machine Vision Conference, pp. 53–62 (1996)

10. Kopf, S., Haenselmann, T., Effelsberg, W.: Shape-Based Posture Recognition in Videos. In: Proc. Electronic Image, vol. 5682, pp. 114–124 (2005)

11. National Library of Tunisia, http://www.bibliotheque.nat.tn

12. Alimi, A.M.: Evolutionary Computation for the Recognition of On-Line Cursive Handwriting. IETE Journal of Research, Special Issue on Evolutionary Computation in Engineering Sciences 48, 385–396 (2002)

13. Boussellaa, W., Zahour, A., Alimi, A.M.: A Methodology for the Separation of Foreground/Background in Arabic Historical Manuscripts using Hybrid Methods. In: Proc. 22nd Annual Symposium on Applied Computing, Document Engineering Track (2007)

14. Zaghden, N., Charfi, M., Alimi, A.M.: Optical Font Recognition Based on Global Texture Analysis. In: Proc. International Conference on Machine Intelligence, pp. 712–717 (2005)

15. Maghrebi, W., Khabou, M.A., Alimi, A.M.: A System for Indexing and Retrieving Historical Arabic Documents Based on Fourier Descriptors. In: Proc. International Conference on Machine Intelligence, pp. 701–704 (2005)

An Ancient Graphic Documents Indexing Method Based on Spatial Similarity

Ali Karray[1,2], Jean-Marc Ogier[1], Slim Kanoun[2], and Mohamed Adel Alimi[2]

[1] Faculty of Sciences and of Sciences and Technology, L3i Research Laboratory,
Université de La Rochelle, Avenue Michel Crépeau,
17042 La Rochelle Cédex 1, France
ali.karray@ieee.org, jean-marc.ogier@univ-lr.fr
http://www.univ-larochelle.fr
[2] National Engineering school of Sfax, REGIM Research Group, University of Sfax
Route Soukra, 3038 Sfax, Tunisia
slim.kanoun@enis.rnu.tn, adel.alimi@ieee.org
http://www.enis.rnu.tn

Abstract. Content based image retrieval using spatial image content
(i.e. using multiple regions and their spatial relationships) is still an
open problem which has received considerable attention in literature. In
this paper we introduce a new representation of image based on the most
general spatial image content representation that is Attributed Relation
Graphs (ARG) representation and also a new method of image indexation. Like all CBIR systems, the one proposed here has two components:
a segmentation component and a matching component using a novel inexact graph matching algorithm. We tested our work in lettrines image
but its also valid in general image.

Keywords: image database, similarity retrieval, attributed relation
graph, ancient documents indexing.

1 Introduction

This paper deals with a project, called MADONNE, aimed at managing various
resources of international inheritance especially books, images collections and
iconographic documents. These numerous documents contain huge amount of
data and decay gradually. One of the goals of MADONNE project is to develop
a set of tools allowing to extract all information as automatically as possible from
digitized ancient images and to index them in order to develop Content-Based
Image Retrieval. This rest of the paper is organized as follows: A short presentation of lettrine is given in section 2. The proposed methodology is presented
in section3. In section 4 we discuss the experimental results and the conclusion
is found in section 5.

2 Lettrine Description

Lettrines are graphical objects which contain a lot of information. A quick see-
ing of several lettrines shows different common points between all those (Fig. 1).

W. Liu, J. Lladós, and J.-M. Ogier (Eds.): GREC 2007, LNCS 5046, pp. 126–134, 2008.

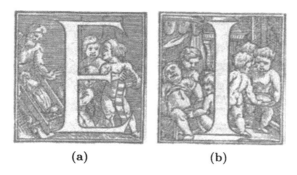

(a)   (b)

Fig. 1. Example of lettrine

For instance, each of them has a framework, parallel groups of lines representing texture, a letter, and the rest, i.e. the illustration itself, which is generally composed of small curves. In our context of Content Based Information Retrieval, the aim is to try to compute an ARG representation allowing to recognize a lettrine, or a part of a lettrine. Many works concerning lettrines can be found in the literature, from an historical point of view [1], [2]. However, from an image analysis dimension, the works concerning the analysis and the indexing of such "graphic objects" are very rare. In our context, our aim is to develop some Content Based Image Retrieval techniques. The history allow us to learn that lettrines are printed on documents using wood plug, and that the same plug may have been re-used on the document until the end of its printing, sometimes the plug can be used for creating other documents. Wood is far from being a matter stable during its use, so it is not rare, when we search about similar lettrine in various documents or between the first and the last page of the same document, to notice that the general shape of the plug changes by an increasing or a decreasing of the thickness of the features.

3 Proposed Method

Nowadays, the literature is very rich in the domain of image retrieval systems. However, most of the classical techniques used in CBIR are not usable in our contest do to the specificity of the features of lettrines. Unfortunately, there are few literatures relating to the retrieval of ancient graphic document [3],[4]. Actually we focus on the spatial organization of the different layers of information extracted thanks to the segmentation stage. To compare lettrine, we use a specific segmentation process developed specially for lettrine [5]. For each lettrine, the segmentation stage provides three layers i.e. homogeneous, texture and contour layers. In this paper, we extract a spatial signature from lettrine based on the well known ARG representation. The overall process of CBIR of lettrines is illustrated in Fig. 2.

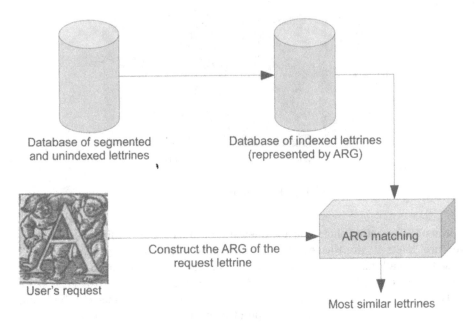

Fig. 2. Lettrine retrieval system

3.1   Lettrine Segmentation

In this paper, our object is to extract a signature from lettrine, in order to permit to find a lettrine among wide set of lettrines. The first stage consists in segmenting the lettrine [5] in different layers of information: different texture layers, uniform areas, outlines. This segmentation stage is a top-down technique based on a strategy inspired from visual perception principles that is summarized in Fig. 3.

As a consequence, for instance, this stage can provide a layer corresponding to textures, a layer corresponding to homogeneous areas, a layer corresponding to outlines. Fig. 4 and Fig. 5 illustrate an example of the homogeneous and textured areas computed from a lettrine image.

The detail concerning the segmentation process that has been implemented was presented in a paper of workshop of GREC 2005 [5].

3.2   Attribute Relational Graphs Representation of Lettrines

Attributed Relational Graph , formally defined as a triple $G=(V,E,\mu,\nu)$, where V is the node set, E is the edge set, μ is the attribute set attached to each node, and ν is the attribute set attached to each edge. In our work, each lettrine is modulated by three ARGs. Every ARG represent one of the three layers obtained after the segmentation process of the corresponding lettrine (See Fig. 6). Nodes correspond to regions and arcs correspond to relationships between regions. There is no overlapping between the regions, thus, the arcs of ARG are

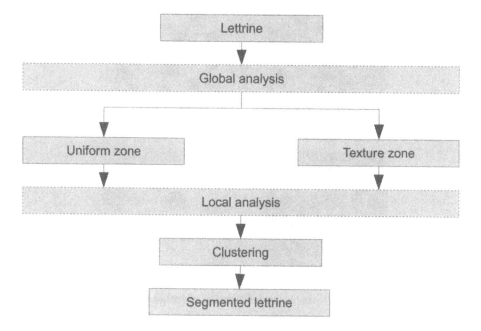

Fig. 3. General schema of lettrine segmentation process

Fig. 4. Original lettrine and its homogeneous areas

undirected. Both nodes and arcs are labeled by the attribute values of the region properties and the relationships properties, respectively. In this work individual regions are described by 2 attributes, namely size and shape. The size computed as the size of the area of a region. The shape descriptor computed as the first three moments of Hu [6]. The chosen shape descriptor present two advantages. Firstly, the shape descriptor is global according to the uncertainty related to the preliminary phase of segmentation. Secondly, the use of only three attributes decrease considerably the algorithm complexity and in consequence the execution time. Spatial relationships between regions are described by 2 attributes, namely distance, computed as the distance between the gravity centers of the two regions. The other is the relative angle between the two regions. In the work described here, all features are automatically computed. It is easy to add more

Fig. 5. Textured background extraction of rather simple lettrine

features as region or relationship attributes. In any case, our proposed method can handle any set of features.

3.3 ARG Matching Algorithm

When image image is described by ARGs, the problem of finding similar image is transformed of a problem of graph matching [7,8,9,10,11,12]. Given two ARGs, the object is to find not only a sequence of error transformations, but also finding the one that leads to the smallest matching error. The smallest matching error represent the distance between the two ARGs. Thus, the smaller the matching error between two ARGs the most similar they are. The A* algorithm [7] guaranteed to find the optimal solution but require exponential time and space due to the NP-completeness. Thus, the A* algorithm is not adapted to the image retrieval and indexing problems due to the high time needed for the execution of this algorithm to find the nearest lettrines that match with the requested lettrine. To resolve this problem, inspired from the work [8],we propose a novel graph matching algorithm that reduces the time execution and gives good results at once. The idea is to match only between the most similar nodes among the two ARGs. So, we introduce a precision K to reduce the complexity of the problem by diminishing the space solution. The algorithm creates the state-space tree. For each level of the tree, node represents a matching of a pair of nodes from the input ARGs. A node is developed at the lower level only if the node of the model ARG belongs to the set of the K-nearest node of the node of the request ARG (see Fig. 7).

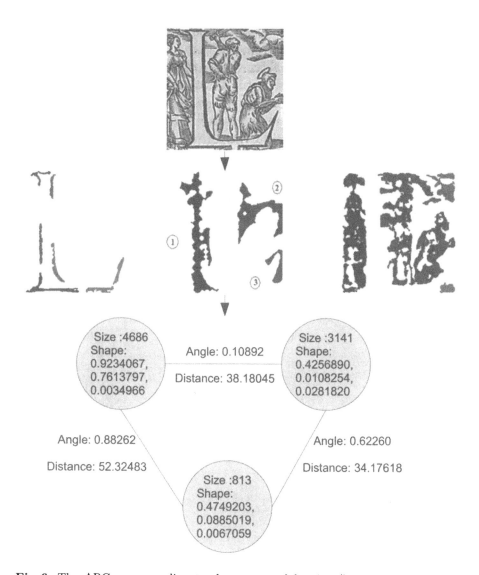

Fig. 6. The ARG corresponding to the segmented lettrine (layer corresponding to homogeneous areas)

We adapt this ARG matching algorithm for ARG of lettrines described in the previous paragraph. The distance between two lettrines L_1, L_2 is defined by: $D(L_1, L_2) = \sum_{i=1}^{3} d_i$ .With d_i is the global error or the error-correcting subgraph isomorphism between ARG representations of the i layer of L_1 and the i layer of L_2.

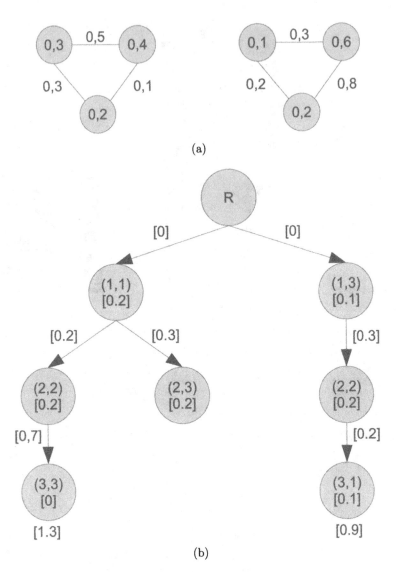

Fig. 7. (a) Input graph and model graph and (b) the space state tree created from the input graph and the output graph of (a) with a precision 2

4   Experimental Result

The experiment was performed by taking into account the lettrines from Centre d'Etudes Supérieures de la Rennaissance de Tours [13]. In this experience, each lettrine was segmented into 3 information layers. After that, each lettrine was represented by 3 ARGs. The corresponding graphs have between 12 and 2 nodes. When a lettrine in the index is queried, it ranks first for the matching

Query   First nearest neighbor

Second nearest neighbor Third nearest neighbor

Fig. 8. Example of a query lettrine (top left) and of its 3 nearest neighbors

score. Also, the other obtained lettrines are close to the query. This strategy is currently evaluated in terms of recall/precision criterion. However, our first experimentation highlights the good quality of the retrieval process. Fig. 8 illustrate the result of the search for a lettrines request. The first and the second lettrines result are very similar to the lettrines request.

5 Conclusion

This paper proposes a novel spatial signature of lettrines with ARGs, in order to build a content based image retrieval system, in the context of ancient document valorization. To compute distance between spatial signatures of lettrines, we propose a novel ARG matching algorithm that reduces the time execution and gives good results at once. The results obtained thus far show that our method can match ARG adequately.

This strategy of graphic image description is currently experimented on a wide set of lettrines that has been provided by one of our partner in the Madonne's project, by using a research engine. Actually, we work to integrate an ARG clustering algorithm; this algorithm will accelerate considerably the retrieval process. We also study to integrate more attributes in ARG to improve the representation of lettrine. These improvements will be the subject of future works.

References

1. Masse de Données issues de la Numérisation du patrimoine,
   `http://l3iexp.univ-lr.fr/madonne`
2. Harris, C., Stephens, M.: A combined corner and edge detector. In: 4th Alvey Vision Conference, pp. 147–151 (1988)
3. Schmid, C., Mohr, R.: Local Grayvalue Invariants for Image Retrieval. IEEE Transactions on Pattern Analysis and Machine Intelligence 19(5), 530–535 (1997)
4. Pareti, R., Uttama, S., Salmon, J., Ogier, J., Tabbone, S., Wendling, L., Vincent, N.: On defining signatures for the retrieval and the classification of graphical dropcaps. In: Conference on Document Image Analysis for Libraries (DIAL), pp. 220–231 (2006)
5. Uttama, S., Ogier, J., Loonis, P.: Top-down segmentation of ancient graphical drop caps: Lettrines. In: Workshop on Graphics Recognition (GREC), pp. 87–96 (2005)
6. Hu, M.K.: Visual pattern Recognition by Moment Invariants. IRE Transactions on Information Theory IT-8, 179–187 (1962)
7. Tsai, W.-H., Fu, K.S.: Error-Correcting Isomorphisms of Attributed Relation Graphs for Pattern Recognition. IEEE Trans. Systems, Man, and Cybernetics 9, 757–768 (1979)
8. Hlaoui, A., Wang, S.: A New Algorithm for Inexact Graph Matching. In: ICPR, vol. (4), pp. 180–183 (2002)
9. Ullmann, J.R.: An Algorithm For Subgraph Isomorphism. J. ACM 23(1), 31–42 (1976)
10. Sanfeliu, A., Fu, K.S.: A distance measure between attributed relational graphs for pattern recognition. IEEE Trans. Syst., Man Cybern SMC-13 (1983)
11. Christmas, W.J., Kittler, J., Petrou, M.: Structural Matching in Computer Vision Using Probabilistic Relaxation. IEEE Trans. Pattern Analysis and Machine Intelligence 17(8), 749–764 (1995)
12. Gori, M., Maggini, M., Sarti, L.: Exact and Approximate Graph Matching Using Random Walks. IEEE Trans. Pattern Analysis and Machine Intelligence 27(7), 1100–1111 (2005)
13. Les Bibliothèques virtuelles humanistes, `http://www.bvh.univ-tours.fr`

A Fast CBIR System of Old Ornamental Letter

Mathieu Delalandre[1], Jean-Marc Ogier[2], and Josep Lladós[1]

[1] CVC, Barcelona, Spain
{mathieu,josep}@cvc.uab.es
[2] L3i, La Rochelle, France
jean-marc.ogier@univ-lr.fr

Abstract. This paper deals with the CBIR of old printed graphics (of XVI° and XVII° centuries) like the headpieces, the pictures and the ornamental letters. These graphical parts are previously segmented from digitized old books in order to constitute image databases for the historians. Today, large databases exist and involves to use automatic retrieval tools able to process large amounts of data. For this purpose, we have developed a fast retrieval system based on a Run Length Encoding (RLE) of images. We use the RLE in an image comparison algorithm using two steps: one of image centering and then a distance computation. Our centering step allows to solve the shifting problems usually met between the segmented images. We present experiments and results about our system concerning the processing time and the retrieval precision.

1 Introduction

This paper deals with the topic of CBIR[1] applied to the document images. During the last years many works have been done for the retrieval of journals, forms, maps, drawings, musical scores, etc. In this paper we are interested on a new retrieval application: the one of old printed graphics. Since the Digital Libraries development in the years 90's numerous works of digitization of historical collections have been done. These old books are composed of text but also of various graphical parts like the headpieces, the pictures and the ornamental letters. From the whole digitized pages these graphical parts are segmented (in a manual or automatic way [1]) in order to constitute image databases for the historians. Some example of famous databases are the Ornamental Letters DataBase[2] and the International Bank of Printers' Ornaments[3].

These databases are composed today of thousands of images that involves to use automatic retrieval tools able to process large amounts of data. Past works have been already proposed on this topic [2] [3] [4] [5]. In [2] the authors propose a system to retrieve similar illustrations composed of strokes using orientation radiograms. The radiograms are computed at a $\frac{\pi}{4}$ interval and used as image signatures for the retrieval.

---

[1] Content Based Image Retrieval.

[2] http://www.bvh.univ-tours.fr/oldb.asp (10 000 images)

[3] http://www2.unil.ch/BCUTodai/app (3000 images)

W. Liu, J. Lladós, and J.-M. Ogier (Eds.): GREC 2007, LNCS 5046, pp. 135–144, 2008.

[3] looks for common sub-parts in the old figures. The segmentation is done by a local computation of the Hausdorff distance using a sliding window. A voting algorithm manages the location process. In [4] a statistical scheme is proposed to retrieve the ornamental letters according a style criterion. Patterns (ie. gray-level configuration of a pixel neighborhood) are extracted from the image and next ranked to build a curve descriptor. The retrieval is done by a curve matching algorithm. In [5] the layout of the ornamental letters is used for the retrieval. A segmentation process extracts the textured and uniform regions from images. Minimum Spanning Tree (MST) are next computed from these regions and used as image signatures for the retrieval.

All the existing systems are dedicated to a specific kind of retrieval (style, layout, etc.). In this paper we focuss on the application of wood plug tracking illustrated in the Figure 1. Indeed, from the 16th to the 17th centuries the plugs, used to print the graphics in the old books, were mainly in wood. These wood plugs could be reused to print several books, be exchanged between printing houses, or duplicated in the case of damage. So to retrieve, in automatic way, printings produced by a same wood plug could be very useful for the historian people. It could solve some dating problems of books as soon as to highlight the existing relations between the printing houses.

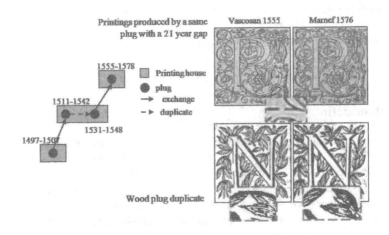

Fig. 1. Wood plug tracking

This retrieval application is an image comparison task [6]. Indeed, the images produced by a same wood plug present similarities at pixel level. However, it raises a complexity problem. First in regard to the amount of data, building a comparison index between thousands of image could require days of computation. Next in regard to the copyright aspects. The images belong to specific Digital Libraries or private collections; in order to allow crossed queries between these databases real-time retrieval processes are required. In regard to these specificities we have developed a system to perform a fast retrieval of images. This one uses two main steps as shown in the Figure 2: one

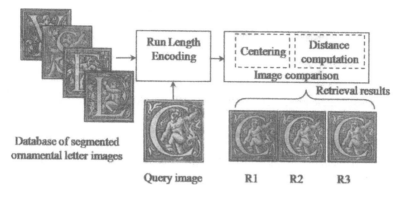

Fig. 2. Our system

of Run Length Encoding of images and the other of image comparison (centering and distance computation). We will present both of them in the next sections 2 and 3. Then, in the section 4 we will present some experiments and results about our system. At last, we will conclude and will give perspectives in the section 5.

2   Run-Length Encoding

Our purpose is to retrieve, in a fast way, the images similar to a query in a large database. To do it it is necessary to decrease the processing time of the retrieval system. One way is to exploit a special hardware architecture [7] like the pipe-line processor or the mesh-connected computer. It is an expensive solution which makes difficult the spreading of the developed systems. The other way is to use a compressed data structure to represent the images in order to decrease their handling times. It exists few works dealing with this topic, some examples are the ones of [8] and [9]. In [8] the authors use a connected-component based representation to perform a fast retrieval of images based on their layout. The system of [9] employs a contour based representation of images in order to perform fast neighboring operations like the erosion, the dilatation or the skeletonization.

For our problematic we have considered the run based representation [10]. The run is well known data structure. As explained in the Definition 1, it encodes successive pixels of same intensity into a single object. The conversion of a raster based representation to

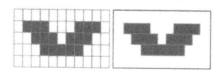

Fig. 3. Example of Run-Length Encoding (RLE)

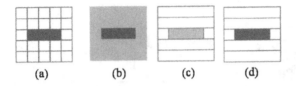

Fig. 4. RLE types (a) raster (b) foreground RLE (c) background RLE (d) foreground/background RLE

a run based representation is called Run-Length Encoding (RLE) in the literature. The Figure 3 gives an example of RLE.

Definition 1. *A run is maximal sequence of pixels defined by o the orientation (either vertical or horizontal), (x,y) the starting point, l the length and c its color.*

In the past the run based systems have been widely used for the document recognition. The first system has been proposed by [11]. Then, several ones have been developed during the years 90's [10] and nowadays they are used for various applications: handwriting recognition [12], symbol recognition [13], structured document segmentation [14], etc. Concerning the use of run for the complexity only the following systems have been proposed up to day: [15] for the contour extraction, [16] for the image comparison and [17] for the morphology operations. Using the RLE, the final sizes of images are reduced. The compression rate defines itself as the ratio between the number of run and the one of pixel. Next, the algorithms have to work on the RLE space to perform faster operations. However, different criteria can influence the compression rate. First, the RLE have to be extracted from binary images and the previous step of color quantization will have a great impact on the encoding results. Next, the RLE can be applied to the foreground and/or to the background of images. In this way, three encodings can be considered as presented in the Figure 4 and each of them will give a different compression result. Finally the RLE can be also done in different ways: vertical, horizontal, zig-zag or others. According to the content of an image this will change the result of the RLE.

For our application we have chosen to use the foreground/background encoding. This seems more adapted for the comparison of ornamental letter images where the objects (letter, character, etc.) appear as soon as on the foreground and the background of the images. Next, we perform this encoding from gray-level images by applying a binarization step with a fixed threshold. Indeed, the processed images by our system have been previously cleaned (lighting correction, filtering, etc.) by the digitalization platforms of old books. Finally, due to the property of visual symmetry of the ornamental letter images (on on both sides of letters) we have chosen to apply a vertical RLE.

3   RLE Based Image Comparison

In this section we present our image comparison algorithm based on the RLE. As presented in the introduction part, the images processed by our system have been

Fig. 5. Vertical and horizontal projection histograms

previously segmented from old books. It introduces shifting problems between the images which makes more harder their comparison. In order to solve this problem our comparison algorithm uses two steps: one of image centering and one other of distance computation. We present each of them in what follows.

Our image centering step exploits horizontal and vertical projection histograms of pixels. The Figure 5 gives examples of such projection histograms. These ones are built during the indexing step (with the RLE) from the black pixels of images. We use the black pixels because the segmentation process adds background areas around the ornamental letter. The centering parameters deduce themselves by the foreground analysis.

We center next two images together by computing the distances between their histograms (vertical and then horizontal). To do it we have chosen the distance presented in the Equation 1. It is a weighted distance between two histograms g and h. We have chosen the weighting because it increases the robustness of the comparison when strong amplitude variations appear in the histograms [18]. Our images could be of different sizes, so we compute our weighted distances using an offset in pixel (from 0 to $l - k$). k and l are the lengths of g and h with l the higher value (h is the largest histogram). Considering two images to center, g and h are chosen when staring the centering step by finding the minimum/maximum widths and heights. The delta to use, ether x or ether y, corresponds to the found minimum weighted distance among the computed offsets (from 0 to $l - k$). The previous Figure 5 corresponds to the deltas $d_x = 1$ and $d_y = 4$.

$$\begin{matrix} g_{1,2,..,k} \\ h_{1,2,...,l} \\ k \le l \end{matrix} \qquad delta = \min \left(\bigcup_{j=0}^{l-k} \sum_{i=1}^{k} \frac{|(h_i - g_{i+j})|}{h_i} \right) \qquad (1)$$

In a second step we compute a distance between our images. This distance is obtained by a "simple" comparison pixel to pixel [6]. However, to compute it our comparison works obviously from the RLE representation of images. We present here the algorithm that we use[4].

---

[4] Presentation based on the LATEX package Pseudocode [19].

Pseudo-algorithm 3.1: DISTANCE(i_1, i_2, d_x, d_y)

$s \leftarrow 0$
$x_1 \leftarrow x_2 \leftarrow 0$
$a_1 \leftarrow a_2 \leftarrow 0$
for each line $L1$ at y of i_1 and $L2$ at $y + d_y$ of i_2

$$
\mathbf{do} \begin{cases}
p_1 \leftarrow \text{NEXT}(L1) \\
x_1 + = p_1.length \\
p_2 \leftarrow \text{NEXT}(L2) \\
x_2 + = p_2.length \\
\mathbf{while}\ (p_1 \neq end) \vee (p_2 \neq end) \\
\qquad \mathbf{do} \begin{cases}
\mathbf{while}\ x_1 \geq (x_2 + d_x) \\
\quad \mathbf{do} \begin{cases}
\mathbf{if}\ p_2.color = p_1.color \\
\quad \mathbf{then}\ s + = p_2.lenght - a_2 \\
p_2 \leftarrow \text{NEXT}(L2) \\
x_2 + = p_2.lenght \\
a_1 + = p_2.lenght - a_2 \\
a_2 = 0
\end{cases} \\
\mathbf{while}\ (x_2 + d_x) \geq x_1 \\
\quad \mathbf{do} \begin{cases}
\mathbf{if}\ p_1.color = p_2.color \\
\quad \mathbf{then}\ s + = p_1.lenght - a_1 \\
p_1 \leftarrow \text{NEXT}(L1) \\
x_1 + = p_1.lenght \\
a_2 + = p_1.lenght - a_1 \\
a_1 = 0
\end{cases}
\end{cases}
\end{cases}
$$

$s \leftarrow s/(\min(i_1.width, i_2.width) \times i_1.height)$

Our algorithm uses the vertical runs to compare two given images i_1 et i_2. It browses all the lines $L1$ and $L2$ of these images at the coordinates y and $y + d_y$. For each couple of line, it browses alternately the runs using two variables $\{p_1, p_2\}$. The Figure 6 explains this run browsing. Two markers $\{x_1, x_2\}$ are used to indicate the current positions of the browsing. The browsed line is already the one of lower position (tacking into account the d_x offset of the centering step). The latest read run of the upper position is used as reference run. The runs of the browsed line are summed using a variable s if they are of the same color than the reference run. During the alternately browsing two stacks $\{a_1, a_2\}$ are used. These last ones allow to deal the browsing switch ($L1 \leftrightarrows L2$). For this purpose, they sum the browsed distances on each line using the reference runs.

4   Experiments and Results

In this section we present experiments and results about our system. For this purpose we have tested our system on the Ornamental Letter Database[2]. In this database we have selected 2048 gray level images digitized from 250 to 350 dpi. The full size of these images (in uncompressed mode) is of 268 Mo (a mean size of 131 Ko per image). We present here experiments and results of our system concerning three criteria: the compression rate, the comparison time and the retrieval precision.

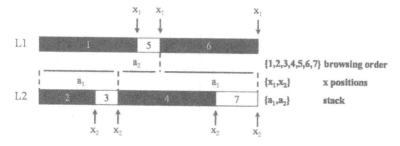

Fig. 6. Run browsing

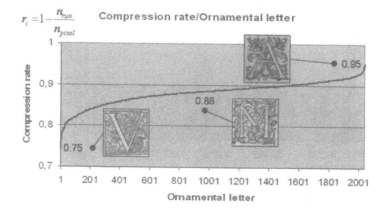

Fig. 7. Compression rates of the ornamental letter images

In a first step we have computed the RLE of images in order to obtain their compression rates. The Figure 7 shows our results. We have obtained a mean rate $\overline{r_c} = 0.88$ on the whole database with minimum and maximum ones of 0.75 and 0.95. These results show that RLE has reduced of 88 % the sizes of images, so from 8 to 9 times. The Figure 7 gives also examples of ornamental letter image corresponding to the characteristic rates min, mean and max. The better rates are obtained for the images composed of strongly homogeneous regions whereas the lower ones correspond to textured images (which produce lot of heterogeneous regions).

We have next evaluated the retrieval times of our system. To do it we have performed a query with each of the images of the database. We have compared each of these queries with all the other images of the database. The comparison is done in two steps, one of image centering and then the distance computation. From the time results we have looked for the min, mean and max ones. In order to compare these values we have also done the same experiments but using a classical image comparison algorithm working from a raster based representation. The both algorithms have been implemented in C++ and tested on a laptop computer using a 2GHz Pentium processor working with a Windows XP operating System. Our results are presented in the Figure 8. We have obtained a mean time less to one minute with a our approach contrary to the several ones needed with the raster based comparison. In any cases, our approach

	Size (k.pixel)	Time (s)
Min	7.74	176.67
Mean	130.8	337.06
Max	600.8	903.62

	Size (k.run)	Time (s)
Min	1.1	22.32
Mean	15.5	41.68
Max	87.8	137.06

Raster based comparison   RLE based comparison

Fig. 8. Time retrieval results

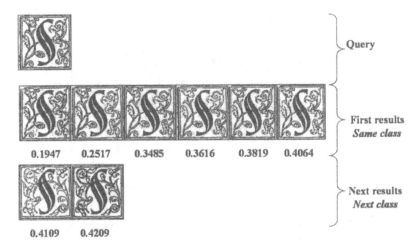

Fig. 9. Example of query result

allows to execute the queries within two minutes whereas the raster based comparison can take until a $\frac{1}{4}$ hour.

At last we have performed, in a random way, some queries in order to evaluate the retrieval precision of our system. The Figure 9 gives an example of query result. In regard to this kind of result our system seems allowing an efficient retrieval of the ornamental letter images. Indeed, as explained previously the image is done at a pixel level. It gives a precise comparison of the images allowing to obtain good retrieval results. The remained retrieval problems concern the very damaged ornamental letter images which appear in the case of broken plugs, ripped parts, darkness papers, bad curvatures, etc.

5   Conclusion and Perspectives

In this paper we have presented a system dealing with the CBIR of old printed graphics (headpieces, pictures and ornamental letters of the XVI° and XVII° centuries). The aim of our system is to process large image databases. For this purpose, we have developed a fast approach based on a Run Length Encoding (RLE) of images. This one allows to reduce the image sizes and then their handling times for the comparison. The core part of our system is an image comparison algorithm. This one uses two steps: one of image centering following by a distance computation. Like this, the centering step allows to solve the shifting problems usually met between segmented images. We have presented different experiments and results about our system. We have shown how our system allows to compress from 8 to 9 times the image sizes, and therefore to reduce the needed retrieval time. We have also illustrated the retrieval precision of our system through an example of query result.

The perspectives concerning this work are of two types. In a first step we work now on a selection process of images based on global features. The key idea to this work is to reduce previously the comparison space by rejecting the images too different from the query one in order to speed-up the retrieval. In a second step we want to evaluate our retrieval results. However, this needs to acquire the ground-truth from the ornamental letter images. Editing the ground-truth in an hand user way could be a long and harder work which could introduce lot of errors. Our key idea to solve this problem is use our system as a ground-truthing one in the way of [20]. It will provide the retrieval results to a user which will valid or will correct them in order to constitute the ground-truth. Like this, the user will be able to edit a ground-truth in the semi-automatic way.

Acknowledgments

This work was funded by the project Madonne[5] of the French ANR program "ACI Masse de Données" 2003 and the Spanish Ministry of Education and Science under grant TIN2006-15694-C02-02. The authors wish to thank Sébastien Busson (CESR, Tours, France) of the BVH project for his collaboration to this work.

References

1. Ramel, J., Busson, S., Demonet, M.: Agora: the interactive document image analysis tool of the bvh project. In: Document Image Analysis for Libraries (DIAL), pp. 145–155 (2006)
2. Bigun, J., Bhattacharjee, S., Michel, S.: Orientation radiograms for image retrieval: An alternative to segmentation. In: International Conference on Pattern Recognition (ICPR), vol. 3, pp. 346–350 (1996)
3. Baudrier, E., Millon, G., Nicolier, F., Seulin, R., Ruan, S.: Hausdorff distance based multiresolution maps applied to an image similarity measure. In: Optical Sensing and Artificial Vision (OSAV), pp. 18–21 (2004)
4. Pareti, R., Vincent, N.: Global discrimination of graphics styles. In: Workshop on Graphics Recognition (GREC), pp. 120–128 (2005)
5. Uttama, S., Hammoud, M., Garrido, C., Franco, P., Ogier, J.: Ancient graphic documents characterization. In: Workshop on Graphics Recognition (GREC), pp. 97–105 (2005)
6. Gesu, V.D., Starovoitov, V.: Distance based function for image comparison. Pattern Recognition Letters (PRL) 20(2), 207–214 (1999)
7. Kumar, V.: Parallel Architectures and Algorithms for Image Understanding. Academic Press, London (1991)

---

[5] http://l3iexp.univ-lr.fr/madonne/

8. Biancardi, A., Mérigot, A.: Connected component support for image analysis programs. In: International Conference on Pattern Recognition (ICPR), vol. 4, pp. 620–624 (1996)

9. van Vliet, L., Verwer, B.: A contour processing method for fast binary neighbourhood operations. Pattern Recognition Letters (PRL) 7(1), 27–36 (1998)

10. Wenyin, L., Dori, D.: From raster to vectors: Extracting visual information from line drawings. Pattern Analysis and Applications (PAA) 2(2), 10–21 (1999)

11. Pavlidis, T.: A minimum storage boundary tracing algorithm and its application to automatic inspection. Transactions on Systems, Man and Cybernetics (TSMC) 8(1), 66–69 (1978)

12. Xue, H., Govindaraju, V.: Building skeletal graphs for structural feature extraction on handwriting images. In: International Conference on Document Analysis and Recognition (IC-DAR), pp. 96–100 (2001)

13. Zhong, D., Yan, H.: Pattern skeletonization using run-length-wise processing for intersection distortion problem. Pattern Recognition Letters (PRL) 20, 833–846 (1999)

14. Shi, Z., Govindaraju, V.: Line separation for complex document images using fuzzy run-length. In: Workshop on Document Image Analysis for Libraries (DIAL), pp. 306–313 (2004)

15. Kim, S., Lee, J., Kim, J.: A new chain-coding algorithm for binary images using run-length codes. Computer Graphics and Image Processing (CGIP) 41, 114–128 (1988)

16. Chan, Y., Chang, C.: Image matching using run-length feature. Pattern Recognition Letters (PRL) 22(5), 447–455 (2001)

17. Breuel, T.: Binary morphology and related operations on run-length representations. In: International Conference on Computer Vision Theory and Applications (VISAPP) (2008)

18. Brunelli, R., Mich, O.: On the use of histograms for image retrieval. In: International Conference on Multimedia Computing and Systems (ICMC), pp. 143–147 (1999)

19. Kreher, D., Stinson, D.: Pseudocode: A LATEX Style File for Displaying Algorithms. Department of Mathematical Sciences, Michigan Technological University, Houghton, USA (2005)

20. Yang, L., Huang, W., Tan, C.: Semi-automatic ground truth generation for chart image recognition. In: Bunke, H., Spitz, A.L. (eds.) DAS 2006. LNCS, vol. 3872, pp. 324–335. Springer, Heidelberg (2006)

Developing Domain-Specific Gesture Recognizers for Smart Diagram Environments

Adrian Bickerstaffe, Aidan Lane, Bernd Meyer, and Kim Marriott

Monash University, Clayton, Victoria 3800, Australia

Abstract. Computer understanding of visual languages in pen-based environments requires a combination of lexical analysis in which the basic tokens are recognized from hand-drawn gestures and syntax analysis in which the structure is recognized. Typically, lexical analysis relies on statistical methods while syntax analysis utilizes grammars. The two stages are not independent: contextual information provided by syntax analysis is required for lexical disambiguation. Previous research into visual language recognition has focussed on syntax analysis while relatively little research has been devoted to lexical analysis and its integration with syntax analysis. This paper describes GestureLab, a tool designed for building domain-specific gesture recognizers, and its integration with Cider, a grammar engine that uses GestureLab recognizers and parses visual languages. Recognizers created with GestureLab perform probabilistic lexical recognition with disambiguation occurring during parsing based on contextual syntactic information. Creating domain-specific gesture recognizers is not a simple task. It requires significant amounts of experimentation and training with large gesture corpora to determine a suitable set of features and classifier algorithm. GestureLab supports such experimentation and facilitates collaboration by allowing corpora to be shared via remote databases.

1 Introduction

There has been considerable research into the automatic recognition of diagrams as the basis for smart diagrammatic environments (SDEs). These SDEs use structured diagrams as a means of visual human-computer interaction [10]. An example SDE is a smart whiteboard that automatically interprets, refines and annotates sketches jotted down in group discussion. Much of this research has focused on generic diagram interpretation engines based on incremental multi-dimensional parsers. Such parsers can automatically be generated from a grammatical specification of the diagrammatic language [5], greatly simplifying the task of implementing SDEs. The inputs to such a parser are lexical tokens such as lines, rectangles, or arrows. Typically the user composes a diagram from these with an object-oriented drawing editor.

Extending the (semi-)automatic generation of diagram interpreters to support sketching in pen-based environments is a challenging task and the focus of this paper. A generic two-stage approach is taken in which syntax analysis (parsing) is

W. Liu, J. Lladós, and J.-M. Ogier (Eds.): GREC 2007, LNCS 5046, pp. 145–156, 2008.

preceded by lexical analysis (gesture recognition). While some previous projects have used parsing techniques for lexical analysis, decades of research into pattern analysis suggests that feature-based and statistical methods are better suited to this problem [7]. The core challenges tackled in our paper are: (1) to automate as far as possible the development of statistical recognizers for stylus-drawn graphical tokens and (2) to integrate statistical lexical recognition with grammar-based syntax analysis.

The main contribution of this paper is to describe GestureLab, a tool for generating probabilistic gesture recognizers. GestureLab is integrated with Cider [5], a multi-dimensional parser generator for diagram analysis: gesture recognizers generated with GestureLab can be interfaced automatically with an incremental parser generated by Cider. Together these two systems provide a suite of generic tools for the construction of interactive sketch interpretation systems. These tools automate the SDE construction process to a high degree. The viability of the GestureLab-Cider approach is demonstrated in the development of a computer algebra system that interprets stylus-drawn mathematical expressions. The tool recognises algebraic and matrix notation from interactive input and demonstrates context-driven disambiguation.

GestureLab uses Support Vector Machines (SVMs) as the default mechanisms for learning new recognizers. SVMs are a popular approach to supervised learning of wide-margin classifiers because they are well-understood, theoretically well-founded and have shown excellent performance across a broad variety of applications [2,13]. However, standard SVMs perform non-probabalistic two-way classification. A second contribution of this paper is to describe an extension to SVMs that allows GestureLab to generate probabilistic k-way recognizers.

2   GestureLab

Given the huge variety of lexical tokens occurring in different types of diagrams, it is clear that generating an interpreter for a new diagram type requires lexical recognition to be tailored to the gestures of interest. GestureLab (see Fig. 1) is a software tool designed to facilitate rapid development and testing of such domain specific gesture recognizers. Recognizers can be developed entirely within GestureLab without any need for a testbed application and can be coupled to Cider without modification.

GestureLab recognizers follow the standard approach to statistical gesture recognition: recognition is performed on digital ink which includes position, timing, pressure, and angle data. Statistical summary features such as the total length of the gesture, initial stroke angle, and maximum curvature are extracted from this data and used by a classifier algorithm to predict class labels (gesture types). A recognizer thus consists of a bundle of feature extractors and a classifier algorithm trained on a particular gesture corpus.

GestureLab supports all phases of the recognizer development process: (a) collecting, manipulating and sharing gesture corpora, and (b) automatic training and cross-validation of feature extraction and recognizer mechanisms. In the

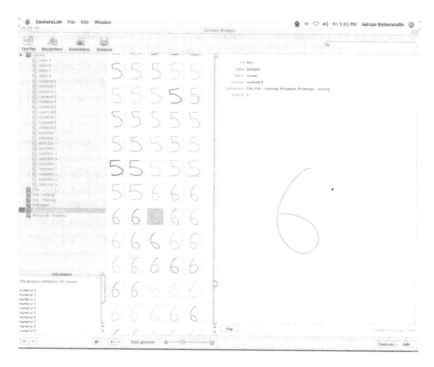

Fig. 1. Main window of GestureLab

event that the built-in feature extraction and recognizer mechanisms are insufficient, GestureLab also allows the developer to readily define (c) new feature extraction mechanisms and (d) new recognizer algorithms.

A core challenge for any two-phase approach which splits lexical and syntactic analysis is that lexical recognition may be ambiguous; contextual information from syntax analysis may be needed to disambiguate the lexical classification. This disambiguation must be delayed until the parsing stage when contextual syntactic information is readily available. To support this, GestureLab generates probabilistic recognizers that return membership probabilities for all possible token classes instead of a single most likely class.

Corpus Management: GestureLab arranges gestures in terms of a library containing named *categories* (or "classes") of gestures and *collections* of gestures selected from these categories. Class membership determines the intended interpretation of a gesture, while collections are named sets for training and testing. Each gesture may belong to any number of named collections and classes. In this way, training and test collections can be created, modified, and deleted without altering the corpora. The library can be accessed and manipulated using an intuitive drag-and-drop interface or via an SQL interface. SQL queries are based on attributes such as collection names, feature data, and experimental results (see below). Gestures can be reviewed visually either as static images or as animations showing the original drawing process.

GestureLab uses a single database to store corpora and experiment data. Remote access to this database is possible using GestureLab clients connected via the Internet. This makes it possible for geographically distributed research groups and for whole research communities to contribute to shared corpora, and to use this data for recognizer development. Corpora and experiment data are also accessible for other software applications via a versatile text-based import/export facility.

Defining New Classifier Algorithms: In the simplest case, a domain-specific recognizer is built by simply training a generic classifier on a domain-specific corpus. GestureLab uses a probabilistic k-way Support Vector Machine as the default recognizer algorithm (see Section 3). When this is insufficient, specialized classifier algorithms can easily be added by the developer. Classifier algorithms are implemented by writing a new C++ class which inherits from a base recognizer and which implements the recognizer interface defined by virtual functions. This interface comprises functions for training and classifying feature vectors, in addition to saving and loading the recognizer to/from file.

Different applications may need different ink pre-processing such as smoothing or hook removal. The responsibility for any pre-processing rests with the individual recognizers so that each recognizer can process stroke data in a manner which is most suitable for the particular application.

Defining New Feature Extractors: The standard GestureLab distribution includes a set of pre-defined feature extractors, following [12]. These include, for example, the initial angle, maxmimum speed, and total duration of the gesture. However, the features required for effective classification of new gesture sets can vary greatly and so GestureLab provides a flexible mechanism for defining new feature extractors. Feature extractors are defined using a plug-in interface and are implemented by writing a C++ class which inherits from a pre-defined feature class. The interface is straight forward: the extractor receives stroke data and returns the feature as a single real-value. In this way, there are no restrictions on the types of features that can be defined or on the algorithms used to compute these. Several feature extractors can be bundled together as a single feature plug-in module.

Automatic Training and Testing: GestureLab offers full support for automatic recognizer training, testing, and experimentation via an intuitive graphical interface. For an experiment, the designer couples specific feature recognizers and classifier algorithms with chosen gesture collections and can then train and validate the thus defined recognizer automatically. This is particularly useful since training times for some classifiers of large alphabets can be extremely long. GestureLab performs automatic cross validation and can automatically create training and test data sets by randomly sampling from a collection of gestures. All experiment data (including feature values, recognition probabilities, parameters settings, etc.) are stored in the central database and are fully accessible so that experiments can be easily repeated and varied. A versatile experiment report facility allows the developer to obtain experiment summaries including the overall recognizer accuracy, the number of gestures correctly/incorrectly classified, and the particular gestures which were misclassified. Results can be filtered to display only correct or only incorrect predictions. Gestures contained in the results table can be displayed graphically and replayed as a temporary collection. This is particularly useful for diagnosing causes of misclassification and developing new feature extractors to address the problems found.

GestureLab also supports quicker, less comprehensive testing of recognizers. A "test pad" allows recognizers to be evaluated on a gesture by gesture basis using interactive input instead of a whole gesture collection. The test pad is particularly useful for investigating unexpected recognizer traits.

3   SVM Gesture Recognition in GestureLab

The default classifier algorithm for GestureLab recognizers is the Support Vector Machine (SVM [2,13]). SVMs are chosen because they are well-understood, theoretically well-founded and have proven performance in a wide range of application areas.

Linear SVMs: Basic SVMs are binary linear wide-margin classifiers with a supervised learning algorithm. Let X be a set of m training samples, $x_i \in \mathbb{R}^n$, with associated class labels $c_i \in \{+1, -1\}$. Assuming linear separability, the goal of SVM learning is to find an $(n-1)$-dimensional hyperplane which separates the classes $\{x_i \in X | c_i = +1\}$ and $\{x_i \in X | c_i = -1\}$. Such a hyperplane fulfils $c_i(w \cdot x_i + b) \geq$ 0 and corresponds to the decision function $c_{pred}(x) = \text{sign}((w \cdot x + b))$. In general, there are an infinite number of such hyperplanes.

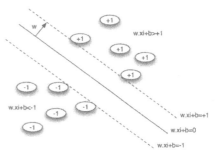

Fig. 2. SVM classification example

SVMs compute the hyperplane that provides the maximum class separation by finding a maximal subset $S \subseteq X$ of so-called support vectors for which w, b can be re-scaled such that $c_i(w \cdot s_i + b) = 1$ for $s_i \in S$. The separation margin perpendicular to the separation hyperplane is $2/||w||$ (see Fig. 2), so that maximizing the margin can be done by solving the quadratic program (QP)

$$\min_{w,b} ||w||^2 \text{ s.t. } \forall i : c_i(w \cdot x_i + b) \geq 1 \qquad (1)$$

or its dual

$$\max_{\alpha_i} \sum_{i=1}^{m} \alpha_i - \sum_{i=1}^{m}\sum_{j=1}^{m} \alpha_i \alpha_j c_i c_j x_i^T x_j \text{ s.t. } \sum_{i=1}^{m} \alpha_i c_i = 0 \wedge \forall i : \alpha_i \geq 0 \qquad (2)$$

where $w = \sum_i \alpha_i x_i c_i$.

Such a hyperplane cannot be found if the two classes are not linearly separable: some x_i will always be on the wrong side of the plane and QP (1,2) is not feasible. In this case, the aim is to minimize the classification error whilst simultaneously maximizing the margin. This is achieved by introducing a penalty term ξ_i for misclassified samples in the corresponding QP (3) or its dual.

$$\min \|w\|^2 + C \sum_i \xi_i \text{ s.t. } \forall i : c_i(w \cdot x_i + b) \geq 1 - \xi_i \tag{3}$$

Non-linear SVMs: The approach described thus far computes only linear classifiers.

In many cases an SVM can, however, separate classes that require non-linear decision surfaces by first transforming the data into some higher-dimensional space in which linear separation is possible. Such transformations can potentially be expensive, but the so-called "Kernel Trick" allows us to side-step the explicit transformation. For a transformation $\phi(\cdot)$, a kernel function $k(\cdot, \cdot)$ computes the dot product of transformed data without explicitly computing the transformation, i.e. $k(x, x') = \phi(x) \cdot \phi(x')$. Of course, kernel functions can only be found for a limited class of transformations $\phi(\cdot)$. Kernel functions provide a general way to apply a linear algorithm (in a limited way) to non-linear problems, provided the crucial computations of the algorithm can be phrased in terms of dot products, as is the case for SVMs.

A non-linear SVM attempts to perform linear separation of the transformed samples $\phi(x_i)$ using the kernel trick [13]. Common kernels include the polynomial kernel $k(x, x') = (x \cdot x')^d$ and the Radial Basis Function (RBF) kernel $k(x, x') = \exp(-\gamma \|x - x'\|^2)$, $\gamma > 0$. Fig. 3 shows a non-linear classification problem.

GestureLab's default classifier uses RBF kernels and performs a two-dimensional grid-search to optimize the kernel parameters. This search is guided by cross-validation results using all training data relevant to the decision node and 5-fold cross-validation.

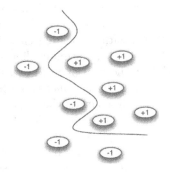

Fig. 3. A non-linear classification problem linearly separable in the transformed feature space

Multiclass SVMs: Standard SVMs are binary classifiers and it is not at all straightforward to use these for multi-way classification. The standard techniques to build k-way SVMs are one-against-all [4], one-against-one [4], and DAGSVM schemes [11]. A one-against-all classifier requires k SVMs for a k-class problem, where the ith SVM is trained using all samples from the ith class versus all other samples. A sample is classified by evaluating all k trained SVMs and the label of the class for which the decision function is maximum is chosen. The one-against-one scheme trains $\frac{k(k-1)}{2}$ classifiers derived from pairwise comparison of target classes. A prediction is made by evaluating each SVM and recording "votes" for the favored class; the class with the most votes is selected as the predicted class. Both methods suffer from very long training times and this issue is further compounded for large data sets such as our corpus of over 10000 gestures. Furthermore, there is no bound on the generalization error of one-against-all schemes, and one-against-one schemes can tend to overfit.

The DAGSVM scheme is more complex. The decision DAG is created by viewing the problem as a series of operations on a list, with each node eliminating one element from the list. Given a list initialized with all class labels, the root node is formed using the training data corresponding to the first and last elements of the list. A decision can now be made which will eliminate one of the two classes being compared. The eliminated class is removed from the list and the DAG proceeds to form a child node again using the first and last list elements. The formation of child nodes in this manner occurs for both decision paths and continues until only one element remains in the list. The DAGSVM will consequently comprise $\frac{k(k-1)}{2}$ nodes and achieve predictions by evaluating $k-1$ of these nodes. Note that each final node can be reached using more than one pathway from the root node and thus, acyclic graph structure is exhibited. The problem of lengthy training times also applies to the DAGSVM schema since, like one-against-one, it requires training $\frac{k(k-1)}{2}$ decision nodes. The performance of a DAGSVM also relies on the order in which classes are processed, and no practical method is known to optimize this order.

We believe a better approach is to reduce the set of possible classes at each decision node and take relative class similarity into account during the construction of the decision tree. We construct the decision tree as a Minimum Cost Spanning Tree (MCST) based on feature distances. Each of the leaves corresponds to a target class and the interior nodes group classes into progressively more disjoint sets. For each internal node in the MCST an SVM is trained to separate all samples belonging to classes in its left subtree from those in the right subtree. Fig. 4 contrasts the DAGSVM and MCST-SVM approaches for a four class example.

The MCST recognizer scales features to $[-1, 1]$ and computes a representative feature vector for each class. The representative feature for a given class is the centroid of all samples belonging to that class. Euclidean distances between all

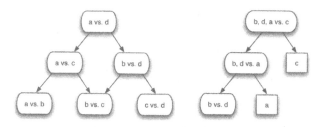

Fig. 4. DAGSVM (left) vs. MCST (right) structure

unique unordered pairs of representative vectors are calculated, and from these distances (or "edge weights") an MCST is constructed (in polynomial time) using Kruskal's algorithm [6]. Average-linkage and complete-linkage versions of the decision tree have also been implemented.

The MSCT recognizer requires $k - 1$ nodes for a k-class problem and a maximum of $k - 1$ decisions for a prediction. MCST recognizers have a core advantage over the other schemas since they discriminate between classes based on class similarity. Furthermore, training time is reduced because only $k - 1$ SVMs must be trained.

Probabilistic SVMs: A standard SVM provides only a non-probabilistic class prediction ("best guess"). As explained earlier, probabilistic predictions are required to perform context-based syntactic disambiguation. The MCST approach facilitates inference of probability distributions for prediction errors during the training phase in a simple manner: after completing the training of all recognizer nodes, a test prediction for each training sample is made and the frequencies of predicting class c_i for a data item of true (known) class c_j are tabulated. From these, maximum likelihood probability distributions are computed for each leaf node of the SVM decision tree.

Coupled with a standard set of feature extractors, the probabilistic MCST-SVM recognizers produce state-of-the-art recognition rates [3].

4   Cider

Syntactic recognition is provided by Cider. Only a quick overview can be given here, for more details see [5]. Fig. 5 shows the components that comprise the Cider toolkit and how these components are used in the creation of an application. The white boxes indicate components of Cider; cross-hatched boxes indicate optional components that can be tailored to extend the capabilities of the toolkit; shading indicates components that must be created by the application developer.

Cider automatically generates a parser for a visual language from a grammatical specification of the visual language's syntax. Parsers produced by Cider are fully incremental which means that users can add, delete, or modify components of a diagram at any time and that the interpretation engine automatically maintains a consistent interpretation of the diagram state. Furthermore, the ability to specify syntactic transformations provides a powerful mechanism for encoding diagram manipulations and user interactions. Cider compiles the grammar and transformation specifications into libraries that can then be used as domain-specific diagram interpretation engines by an application.

Both the syntax and transformation rules are specified using Constraint Multiset Grammars [9], a kind of attributed multi-set grammar. As a simple example, consider the following production which defines a division term t as composed of two numerals a and b with a horizontal division line:

```
t:Term      ::= l:Line
                exist a:Numeral, b:Numeral
                  where immediately_above(a.bbox, l.bbox) and
                    immediately_below(b.bbox, l.bbox) and
                    horiziontally_centered(l.bbox, a.bbox) and
                    horizontally_centered(l.bbox, b.bbox)
                { t.value = a.value / b.value }
```

Importantly, Cider supports structure preserving diagram manipulation. This means that specifications can be written so that once a syntactic diagram component has been recognized, the syntactic structure is automatically maintained when the user manipulates one of the component constituents. For example, when a fraction has been recognized in a mathematical expression and the user extends the denominator, the fraction line can automatically be extended; when the fraction line is dragged, numerator and denominator terms can be dragged with it, etc. This is achieved by using a constraint solver in the diagram processor to automatically update attribute values of tokens so that the specification remains consistent with the visual state of the diagram.

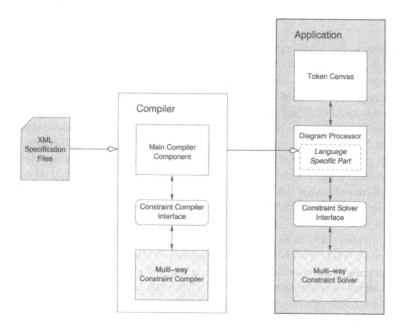

Fig. 5. Cider Architecture

5   System Integration

Building an SDE with GestureLab and Cider: Cider and GestureLab provide a powerful tool suite for building pen-based SDEs. An SDE created with GestureLab and Cider has three main components: a graphical front-end where

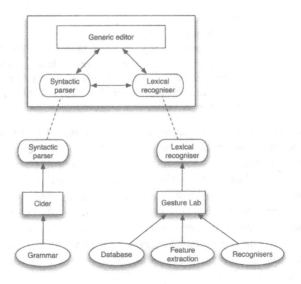

Fig. 6. System components

input is drawn, the Cider runtime environment, and the recognition engine. The recognition engine consists of the lexical recognizer generated by GestureLab and the syntax recognizer (i.e. parser) generated by Cider (see Fig. 6). The graphical front-end is either provided by the application or a generic graphical editor.

The first step in building such an SDE is to bundle the gesture recognizers developed in GestureLab as a single static library. The second step is to develop a Constraint Multiset Grammar that parses the targeted diagram language. Using Cider, this grammar is compiled into another static library that encapsulates the syntactic recognition engine. The application must then be set up to communicate with Cider through the Cider controller API. Primarily, the controller handles the addition, modification and deletion of gestures on the front-end side, as well as the addition of non-terminals (recognition of syntactic components), and their modification and deletion. Non-terminals are added, modified and deleted by Cider as the consequence of a production application or in reaction to a structure preserving manipulation. Communication between the front-end and Cider occurs in terms of requests made to Cider from the front-end. Asynchronous responses from Cider are handled by registering callback functions for events such as the creation or removal of a non-terminal symbol, or attribute changes.

Handling Lexical Ambiguity: Interpretation by the recognition engine is a two-phase process: first, the lexical recognizer provides a probabilistic classification of the gesture; second, resolution of lexical ambiguity is delayed until parsing so as to allow the use of contextual information. For instance, consider a toy problem in which we want to recognize divisions written in TeX in-line style, such as "4 / 5" where numbers always comprise a single digit (the extension to multi-digit numbers is simple).

The grammar contains a "Gesture" terminal symbol in addition to "Numeral", "Operator", and "Term" non-terminal symbols. All symbols have bounding box attributes. The "Numeral" and "Term" non-terminals require a further integer attribute to store their numerical value.

The problem that we encounter is that there can be ambiguity on the lexical level when classifying a vertical line: depending on the angle of the line, it may represent the numeral "one" or a division operator. At some angles the interpretation will be ambiguous and the classification must be delayed until sufficient syntactic context is available to disambiguate in the parser. A vertical line must be a division operator if there are numeric operands to its left and right, whereas the line must be a numeral with value "one" if there is no numeral immediately to the left or right of it. This syntactic disambiguation is taken into account in the following CMG grammar fragment:

```
n:Numeral   ::= g:Gesture where ( most_likely(g, zero) )
                { n.value := 0, n.bbox = Gesture.bbox }
n:Numeral   ::= g:Gesture where ( most_likely(g, line) )
                not exists ( n:numeral )
                    where ( immediately_left_of(n.bbox, g.bbox) or
                            immediately_right_of(n.bbox, g.bbox) )
                { n.value := 1, n.bbox = Gesture.bbox }
...
n:Numeral   ::= g:Gesture where ( most_likely(g, nine) )
                { n.value := 9, n.bbox = Gesture.bbox }

d:Operator     ::= g:Gesture where (most_likely(g, line) )
                exists ( n:numeral )
                    where ( immediately_left_of(n.bbox, g.bbox) or
                            immediately_right_of(n.bbox, g.bbox) )
                { d.bbox = g.bbox }
t:Term      ::= o:Operator
                exist a:Numeral, b:Numeral
                    where immediately_left_of(a.bbox, o.bbox) and
                        immediately_right_of(b.bbox, o.bbox) and
                        vertically_centered(o.bbox, a.bbox) and
                        vertically_centered(o.bbox, b.bbox)
                { t.value = a.value / b.value }
```

The attribute constraints will be automatically processed with default tolerances, however tolerances can be set explicitly and arbitrary attribute tests can be implemented as user-defined functions.

Case study: The GestureLab-Cider pair has been used to create a pen-based front-end for a computer algebra system. This system interprets stylus-drawn mathematical expressions and handles fractions, exponentials, basic arithmetic, and matrices. The case study demonstrates the viability of the GestureLab-Cider approach for generating domain-specific SDEs. Note that due to the incremental nature of Cider parsers, expressions can be written in any order of symbols and can arbitrarily be modified; a consistent interpretation will automatically be maintained at all times.

6 Conclusions

This paper has described GestureLab, a tool designed for building domain-specific gesture recognizers, and its integration with Cider, a grammar engine for parsing visual languages that use GestureLab recognizers. Together these two systems form a suite of generic tools for the construction of interactive sketch interpretation systems. These tools automate the SDE construction process to a high degree.

GestureLab has been specifically designed to facilitate collaboration between researchers, allowing gesture corpora to be stored and shared via remote databases either locally or via the Internet. The software has been released into the public domain at http://www.csse.monash.edu.au/~adrianb/GL/Home.html. GestureLab aims to provide synergy between different research efforts by facilitating the sharing of corpora and recognizer reference implementations. Cider is also available upon request.

References

1. Chang, C., Lin, C.: LIBSVM: a library for support vector machines (2001), http://www.csie.ntu.edu.tw/~cjlin/libsvm
2. Cortes, C., Vapnik, V.: Support-vector network. Machine Learning 20, 273–297 (1995)
3. Garain, U., Chaudhuri, B.B.: Recognition of online handwritten mathematical expressions. IEEE Transactions on Systems, Man, and Cybernetics - Part B 34(6), 2366–2376 (2004)
4. Hsu, C.W., Lin, C.J.: A comparison of methods for multi-class support vector machines. IEEE Transactions on Neural Networks 13 (2002)
5. Jansen, A.R., Marriott, K., Meyer, B.: Cider: A component-based toolkit for creating smart diagram environments. In: International Conference on Distributed and Multimedia Systems, Miami (September 2003)
6. Kruskal, J.B.: On the shortest spanning subtree and the traveling salesman problem. Proceedings of the American Mathematical Society 7, 48–50 (1956)
7. Liu, W.: On-line graphics recognition: state-of-the-art. In: Lladós, J., Kwon, Y.-B. (eds.) GREC 2003. LNCS, vol. 3088, pp. 291–304. Springer, Heidelberg (2004)
8. Lorena, A.C., de Carvalho, A.C.P.L.F.: Minimum spanning trees in hierarchical multiclass support vector machines generation. In: Ali, M., Esposito, F. (eds.) IEA/AIE 2005. LNCS (LNAI), vol. 3533, pp. 422–431. Springer, Heidelberg (2005)
9. Marriott, K., Meyer, B.: On the classification of visual languages by grammar hierarchies. Journal of Visual Languages and Computing 8(4), 374–402 (1997)
10. Meyer, B., Marriott, K., Allwein, G.: Intelligent diagrammatic interfaces: state of the art. In: Diagrammatic Representation and Reasoning, pp. 411–430. Springer, London (2001)
11. Platt, J.C., Cristinini, N., Shawe-Taylor, J.: Large margin DAGs for multiclass classification. Advances in Neural Information Processing Systems 12, 547–553 (2000)
12. Rubine, D.: Specifying gestures by example. Computer Graphics 25(4), 329–337 (1991)
13. Schölkopf, B., Smola, A.: Learning with kernels. MIT Press, Cambridge (2002)

Using Error Recovery Techniques to Improve Sketch Recognition Accuracy

Gennaro Costagliola, Vincenzo Deufemia, and Michele Risi

Dipartimento di Matematica e Informatica – Università di Salerno
Via Ponte don Melillo, 84084 Fisciano (SA), Italy
{gcostagliola,deufemia,mrisi}@unisa.it

Abstract. Sketching is an activity that produces informal documents containing hand-drawn shapes highly variable and ambiguous. In this paper we present a diagrammatic sketch recognizer that is able to cope with the recognition of in-accurate hand-drawn symbols by exploiting error recovery techniques as developed for programming language compilers. The error recovery algorithms are able to interact with recognizers automatically generated from grammar specifications in order to obtain the information on missing or misrecognized strokes.

Keywords: Sketch Recognition, LR parsing, Error Recovery.

1   Introduction

Sketches represent an effective medium for facilitating conceptual design activities by enabling designers to focus on critical issues rather than on intricate details such as precise size, shape, location, and color. In the last decade, many efforts have been put into developing software capable of understanding sketches with objects that can be represented using structural descriptions [1,2]. These recognizers enable users to create sketches using pen-based devices and to transform the edited sketches into input for more powerful design systems. However, existing sketch recognition techniques are error-prone or severely limit the user's drawing style.

In this paper we present a sketch recognition technique able to identify hand-drawn messy symbols of diagrammatic languages. In particular, it extends the approach introduced in [3] with two important features: recognizing hand-drawn symbols with missing strokes and automatically correcting errors of stroke misrecognition. The approach is based on Sketch Grammars [4] for modeling diagrammatic sketch notations and for the automatic generation of efficient recognizers whose parsing technique is based on LR parsing techniques [5]. The recognition of inaccurate hand-drawn symbols is faced by using error recovery techniques as developed for programming language compilers [6].

The sketch recognition system performs an on-line interpretation of the user strokes using the eager modality [7]. This means that the recognizers interpret the strokes immediately after they have been drawn, and provide a user with feedback of the recognized symbols as soon as possible. This approach is more

W. Liu, J. Lladós, and J.-M. Ogier (Eds.): GREC 2007, LNCS 5046, pp. 157–168, 2008.
© Springer-Verlag Berlin Heidelberg 2008

robust and efficient, and consequently more usable, than the lazy one, where the recognition occurs only when explicitly requested by the user and it involves all strokes previously drawn. During the recognition process, the user validates or rejects the symbol interpretations progressively.

The paper is organized as follows. The related work is discussed in Section 2. In Section 3 we describe the proposed approach for the recognition of hand drawn diagrammatic symbols, whereas in Section 4 the results of a preliminary study are presented. Finally, the conclusion and further research are discussed in Section 5.

2   Related Work

A large body of work has been proposed for the recognition of freehand drawings using structural, syntactic, and temporal methods.

In [8] Kara and Stahovich present a multi-level parsing scheme that uses contextual knowledge to both improve accuracy and reduce recognition times. However, the recognition process is guided from "marker symbols", which are symbols easy to recognize, that they assume to exist always in the sketch. Moreover, the approach assumes that the hand-drawn diagram consists of shapes linked by arrows.

In [9] Hammond and Davis present LADDER, a language that allows designers to specify how shapes are drawn, displayed and edited in a certain domain. The language consists of a set of predefined shapes, constraints, editing behaviors and a syntax for combining them. New domain objects are created by specifying the shape descriptions. They do not consider the problem of stroke segmentation in case of multi-symbol strokes.

A strategy quite similar to that proposed in this paper has been developed by Alvarado and Davis [10]. They describe a blackboard-based architecture with a top-down recognition component based on dynamically constructed Bayesian networks that allows recovery from bottom-up recognition errors. In particular, the approach allows to model low-level errors explicitly and to use top-down and bottom-up information together to fix errors that can be avoided using context. However, the high computational cost of the whole method makes the system unsuitable for real-time recognition of realistic sketches. To keep the search tractable, the spatial recognition method for text and graphics proposed in [11] makes some assumptions about the objects in the domain, such as, the objects have no more than eight strokes. However, these assumptions limit the applicability of the method to domains where objects vary in size and shape or where assumptions on the object size and scale might not hold.

Another approach to reduce the computational cost of sketch recognition is to exploit the preferred stroke orders. By the observation that when asked to draw a symbol more than once, people tended to draw it in the same order and that certain regularities exist between individuals, Sezgin and Davis construct a Hidden Markov Model based on these orders for recognizing each symbol [12]. The HMM-based approach exploits the regularities to perform very efficient segmentation and recognition. However, the recognition algorithm requires each object to be completed before the next one is drawn.

3 The Proposed Approach

As shown in Fig. 1, the proposed sketch recognition system interacts with the sketch interface to obtain the edited strokes, to provide the results of its interpretation process, and to receive user's feedback on the recognized symbols.

The sketch recognition module works in the eager mode and is composed by three sub-modules. The *domain independent recognizer* interprets the strokes as primitive shapes, such as lines, arcs, ellipses, etc. Moreover, to support the recognition of multi-stroke symbols the strokes to be classified are suitably split into single-stroke segments by exploiting stroke information such as curvature, speed and direction.

The *symbol recognizers* cluster the primitive shapes in order to identify possible domain symbols. In particular, when a symbol recognizer is able to parse a new stroke, it gives as output the new status of the symbol, which can be partially or completely recognized. The strokes not parsed by a symbol recognizer are temporarily stored in its unmatched strokes repository. This repository contains both graphical and classification information of each unparsed stroke.

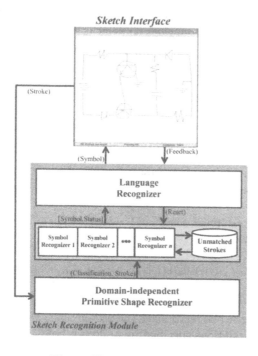

Fig. 1. The recognition process

The interpretations produced by the symbol recognizers are analyzed by the *language recognizer*, which applies its recognition context for selecting the interpretations to be forwarded to the sketch interface.

The *sketch interface* visualizes as feedback the information obtained by the language recognizer. In particular, the interface requires users to accept or reject the recognized symbols. When a symbol is accepted, the language recognizer updates its parsing status and discards the interpretations of symbol recognizers that are in conflict with the accepted symbol.

In the following, we first describe a grammar-based implementation of a symbol and language recognizer and then show how to integrate error recovery techniques to improve the robustness of the recognition.

3.1 Describing and Recognizing Sketched Symbols

The formalism used to specify the sketched symbols of a domain language is the *Sketch Grammar* [4]. A Sketch Grammar represents an extension of string grammars, where also geometric and topologic relations are allowed. The grammar productions alternate symbols and relations, and cluster the input strokes into shapes of a domain language. As an example, the following symbol grammar productions specify a transistor symbol of electric circuit diagrams, represented in Fig. 2:

(1) $npnTransistor \rightarrow$ ELLIPSE $\langle joint_{1\_1}(t_1), joint_{1\_2}(t_1), contains(t_3) \rangle^{r_1}$ LINE$_1$
$\qquad \langle joint^1_{1\_1}(t_1), rotate(90, t_4), contains^1(t_3), near_2(t_5) \rangle^{r_2}$ LINE$_2$
$\qquad \langle joint^2_{1\_1}(t_1), rotate^1(45, t_4), contains^2(t_3), near^1_2(t_5) \rangle^{r_3}$ $wireUp$

(2) $npnTransistor \rightarrow$ ELLIPSE $\langle r_1 \rangle$ LINE$_1$ $\langle r_2 \rangle$ LINE$_2$
$\qquad \langle joint^2_{1\_1}(t_1), rotate^1(-45, t_4), contains^2(t_3), near^1_2(t_5) \rangle^{r_4}$ $wireDown$

(3) $wireUp \rightarrow$ LINE$_3$ $\langle joint^3_{1\_1}(t_1), rotate^2(-45, t_4), contains^3(t_3), near^2_2(t_5) \rangle^{r_5}$ LINE$_4$
$\qquad \langle contains^4(t_3), length(0.33, t_6), rotate(-45, t_4) \rangle^{r_6}$ $Arrow$

(4) $wireDown \rightarrow$ LINE$_4$ $\langle \overline{r_6}, \overline{r_8}, joint^3_{1\_1}(t_1), rotate^2(45, t_4), contains^3(t_3), near^2_2(t_5) \rangle^{r_7}$ LINE$_3$
$\qquad \langle joint^1_{2\_1}(t_1), length(0.33, t_6), contains^4(t_3), rotate^1(-45, t_4) \rangle^{r_8}$ $Arrow$

(5) $wireDown \rightarrow$ LINE$_4$ $\langle r_6 \rangle$ $Arrow$ $\langle joint^4_{1\_1}(t_1), rotate^3(45, t_4), contains^5(t_3), near^3_2(t_5) \rangle^{r_9}$
$\qquad$ LINE$_3$

(6) $Arrow \rightarrow$ LINE$_5$ $\langle joint_{1\_1}(t_1), length(1, t_6), rotate(90, t_4) \rangle^{r_{10}}$ LINE$_6$

Notice that in order to simply the description of the productions we associate a label r_i to each sequence of relations in the productions. Moreover, the notation $\overline{r_i}$ indicates that the relations specified by r_i do not hold.

The transistor symbol is composed by an ellipse and several lines as shown in Fig. 2(a-b). Each primitive shape has associated a set of attributes, which are used to relate a shape to the others, and their values depend on the "position"

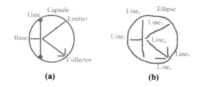

Fig. 2. The npn Transistor symbol (a) and a hand-drawn sketched version (b)

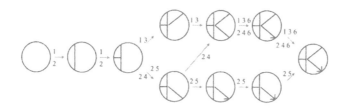

Fig. 3. The three main drawing sequences of the npn Transistor

and "size" of the shape in the sketch. The relations specified in the productions are such that the strokes in the sketch in Fig. 2(b) satisfy them. The values $t_1, \ldots, t_6$ are specified by the grammar designer and indicate the error margin in the satisfaction of the relations. The rotation direction is meant to be counter-clock.

As an example, the first relation in production 1 (labeled with r_1) relates through two *joint* relations attribute 1 of ELLIPSE, which represents its borderline, to attribute 1 and 2 of a LINE, which represent its end points, highlighted by bullets in Fig. 2(a). Moreover, relation *contains* specifies that LINE is contained into the borderline of the ELLIPSE.

Apex 1 in $joint^1_{1-1}$ of the first production (labeled with r_2) indicates that the simple relation *joint* must hold between ELLIPSE (1 position behind LINE$_1$) and LINE$_2$. Note that the apex may also refer to symbols in previously applied productions.

Given a sketch grammar for a symbol of a diagrammatic language it is possible to automatically generate the corresponding symbol recognizer. This recognizer parses a symbol scanning the strokes in the order defined by the productions. As an example, the previous productions describe an npn Transistor symbol with the sequence: capsule, gate, base, collector, and emitter. In order to perform a robust symbol recognition process, the recognizers should be able to parse symbols using more than one sequence. Since when drawing symbols the users employ only a subset of all possible sequences [12], the sketch grammar for a symbol should only include the stroke temporal patterns mainly used. As an example, the npn Transistor is characterized by the three main sequences shown in Fig. 3. Here the number lists indicate the active productions for each sequence.

The parser built from the grammar is based on an extension of LR-parsing. The parsing algorithm analyzes the input exploiting the information contained in

Table 1. The parsing table for the npn Transistor grammar

State		Action			Goto				Next
		LINE	ELLIPSE	$	npnTransistor	wireUp	wireDown	Arrow	
0			:sh 1		:17				(start,ELLIPSE)
1		:sh 2							(r_1, LINE)
2		:sh 3							(r_2, LINE)
3	1	:sh 4				:6	:6		(r_3, LINE)
	2	:sh 5				:7	:7		(r_4, LINE)
4		:sh 8							(r_5, LINE)
5	1	$\overline{r_6}, \overline{r_8}$:sh 9						:12	(r_7, LINE)
	2	:sh 10						:12	(r_6, LINE)
	3	:sh 11						:12	(r_8, LINE)
6			Production 1						-
7			Production 2						-
8	1	:sh 10						:13	(r_6, LINE)
	2	:sh 11						:13	(r_8, LINE)
9	1	:sh 10						:14	(r_6, LINE)
	2	:sh 11						:14	(r_8, LINE)
10		:sh 15							(r_{10}, LINE)
11		:sh 15							(r_{10}, LINE)
12		:sh 16							(r_9, LINE)
13			Production 3						-
14			Production 4						-
15			Production 6						-
16			Production 5						-
17			accept						-

a parsing table. The parsing table for the previous grammar of the npn Transistor symbol is shown in Table 1.

The parsing table is composed by a set of rows and is divided into three main sections: *Action*, *Goto*, and *Next*. Each row is composed of a set of one or more sub-rows each corresponding to a parser state. The *Action* and *Goto* sections are similar to the ones used in the LR parsing tables for string languages [5], while the *Next* section is used by the parser to select the next stroke to be processed. In particular, an entry $Next[k]$ for a state k contains the couple (*relations*, x), which drives the parser in selecting a symbol x satisfying relations.

The entries **sh** x are used to perform both a *shift* operation, i.e., the current symbol is recognized and pushed onto the stack, and a *goto* operation on x, i.e., the current state of the parser is changed to state x. The entries r_i refer to the relation labels introduced in the grammar productions.

Sketch grammars are used also to specify the *language grammars*, which define the sentences of the language as composition of the shapes defined by symbol grammars through spatial relations [3]. As an example, the following productions represent some of the language grammar productions for the circuit diagrams.

(1) $Circuit \rightarrow SubCircuit$

(2) $SubCircuit \rightarrow SubCircuit \langle \textbf{any} \rangle SubCircuit$

(3) $SubCircuit \rightarrow SubCircuit \langle \textbf{joint}_{1\_1}(\textbf{t}_1) \rangle Wire \langle \textbf{joint}_{2\_1}(\textbf{t}_1) \rangle Component$

(4) $SubCircuit \rightarrow SubCircuit \; \langle \mathbf{joint_{1\_1}(t_1)}, \mathbf{joint_{1\_2}(t_1)} \rangle \; Wire$

(5) $SubCircuit \rightarrow Component$

(6) $Component \rightarrow npnTransistor$

(7) $Component \rightarrow pnpTransistor$

(8) $Component \rightarrow Diode$

(9) $Component \rightarrow Lamp$

(10) $Component \rightarrow Battery$

(11) $Component \rightarrow Capacitor$

(12) $Component \rightarrow Resistor$

(13) $Component \rightarrow Ground$

(14) $Component \rightarrow Switch$

Production 2 states that a circuit can be composed of unconnected *SubCircuits*. Production 3 recursively defines a *SubCircuit* as the composition of three language symbols: *SubCircuit*, *Wire*, and *Component*. Production 4 specifies that a *SubCircuit* can be connected through a *Wire* to itself. Finally, Productions 6–14 list the possible components compounding a circuit.

Fig. 4 shows how to recognize a sketched switch circuit through the previous productions. In particular, the recognized circuit uses the npn Transistor in place of the switch to control the flow of electrons from battery through the lamp.

Fig. 4(a) shows the initial switch circuit and two dotted ovals indicating the handles to be reduced. By applying production 14 and then production 5, the initial node *Switch* is reduced to the *SubCircuit* symbol (represented with a dashed oval in Fig. 4(b)). Similarly, the application of production 6 substitutes the *npnTransistor* with the *Component* symbol. Fig. 4(b) shows the handle for application of production 3. In Fig. 4(c), *SubCircuit* and a *Wire* are reduced applying production 4, whereas production 9 substitutes the *Lamp* with a *Component* symbol. Similarly, productions 3 and 10 are applied in Fig. 4(d). In Fig. 4(e) production 3 reduces a *SubCircuit* and a *Component* connected through a *Wire* into a *SubCircuit* symbol. Finally, the subsequent application of productions 4 and 1 reduces the original switch circuit to the starting *Circuit* symbol in Fig. 4(f).

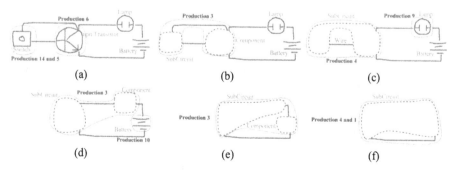

Fig. 4. The reduction process of a switch circuit

Similarly to symbol recognizers, the language recognizer is automatically generated from the language grammar.

3.2   Sketch Symbol Recognition with Error Recovery

The parsing approach described in previous works [3] and based on the shown grammars correctly works if the users completely drawn all the strokes composing a symbol. However, inaccuracy is intrinsic in hand-drawn sketches, thus the user could miss to drawn some symbol strokes, or could draw strokes that result to be difficult to identify, yielding the recognition approach ineffective. In order to cope with these difficulties we will show how to introduce error recovery techniques in the previous parsing algorithm.

The goal of the error recovery algorithm is to overcome the problem of stroke linearization performed by the parsing algorithm, which prevents the recognition of incomplete symbols. To this aim, when a stroke of a symbol is missing (or misrecognized) the parsing algorithm activates an error recovery procedure that allows the parser to proceed in the recognition of the symbol. Nevertheless, the error recovery process should not be applied if it does not lead to the recognition of a symbol, i.e., many strokes of the symbol are still missing. In order to face with this requirement the grammar developer associates to the terminals of the symbol grammars a discriminant value indicating the importance of the stroke in the described symbol. As an example, for the terminals of the previous npn Transistor symbol we can associate the following discriminant values, specified within parenthesis.

(1) $npnTransistor \rightarrow$ ELLIPSE(20) $\langle r_1 \rangle$ LINE$_1$(15) $\langle r_2 \rangle$ LINE$_2$(15) $\langle r_3 \rangle$ $wireUp$

. . .

(3) $wireUp \rightarrow$ LINE$_3$(10) $\langle r_5 \rangle$ LINE$_4$(5) $\langle r_6 \rangle$ $Arrow$

(4) $wireDown \rightarrow$ LINE$_4$(5) $\langle r_7 \rangle$ LINE$_3$(10) $\langle r_8 \rangle$ $Arrow$

. . .

(6) $Arrow \rightarrow$ LINE$_5$(25) $\langle r_{10} \rangle$ LINE$_6$(10)

The discriminant values indicate that the left head segment of the Collector wire has a weight (25) greater than the other segments. This allows the parser to discriminate npn Transistors against pnp Transistors, since the latter represents the Collector symbol in a inverse direction with respect to the first, as shown in Fig. 5. Thus, when the strokes that allow the parser to distinguish a symbol from the others are drawn, the partially recognized symbols will have associated a high discriminant value.

The discriminant values are stored in the *Next* section of the generated parsing table, associated to each next symbol to be processed. As an example, the *Next* entry of state 1 of the npn Transistor parsing table contains the triple (r_1, LINE, 15). Thus, the discriminant values associated to the complex symbols allow the generated recognizers to associate to the partially recognized symbols a value that can be used both to disambiguate the recognition of quite similar symbols and to pose a limit in the number of missing strokes admissible in a symbol. In

Fig. 5. The pnp Transistor symbol

particular, let t be a value such that if the sum of discriminant values associated to the recognized strokes of a symbol S is greater than t then S is (partially or completely) recognized. The error recovery process should terminate when the discriminant values associated to the missing strokes exceed the value $1 - t$. Indeed, in this case the recognized symbol will never exceed threshold t even if all the remaining symbol strokes have been drawn.

In the following we provide the algorithms implementing the error recovery technique.

```
Recovery() {
        PT = parsingtable[parser];
        state = stack[parser].currentState;
        while (state != null) {
          (r, s, d_value) = PT.Next[state];          // for multiple instances the triple with
                                                      //   lowest discriminant value is selected
          if (d_value > 100-threshold[parser]) exit;
          threshold[parser]= threshold[parser]+ d_value;
          newStroke = Fit(r, s);
          state = ContinueParsing(newStroke);
        }
}

Fit(r, s) {
        rep = repository[parser];
        foreach(x in rep) {
          if (r is a valid relation for x) {  // using complete classification
                delete x from rep;
                return x;
          }
        }
        simulatedStroke = use the constraint solver to calculate a shape of type s
                        and compatible with the stack and relation r;
        return simulatedStroke;
}

ContinueParsing(s) {
        action[parser].shift(s);
        rep = repository[parser];
        input[parser]=rep;
        action[parser].continue;                    // reactivate the parser on rep
        input[parser]=PrimitiveShapeRecognizer;     // restore the input
        if (state[parser] != accept and rep.empty == true) return null;
        if (state[parser] == accept) return null;
        return stack[parser].currentState;
}
```

The *Recovery* function is automatically invoked by a symbol recognizer when a syntax error occurs. In particular, if the parser can still recognize a symbol (i.e., the sum of discriminant values of the recognized strokes is lower than the value $1 - t$) then the *Fit* and *ContinueParsing* functions are invoked, respectively, to look for a stroke able to reactivate the parser, and to continue the parsing of the symbol from the stroke in the symbol stroke sequence that follows the missing stroke.

The *Fit* function first analyzes the repository of unmatched strokes to find a stroke, with low accuracy, able to solve the syntax error. If it is found, it is used as the next input for the parser. Otherwise, the function computes the information to simulate the missing stroke. The stroke given in output by *Fit* function is then used by *ContinueParsing* function for updating the parsing state and then reactivating the parser on the unmatched strokes repository to recover from unparsed strokes.

After the recovery process, the symbol recognizer checks the acceptance state of the parser to verify if the symbol is completely or partially recognized.

3.3   An Example

Let us suppose that the user has drawn a pnp Transistor without following any of the stroke sequences of Fig. 3 as shown in Fig. 6(a-e). Moreover, let us consider a threshold value of 70 for the recognition of this kind of symbol.

After the recognition of the Ellipse stroke in Fig. 6(a), the npn Transistor parser reaches state 1, but fails because the next symbol in the input is an Arc against an expected Line. The parser stores the unparsed stroke in the repository associated to the npn Transistor recognizer and then invokes the error recovery procedure. The *Recovery* algorithm invoked by the parser uses the triple $(r_1, LINE, 15)$ to check if the discriminant value 15 exceeds 30 (i.e., 100 less the threshold), and then finds in the repository a valid stroke compatible with relation r_1 and primitive shape LINE. Thus, thanks to the recovery algorithm the stroke is correctly interpreted as Line solving a misrecognition error, as shown in Fig. 6(b). The parser reaches state 2 and waits for the next stroke.

Fig. 6(c) shows the sketch after the drawing of a horizontal line. The new stroke is not compatible with the triple $(r_2, LINE, 15)$ associated to state 2. Thus, the error recovery algorithm simulates the Base wire of the transistor, but cannot continue since will never reach the maximum admissible discriminant value (i.e., 30). In this case, a backtracking process restores the parser state to the one preceding the error recovery invocation. After the drawing of the Base wire stroke (see Fig. 6(d)), the parser moves from state 2 to state 3, and simulates the Collector Wire with the triple $(r_4, LINE, 5)$, considering all the strokes provided as input and including the horizontal line previously sketched.

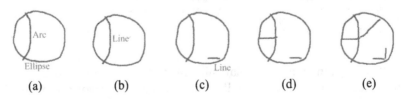

(a)   (b)   (c)   (d)   (e)

Fig. 6. The pnp Transistor symbol

However, the threshold value of 70 for the recognition of the npn Transistor symbol is not reached.

When the user draws the symbol in Fig. 6(e), the parser simulates the Collector Wire and reaches the acceptance state, and the symbol with missing strokes is completely recognized.

4   Preliminary Evaluation

We have conducted a preliminary user study of the proposed approach by implementing a recognizer for electric circuit diagrams. The study included ten subjects who were asked to sketch in an unconstrained fashion the circuit diagram in Fig. 4(a).

Statistical results about the experiment are summarized in Table 2. The results show that the error recovery procedure is particularly useful for improving the recognition performances of complex symbols. In fact, npn Transistor and Lamp symbols can be drawn by using several stroke sequences out of the ones defined by the grammar productions. Thus, the error recovery process is applied six times for npn Transistors and three times for Lamps, obtaining an improvement of the overall precision of 17,56%. On the contrary, the Wire symbol is formed by a simple sequence of connected line strokes which are difficult to be misunderstood. Moreover, Battery and Switch symbols are composed of few strokes and the corresponding symbol recognizers cover all possible stroke orders.

Table 2. Recognition performances

		#Instances	Correctly Recognized	Recognized by using Error Recovery	%Precision
Symbols	Switch	10	10	0	100
	npn Transition	10	7	6	70
	Lamp	10	9	3	90
	Battery	10	10	0	100
	Wire	50	50	0	100
	Total	90	86	9	95,56

5   Conclusions and Future Work

The paper introduces a recognition technique for diagrammatic sketches, which is able to recognize hand-drawn symbols with missing strokes and to correct stroke misrecognition errors automatically. The recognition system is composed of hierarchically arranged recognizers automatically generated from grammar specifications, which integrates error recovery techniques to improve the robustness of the recognition.

In order to reduce the number of active error recovery processes, and consequently the number of user feedback requests, in the future we intend to exploit the feed-backs provided by the users to adapt the behaviour of the recognizers.

In particular, we plan to modify the discriminant values associated to the terminals of the symbol grammar based on the acceptance/reject decisions of incomplete symbols.

Finally, in order to reduce the effort to construct the sketch grammar for a given domain language we plan to investigate how to infer the grammar from a set of sketch samples automatically, similarly to the work proposed in [13].

References

1. Alvarado, C., Davis, R.: A framework for multi-domain sketch recognition. In: Proc. of AAAI Spring Symposium on Sketch Understanding, pp. 1–8 (2002)
2. Caetano, A., Goulart, N., Fonseca, M., Jorge, J.: JavaSketchIT: Issues in sketching the look of user interfaces. In: Proc. of AAAI Spring Symposium on Sketch Understanding, pp. 9–14 (2002)
3. Costagliola, G., Deufemia, V., Risi, M.: A multi-layer parsing strategy for on-line recognition of hand-drawn diagrams. In: Proc. of IEEE Symposium Visual Languages and Human Centric Computing (VL/HCC 2006), pp. 103–110 (2006)
4. Costagliola, G., Deufemia, V., Risi, M.: Sketch Grammars: A formalism for describing and recognizing diagrammatic sketch languages. In: Proc. of International Conference on Document Analysis and Recognition (ICDAR 2005), pp. 1226–1230. IEEE Press, Los Alamitos (2005)
5. Aho, A.V., Sethi, R., Ullman, J.D.: Compilers Principles, Techniques, and Tools. Addison-Wesley, Reading (1987)
6. Snelting, G.: How to build LR parsers which accept incomplete input. SIGPLAN Notices 25(4), 51–58 (1990)
7. Blostein, D., Lank, A., Rose, A., Zanibbi, R.: User interfaces for on-line diagram recognition. In: Blostein, D., Kwon, Y.-B. (eds.) GREC 2001. LNCS, vol. 2390, pp. 92–103. Springer, Heidelberg (2002)
8. Kara, L.B., Stahovich, T.F.: Hierarchical parsing and recognition of hand-sketched diagrams. In: Proc. of ACM Conference on User Interface and Software Technologies (UIST 2004), pp. 13–22 (2004)
9. Hammond, T., Davis, R.: LADDER, A sketching language for user interface developers. Computers & Graphics 29(4), 518–532 (2005)
10. Alvarado, C., Davis, R.: SketchREAD: A multi-domain sketch recognition engine. In: Proc. of User Interface Software and Technology (UIST 2004), pp. 23–32 (2004)
11. Shilman, M., Viola, P.: Spatial recognition and grouping of text and graphics. In: Proc. of Workshop Sketch-Based Interfaces and Modeling, pp. 91–95 (2004)
12. Sezgin, T.M., Davis, R.: HMM-based efficient sketch recognition. In: Proc. of 10th Int. Conf. Intelligent User Interfaces (IUI 2005), pp. 281–283 (2005)
13. Romeu, J.M., Lamiroy, B., Sanchez, G., Llados, J.: Automatic adjacency grammar generation from user drawn sketches. In: Proc. of International Conference on Pattern Recognition (ICPR 2006), pp. 1026–1029. IEEE Press, Los Alamitos (2006)

Representing and Parsing Sketched Symbols Using Adjacency Grammars and a Grid-Directed Parser

Joan Mas[1], Joaquim A. Jorge[2], Gemma Sanchez[1], and Josep Llados[1]

[1] Computer Vision Center, Computer Science Department, Edifici O Campus UAB,
Bellaterra, Spain
{jmas,gemma,josep}@cvc.uab.es

[2] Departamento de Engenharia Informática, INESC, Rua Alves Redol, 9, Lisboa, Portugal
jaj@rtr.inesc-id.pt

Abstract. While much work has been done in Structural and Syntactical Pattern Recognition applied to drawings, most approaches are non-interactive. However, the recent emergence of viable pen-computers makes it desirable to handle pen-input such as sketches and drawings interactively. This paper presents a syntax-directed approach to parse sketches based on Relational Adjacency Grammars, which describe spatial and topological relations among parts of a sketch. Our approach uses a 2D grid to avoid re-scanning all the previous input whenever new strokes entered into the system, thus speeding up parsing considerably. To evaluate the performance of our approach we have tested the system using non-trivial inputs analyzed with two different grammars, one to design user interfaces and the other to describe floor-plans. The results clearly show the effectiveness of our approach and demonstrate good scalability to larger drawings.

1 Introduction

Sketching interfaces are a useful and natural way for people to communicate with computers. By using a digital pen, users can input information such as cursive script annotations or, in a graphical domain, freehand diagrams or graphical gestures. Sketch recognition is therefore a powerful tool in disciplines such as architecture or engineering. In the graphical domain, sketches have an important value. Indeed, sketches provide the ability of expressing complex ideas with simple visual notations, and are a fluent way of human-computer interaction. From a technical point of view, according to Liu's [4] interesting survey, on-line graphics recognition processes may be divided into three main parts: Primitive Shape Detection, Composite Object Recognition and Sketch Understanding. In this work we focus on symbol recognition in sketching diagrams.

Sketches are collections of strokes, i.e. line drawings where basic primitives are the sequences of points captured between a stylus' pen-up and pen-down events. Roughly speaking, sketched symbols are sets of line primitives organized spatially and sometimes in a temporal sequence. These characteristics make desirable to use structural approaches to recognize drawings. In this paper we focus on a syntactic approach to describe and recognize graphical symbols in an on-line framework. A syntactic approach addresses two relevant issues: the description of the graphical entity and the recognition process. The former is based on the theory of formal languages, where a grammar

W. Liu, J. Lladós, and J.-M. Ogier (Eds.): GREC 2007, LNCS 5046, pp. 169–180, 2008.

describes the recognized shapes and its productions represent the relations among composition elements. The latter requires a parsing approach. A parser is a process that, given an input and a grammar G, says if the input belongs to the language generated by the grammar, L(G).

A number of grammatical formalisms exist in the literature to describe bi-dimensional graphical objects. Early attempts augmented linear languages with 2D operators to express the spatial relations among the primitives. Picture Description Languages or Plex grammars [7] are two examples of this approach. Over the last two decades new paradigms of languages have been studied. These languages are inherently bi-dimensional, and thus more apt to describe 2D symbols. They are referred as Visual Languages (VLs). Among the different approaches to VLs we find Relational Grammars [6], where productions describe relations among the different primitive symbols in terms of a set of attributes defined as join points. Adjacency Grammars [2] define a set of constraints denoting spatial, temporal or logic relations. Graph Grammars [19] define productions in terms of graph-based rewriting rules but require complex rules.

Together with grammatical formalisms parsing paradigms were devised to validate whether visual languages belonged to the language generated by those grammars. Most of these parsing methodologies are tailored to a specific grammatical formalism, such as the parsers presented in [16], [18]. Despite these specific methodologies other works extend traditional parsing techniques to try and develop more general methods such as the work of Costagliola et al. [21] which extends conventional LR-parsing techniques to the realm of visual languages.

As described above, a syntactic approach to sketched symbol recognition requires first a grammatical model and second a parsing engine to perform the proper recognition. To this end, we adopt an Adjacency Grammar to describe 2D shapes using a linear language by defining constraints to describe the different relations among the parts composing a sketch. Then we use an incremental parser, to analyze visual sketched sentences. This is done by constructing a parsing-tree each time a new token is drawn. Differently from traditional parsers, our parsing algorithm is able to cope with the main issue of VLs, that is, parsing the input in an order free manner. In this way, the relations the parse tree is built according to spatial or logical relations among the different symbols composing a sketch rather than relying on their temporal sequence as happens with conventional textual languages. This is because, our parser uses a spatial data structure to allocate the different tokens as they are analyzed and, to search the set of neighbouring symbols to match grammatical rules when a new token is drawn and recognized.

This paper is organized as follows: section 2 presents related work on sketch recognition systems and syntactic approaches to describe 2D patterns. In section 3 we present the syntactic approach used to describe and interpret sketches. Section 4 presents experimental evaluation of our work. Finally, section 5 discusses these results and points to future directions in our research.

2   Related Work

This section discusses related work developed in the field of sketch recognition and compares syntactic approaches to describing 2D symbols.

2.1   Sketch Recognition

Sketch recognition is a field of increasing interest due to the progress of digital pen devices allowing interaction between Humans and Computers. This requires designing new applications to analyze sketched inputs, either statically, as in document analysis and recognition or more recently in interactive settings, including calligraphic and pen-based interfaces. In the literature, we find different work that describes sketch recognition as applied to different fields. Landay and Myers presented SILK [13] a system to describe User Interfaces. SILK system attempts to recognize basic primitives forming the sketch using Rubine's algorithm [14] to recognize gestures. Once a primitive is detected the system tries to detect the spatial relations between the primitive and other components or widgets. The system then returns the recognized widget to the user, who is able to correct this output if an error occurs. However, using Rubine's recognizer limited the system to single-stroke basic primitives.

Hammond and Davis [15] developed a system to recognize UML Class Diagrams in four steps: pre-processing, selection, recognition and identification. The pre-processing step classifies the most recent stroke in one of four categories an ellipse, a line, a poly-line or a complex shape. Next, the system attempts to match this to a set of (previously drawn) unrecognized strokes. The authors limit the match candidates to at most nine trying to avoid an exponential time search on the number of strokes required to identify a symbol. This task is ascribed to specific symbol recognizers which identify each different symbol allowed in UML diagrams. The identification process combines the probability of each recognizer with other criteria, such as the number of strokes that compose the symbol recognized.

Kara and Stahovic [11] developed a sketch recognition system to describe engineering circuits that can be used as input to Simulink, using a hierarchical recognition approach. First, the system tries to identify specific symbols as markers. These symbols should be easy to recognize and serve as anchors to help recognizing the remainder of the sketch. In this case, the markers describe arrows which are recognized according to features based on drawing speed. Then the system generates a set of symbol candidates, by taking into account the number of input and output arrows for a given cluster. To recognize a symbol among the different candidates the authors use the combination of four different recognizers by choosing the answer with the best score.

Alvarado and Davis in [10] present a multi-domain sketch recognition system, based on a dynamically constructed bayesian network. The network is constructed using LADDER [8] a language that is able to describe and draw shapes for a specific domain. For each new stroke is drawn the system classifies it in one of the basic categories. Then a hypothesis is generated in three steps: a bottom-up step generates the hypothesis from the new stroke, followed by a top-down step that attempts to find subshapes missing on the partial hypothesis created by the previous step and finally, a pruning step that keeps the number of hypothesis manageable to be analyzed in real time.

2.2   Grammatical Symbol Recognition

Different grammatical formalisms have been proposed to describe visual constructs (symbols). These formalisms describe symbols as a collection of basic primitives and a set of relations connecting those shapes.

Linear grammatical formalisms are presented in [16], [17] and [18]. These three grammars describe productions as a set of symbols and a set of constraints among those symbols. While *Relation Grammars* [17] define their productions over a set of un-attributed tokens, *Constraint Multiset Grammars* [16] and *Picture Layout Grammars* [18] describe their productions using sets of attributed tokens. *Constraint Multiset Grammars* are context-sensitive grammars that define a set of tokens that may exist to produce a valid rule. In this latter formalism, contextual tokens may be specified in spatial or logical constraints, while not being considered part of the production proper.

Coüasnon developed a language named EPF (Enhanced Position Formalism) [20] using an operator based grammatical formalism. In EPF the productions are concatenations of symbols and operators between the symbols. The operators may describe positional relations, factorization of symbols, etc.

While the grammatical formalisms presented before are linear grammars, other researchers focus on visual languages defined via high-dimensional grammars. Wittenburg and Weitzmann [6] present Relational Grammars. These grammars are high-dimensional context-free grammars. Grammar productions are defined over ordered sets of symbols and a set of constraints among the symbols. Other high-dimensional formalisms include Layered Graph Grammars [19]. These grammars are context-sensitive, restricting the size on the left-hand of the production to be smaller than the right-hand.

More recent work that combines sketch recognition with a syntactic approach. Sketch Grammars [21] extend context-free string grammars, by defining relations other than concatenation relations. These relations include temporal or spatial constraints among symbols. Productions in these grammars also define a set of actions that may include drawing constraints and semantic context. LADDER [8] is a closely related technique that describes relations among complex symbols defined in terms of basic elements and specifies drawing, editing and semantic actions as part of a production.

3   A Syntactic Approach to Recognize Hand-Drawn Sketches

The syntactic formalism presented in this paper is a one-dimensional grammar based on Adjacency Grammars [2]. The symbols in the right-hand-side of a production are described as an unordered set of tokens that should obey a set of constraints, which allows these grammars to describe drawings in an order-free manner.

Adjacency Grammars are formally defined as a 5-tuple $G = \{V_t, V_n, S, P, C\}$ where:

- V_t represents the terminal vocabulary.
- V_n represents the non-terminal vocabulary.
  With $V_t \cap V_n = \emptyset$ and $V_t \cup V_n = \Sigma$ being Σ the alphabet of the language generated by the grammar L(G).
- S is the start symbol.
- P is the set of productions of the grammar defined as:

$$\alpha \rightarrow \{\beta_1, \ldots, \beta_j\} \text{ if } \Gamma_1(\Phi_1, c_1), \ldots, \Gamma_n(\Phi_n, c_n)$$

Where $\alpha \in V_n$ and $\forall i \in [1, \ldots, j]\beta_i \in \{V_t \cup V_n\}$, constitute the possible empty multiset of terminal and non-terminal symbols. $\forall k \in [1, \ldots, n]$ Γ_k are the adjacency constraints defined on the attributes of the subsets $\Phi_k \subset \{\beta_1, \dots, \beta_j\}$ and c_k are the cost functions associated to each constraint.

- C represents a set of constraints. This set represents the different spatial relations that we may find between two different grammatical symbols.

$f_\alpha\colon R_1^d \times \dots \times R_j^d \to R_\alpha^d$ is a function that calculates the attributes of the new token from the attributes of the tokens of the right hand of the grammatical productions. Being d the cardinality of the attributes of the token α and j the number of tokens on the right-hand-side. This function is used when a symbol is reduced from a grammatical production during the parsing process.

Concerning the parser required in this approach, it is an incremental on-line parser to recognize sketches, which analyzes each new input token drawn until all inputs are processed. Then the parser either recognizes the whole sketch, or signals an error due to a invalid input.

While conventional parsers analyze input tokens with these grammars according to a predefined input order, in a sketching framework we can not expect such an ordered list of tokens when a user is drawing a symbol. On the contrary, each user may draw the constituents of a symbol in a different order. There are two solutions to this problem. The first establishes a predefined input order, and the user has no freedom to draw a sketch which allows a conventional linear parsing to be applied. The second works with no predefined order but entails a high computational cost because, for each new token, the parser has to look among all the previously drawn symbols for those that may be combined to yield a valid rule.

The parser presented in this paper requires no predefined order in the input but uses a uniform grid to avoid re-scanning all the input as each new token is drawn. When a new token is read by the parser, it is placed into an array of rectangular cells. Then the parser searches the neighbouring cells for symbols that may produce a valid rule, instead of searching all symbols seen so far. For well-behaved languages this allows polynomial-time search while making possible to provide an analysis in real time for reasonable input sentences.

Constructing the grid entails some decisions: whether it should be a *static* or *dynamic* structure, the size of the cells, whether cells should be of fixed size or use adaptive dimensions, and how to place the symbols into the grid. Using a *dynamic* grid requires recalculating the regions each time that a new item is inserted. Regarding the size of cells, if we choose too small a size, each inserted symbol will be stored in many small cells, thus taking up more space, and the parser will need to analyze more cells to find candidate tokens, taking up more time. Using larger cells means that a major number of symbols could be stored inside any given cell. This may mean a large number of "false candidates" showing up in neighbor queries, thus wasting computational resources. Finally, we need to take into account the method used to calculate the cells each symbol belongs to. Using a bounding-box is not always an adequate heuristic to find the cells spanned by a given symbol. I.e. if we have a token that describes a diagonal line its bounding box will intersect many cells that do not really belong the symbol. This will then waste computational resources, by attempting to match the symbol against non-neighbor terms.

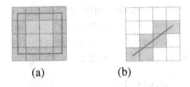

(a)   (b)

Fig. 1. Different placements algorithms used on the grid based parser:(a) Bounding-Box and (b) Bresenham's Algorithm

---

Being:
$Gr = \{Cx_1, \ldots, Cx_{n \times m}\}$ the Grid over the space, with each of its Cx_i cells.
$BB(Cx_i)$: The bounded box of the cell Cx_i $(x_1, y_1), (x_2, y_2)$
$N(Cx_i)$: The set of Neighbouring cells of Cx_i
$E(Cx_i)$: The elements (tokens, completed or unfinished rules) inside a cell Cx_i
$E(Cx_{i_1}, \ldots, Cx_{i_m})$: The elements (tokens, completed or unfinished rules) inside a set of i_m cells

t is a token
$Cells(t)$ is the set of cells that belongs to the token t following fig. 1.
$r' = Validate(r, e_1, e_2)$ where r' is the modification of the rule r if e_1 and e_2 satisfy the constraints defined in r.
$r' = Finalize(r, e_1)$ add the element e_1 if it satisfies the constraints defined in r.
$Valid(e)$ returns true if the element e has all of its composing elements and they have produced a valid reduction. Given a token t

```
Parser(t)
    C = Cells(t)
    N = {Cx_j, ∀Cx_j ∈ C && Cx_i ∈ N(Cx_j)}
        For each t' ∈ E(N)
            If Valid(t') then
                Find productions Pr with t and t'
                For each p' ∈ Pr
                    t'' = Validate(p', t, t')
                    insert t'' into Cells(t'')
                    If Valid(t'') then
                        Parser(t'')
                    End If
                End For
            Else t'' = Finalize(t', t)
            If Valid(t'') then
                Parser(t'')
            End If
        End For
```

Fig. 2. Incremental Grid-Based On-line Algorithm

In our method after some experiments we have defined a static grid with a fixed cell size of $1/2 \times 1/2$ inch. Using larger cells involves covering a more extensive area and therefore taking more primitives into account. On the contrary, smaller cells entail a larger number of memory accesses. To place the symbols into the cells we decide to use two different methods depending on whether the token is a line or not. If the token is a line we use Bresenham's algorithm [9] to compute the cells that contain the line. On the contrary, the bounding box determines the set of cells belonging to it, see fig. 1.

Our parser works as described in fig. 2: When a new primitive p, is input, the parser finds out which grid cells it overlaps. Then the parser searches among the neighbouring cells which contains graphical elements that, together with the new primitive, can match a production in the grammar. If no such elements are found and p does not match the right-hand-side of a production by itself, the parser looks for the next primitive input. If a matching production is found, the parser will check to see if the constraints on the right-hand-side are met. If all the constraints match, the symbol s_1 on the left-hand-side of this rule, generates a parse item, which is placed into the grid. Otherwise, if part of the production rule is valid but there are still missing components s_1 is marked as an unfinished parse item and it is placed also into the grid. Note that incomplete parse items can match incoming primitives as a production would do. If a complete parse item has been produced, we check to see if additional productions can match s_1 (the non-terminal labelling the new item). We recurse on new (completed) items until no more production rules can fire.

4 Experimental Evaluation and Discussion

To evaluate our syntactic approach, we have defined two example grammars, as shown in fig. 3. The first grammar, in fig. 3.a, describes architectural floor-plans. Here, a room is defined as a *rectangle* which has a *door* and a *window* that intersects with it. The second grammar describes a graphical user interface GUI as shown in fig. 3.b. For example, we define a MenuBar as a *rectangle* that contains two or more Menu tokens. The basic primitives of these grammars are described by Adjacency constraints. To simplify writing these grammars, the definition of these basic primitives is automatically done using a grammatical inference method (see [5]).

The experiments show how our grammatical formalism is able to describe hand-drawn sketches using two grammars described above. The first experiment highlights the resource savings due to the adoption of a grid and which is the correct size of its cells. The second one shows how our method is able to cope with distortion, errors and uneven spatial distribution of tokens. Finally, we have tested the syntactic approach with sketches drawn by different users.

a)   b)

Fig. 3. Example of a grammar specification

4.1 Experiment 1: Grid Tuning

This experiment evaluates the computational resource savings afforded by the grid and allows us to estimate a good cell size. Table 1.a shows number of memory accesses by our parsing algorithm. As we can see a small cell size increases the number of memory accesses. Such grids involves small size on the cells that may introduce errors at the time to search symbols in the neighbouring cells of the new token, as two real neighbouring tokens may be placed in not neighbouring cells. This requires expanding the search area to encompass more cells, which in turn increases again the number of memory accesses. On the other hand, large cells yield more "false positive" tests, as more elements which

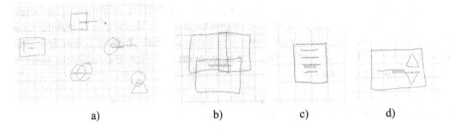

a)   b)   c)   d)

Fig. 4. Examples used on grid tuning: a) Example1, b) Example2, c) Example3, and d) Example4

are placed too far apart to match adjacency relations show up in neighbor queries. As can bee seen from Table 1, the smallest number of memory accesses seem to occur around 1/2 in x 1/2 in cells. This, of course is dependent on input data average size and primitive types. Different input tokens would probably require cells of a different size. Further more comprehensive testing would be required to establish better heuristics for cell size which would take into account grammar tokens and input characteristics.

Table 1.b shows a comparison in terms of finished and unfinished parse items between our method and a non-constrained (i.e. griddles) parser. This parser is somewhat similar to the parser defined by Golin in [18]. We consider as unfinished parse items those productions for which the parser has not been able to match all symbols on the right-hand-side. As we can see, the difference between the two methods lies in the number of unfinished productions. When the user draws a new stroke, unconstrained parsers have to re-analyze all the primitives previously seen, even if they bear no relation to the symbol being drawn. Our method takes into account less primitives, thus it reduces the number of feasible productions tested while generating less temporary (unfinished) parse items. Finally, table 1.c shows the recognition times for each of the samples. As we can see, the time values are reasonable to analyze an input in real time.

Table 1. Parser performance a) Number of memory accesses as a function of cell size, b) Number of Finished and Unfinished parse items created and c) Recognition Time

Cell Size / Sample	1 inch	1/2 inch	1 cm	1/2 cm
Example1	360	504	774	2529
Example2	189	333	504	1944
Example3	108	108	180	450
Example4	126	144	378	945

a)

Sample	With Grid Finish./Unfinish.	Without Grid Finish./Unfinish.
Example1	15/1	17/125
Example2	1/2	1/65
Example3	8/4	8/86
Example4	2/1	2/67

b)

Sample	#tokens	#symbols	Time(ms)	Average(ms)
Example1	19	15	404	26.93
Example2	12	1	155	155
Example3	8	8	220	27.5
Example4	11	3	203	67.3

c)

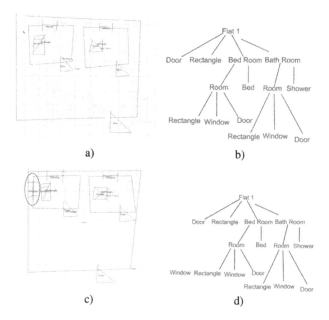

Fig. 5. Architectural floor-plan a) and corresponding parse tree b) c)Another Instance and its parse tree d)

4.2 Experiment 2: Distortion Tolerance

Contrary to document recognition, error tolerance and the ability to cope with sizeable variations in input tokens are very desirable features to have in a parsing algorithm tailored to interactive use. This experiment shows how tolerant our method is to distortion and input errors such as added (extraneous) elements and changes in spatial relationships among the strokes on a given sketch.

Figure 5 shows two samples representing sketched floor-plans and their corresponding parse trees. As we can see, at the time that each new token of the input of the left-hand is drawn by the user, the parsing algorithm constructs the parse tree on the right side where the leaves corresponds to the terminal symbols of the alphabet and each non-leaf node represents a non-terminal symbol labelling the corresponding production in the grammar. Figure 5.c shows the same visual sentence where an additional element circled in red has been added. This element has been recognized as a window (as described by the grammar in Fig. 3.a), it does not satisfy the relational constraint to the other elements of the production. Although this additional symbol appears to be spurious input, we can see that the parser is able to describe the whole sketch, and with some extra work, it would be possible to identify which strokes do not participate on a complete parse tree (i.e. one labelled by the start symbol at the root). This is accomplished through cover sets as described in [1] .

4.3 Experiment 3: User Testing

This experiment evaluates the ability of our methodology to accommodate the variability in drawing styles produced by different users. To perform the experiment we

Table 2. Experimental test based on a set of users

Sample	# Recognized tokens	# Unrecognized tokens	Error Description
1	13	1	Constraints Failure on the token window.
2	15	0	No problems into the sketch.
3	15	0	No problems into the sketch.
4	15	0	No problems into the sketch.
5	12	3	Window recognized as a Rectangle.
6	15	0	The sketch does not present problems
7	15	0	The sketch does not present problems
8	15	0	The sketch does not present problems
9	15	0	The Sketch does not present problems
10	19	0	The sketch presents some extra tokens forming a bed-room that has not been recognized.

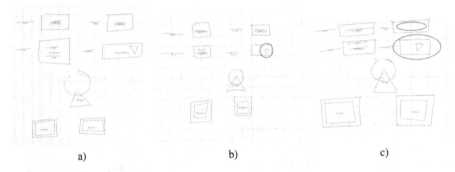

a)   b)   c)

Fig. 6. Three different samples from Sketched User Interfaces a) Complete Recognition, b) token mis-recognized and c) missing token

showed a sketched floor-plan, see fig. 5.a, to ten users who were then asked to draw a similar sketch. Each floor plan requires drawing roughly 40 strokes to depict 15 visual elements. We ran our parsing algorithm on these sketches. Out of the ten sketches, two were not correctly interpreted. Seven sketches contain the forty recognized strokes with a correct interpretation. On one case the user repeated a token, due to a mis-recognized production (constraint failure). Table 2 summarizes the experimental outcome. From the results obtained by the parsing methodology we can see that our approach is able to interpret floor plans interactively. Indeed, some of the failures can be corrected by adding interactive correction capabilities to our method. This will be done in future versions of the system.

We also tested the syntactic approach presented in this paper with a grammar that describes User Interfaces. Figure 6 shows three samples of the same sketch drawn by three different users. Figure 6.a shows a sketch where all the tokens are well recognized comparing it with the sketches in fig. 6.b and fig. 6.c we can see that in the first only one token is not recognized. In this case the unrecognized token is due to a stroke preprocessing error. The system has recognized an arc instead of two segments. The sketch in fig. 6.c has one token missing on its top part and there is a token that has not been well recognized. This is due to the fact the token was highly distorted.

5 Conclusions

In this paper we have presented a syntactic approach for sketch recognition including two components: First, we have described a grammatical formalism based on an Adjacency grammar that allows describing spatial relations among the different symbols that compose a hand-drawn sketch. Second, a parsing method based on a regular grid has been introduced. This parsing method allows partially re-scanning the input when new strokes are entered by users, which makes it suitable for interactive use. Preliminary experimental evaluation of the system shows that using a uniform grid with cells of adequate size reduces the spatial and temporal complexity of the parsing algorithm, thus making it suitable to be used in an interactive sketch recognition framework. Moreover, the syntactic approach presented is flexible enough to describe and interpret sketches from different problem domains, in this case architectural floor-plans and User Interfaces. Finally, the results obtained from the evaluation suggest a good scalability to larger drawings. As future work we plan to expand the system to make it more flexible and support interactive user correction and modification of input drawings.

Acknowledgements

This work has been partially supported by the Spanish project TIN2006-15694-C02-02, the consolider project CONSOLIDER-INGENIO 2010 (CSD2007-00018) and by Portuguese Science Foundation grant POSC / EIA / 59938 / 2004 (DECORAR).

References

1. Jorge, J.A.P.: Parsing Adjacency Grammars For Calligraphic Interfaces. Rensselaer Polytechnic Institute, New York (1995)
2. Jorge, J.A.P., Glinert, E.P.: Online Parsing of Visual Languages Using Adjacency Grammars. In: 11th International IEEE Symposium on Visual Languages, pp. 250–257 (1995)
3. Lladós, J., Valveny, E., Sánchez, G., Martí, E.: Symbol Recognition: Current Advances and Perspectives. In: Blostein, D., Kwon, Y.B. (eds.) GREC 2001. LNCS, vol. 2390, pp. 104–127. Springer, Heidelberg (2002)
4. Wenyin, L.: On-line Graphics Recognition: State-of-the-Art. In: 5th IAPR Workshop on Graphics Recognition, Barcelona, pp. 291–304 (2003)
5. Mas Romeu, J., Lamiroy, B., Sanchez, G., Llados, J.: Automatic Adjacency Grammar Generator from User Drawn Sketches. In: 18th International Conference on Pattern Recognition, Hong-Kong, pp. 1026–1029 (2006)
6. Wittenburg, K., Weitzman, L.: Relational Grammars: Theory and Practice in a Visual Language Interface for Process Modelling. In: International Workshop on Theory of Visual Languages, Italy (1996)
7. Bunke, H.: Hybrid Pattern Recognition Methods. In: Bunke, H., Sanfeliu, A. (eds.) Syntactic and Structural Pattern Recognition.Theory and Applications, pp. 307–347. World Scientific Publishing Company, Singapore (1990)

8. Hammond, T., Davis, R.: LADDER: A Language to Describe Drawing, Display and Editing in Sketch Recognition. In: Internaltional Joint Conference on Artificial Intelligence, Acapulco, pp. 461–467 (2003)
9. Bresenham, J.: Algorithm for Computer Control of a Digital Plotter. IBM System Journal 4(1), 25–30 (1965)
10. Alvarado, C., Davis, R.: Dinamically Constructed Bayes Nets for Multi-Domain Sketch Undrestanding. In: Internaltional Joint Conference on Artificial Intelligence, San Francisco, pp. 1407–1412 (2004)
11. Kara, L.B., Stahovich, T.F.: Hierarchical Parsing and recognition of hand-sketched diagrams. In: 17th Annual ACM Symposium on User Interface Software and Technology, pp. 13–22 (2004)
12. Costagliola, G., Deufemia, V., Risi, M.: A Multi-layer Parsing Strategy for On-line Recognition of Hand-Drawn Diagrams. In: IEEE Symposium on Visual Languages and Human-Centric Computing, pp. 103–110 (2006)
13. Landay, J.A., Myers, B.A.: Sketching Interfaces: Toward More Human Interface Design. IEEE Computer 34(3), 56–64 (2001)
14. Rubine, D.: Specifying Gestures by Example. In: 18th Annual Conference on Computer Graphics and Interactive Techniques, pp. 329–337 (2001)
15. Hammond, T., Davis, R.: Tahuti: A Geometrical Sketch Recognition System for UML Class Diagrams. In: AAAI Spring Symposium on Sketch Understanding, pp. 59–68. Palo Alto, Menlo Park (2002)
16. Marriot, K.: Constraint Multiset Grammars. In: IEEE Symposium on Visual Languages, St. Louis, pp. 118–125 (1994)
17. Crimi, C., Guercio, A., Nota, G., Pacini, G., Tortora, G., Tucci, M.: Relation grammars and their application to multi-dimensional languages. Journal of Visual Languages and Computing 2(4), 333–346 (1991)
18. Golin, E.J.: Parsing Visual Languages with Picture Layout Grammars. Journal of Visual Languages and Computing 2(4), 371–394 (1991)
19. Rekers, J., Schurr, A.: Defining and Parsing Visual Languages with Layered Graph Grammars. Journal of Visual Languages and Computing 8(1), 27–55 (1997)
20. Coüasnon, B.: DMOS, a generic document recognition method: application to table structure analysis in a general and in a specific way. International Journal on Document Analysis and Recognition 8(2-3), 111–122 (2006)
21. Costagliola, G., Deufemia, V., Risi, M.: Sketch Grammars: A Formalism for Describing and Recognizing Diagrammatic Sketch Languages. In: International Conference on Document Analysis and Recognition, Hong-Kong, pp. 1226–1230 (2005)

Categorization of Digital Ink Elements Using Spectral Features

José A. Rodríguez, Gemma Sánchez, and Josep Lladós

Computer Vision Center (Computer Science Department)
Edifici O, Campus Bellaterra, 08913 Bellaterra, Spain
{jrodriguez,gemma,josep}@cvc.uab.es

Abstract. In sketch-based interfaces, the separation of text and graphic elements can be essential when a system has to react to different kinds of input. Even if the interaction with the interface consists in drawing graphic elements, text input may be considered for some purposes, such as annotation, labelling, or input of recognizable text. This work deals with the detection of textual patterns in a set of digital ink elements. The main idea is that text needs a special hand behaviour to be produced, different from the behaviour employed to draw symbols or other graphic elements. Inspired by the models that describe handwriting as a system of coupled oscillations, we believe that the frequencies of these oscillations contain some information about the symbol nature. Therefore, we employ a descriptor that works in the Fourier space. Results show that this representation leads to distinguished patterns for text and graphic elements. The performance of our system is close to the performance one would obtain by using a handwriting recognition engine tuned for this task, while being much faster. Some benefits are also present when both approaches - the proposed and the engine - are combined.

1 Introduction

Digital ink applications can be classified into three levels according to the nature of the interaction [1]. In **Level 0**, the exclusive objective of a pen is to substitute the mouse, and the interaction takes place through usual elements of user interfaces - buttons, menus, etc. **Level 1** applications introduce interaction capabilities using the pen, such as recognition of certain gestures or text recognition inside widgets, but still rely on buttons and other conventional components for a subset of the input actions. Eventually, in **Level 2** applications, the full functionality can be reached using pen movements.

Most commercial applications working on TabletPCs are exactly the same versions that run on non-tablet computers, and therefore can be placed in the Level 0. Only a small portion of TabletPC applications integrate special functionalities that promote them to Level 1. Finally, the development of Level 2 applications is mainly reserved to research.

One distinctive feature of Level 2 applications is that the input can present different modes or (from now on) categories. For instance, consider a system with

W. Liu, J. Lladós, and J.-M. Ogier (Eds.): GREC 2007, LNCS 5046, pp. 181–190, 2008.
© Springer-Verlag Berlin Heidelberg 2008

textual and non-textual (e.g. graphic) elements. It is important to distinguish them since they may need to be processed differently. Therefore, one fundamental problem to be addressed in Level 2 recognition systems is to distinguish between different kinds of inputs.

In this article, the focus of attention is set on text detection, a task that can be useful in applications such as interpretation of architectural sketches [2][3], recognition of mathematical formulae/graphics [4], free-form ink input [5] [6] or automated proofreading [7][8]. Although these applications are mainly concerned with graphics recognition, they involve some step in which text can be present. For example, in architectural design a word can be employed to label a room. In such situations, it is important to face the problem of text/graphics separation, since these two types of information need to be processed by different modules of the system.

Indeed, in [7] [8] we presented applications for interpretation of proofediting sketches containing text and graphics. An example is given in Fig. 1. The system displays a document on screen (it can also be printed on a digital paper) and the user is supposed to proofread it using a pen. The user can draw correction marks and input words using specific rules to indicate what should be changed. A graphics recognition module interprets the graphical symbols and a text recognition module translates the text to ASCII format. This example illustrates the relevance of the text/graphics separation problem. But actually, in the mentioned application the interpretation is done without an explicit text/graphics separation module, being this task performed indirectly. Therefore, we believe a previous stage in which each input element is detected as text or graphics (or at least associated to a confidence measure) would bring valuable information that the system could exploit to increase performance and usability.

Until now, text detection in digital ink or its categorization is a not very popular subject, and it is solved using a variety of approaches. For instance [2], presents a multi-agent system for interpretation of architectural sketches, where the text detection module is one of the multiple agents. Its behaviour depends on

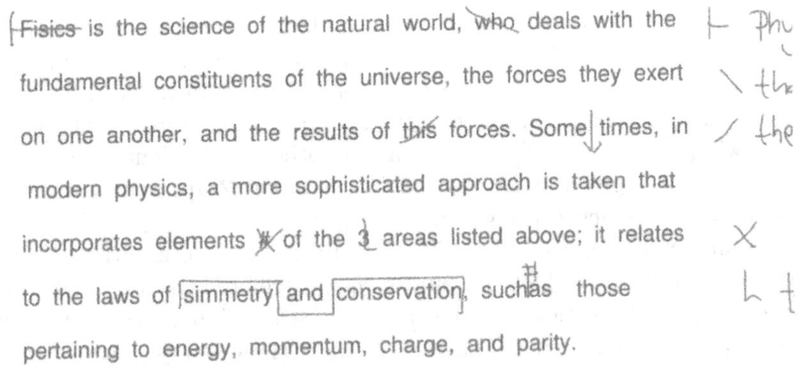

Fig. 1. Example of input in the proofreading recognition application

the confidence value obtained by a character recognizer and the feedback from other agents. Besides that, direct categorization of ink annotations is faced in [9], based on different detectors and heuristics for each different type of annotation. Probably, a more interesting work is [10], where a system for distinguishing text and graphics in digital ink is presented. It works using a hidden Markov model and common features. However, we are not aware of any other work using spectral features for ink categorization purposes.

The approach that is used in this work is inspired by the so-called motor models of handwriting [11]. It should be remarked that only the general idea is imported but a motor model itself is not applied in the current work. Briefly, motor models are one approach for on-line handwriting recognition that assume that the glyph, the image of what is written, is produced as a result of an oscillatory process of hand movement(s). The main idea in this work is that different categories of inputs, such as text and graphics, require very different types of hand movements to be produced. Therefore, a concrete analysis of these oscillations could reveal whether the input is a text, a graphic element, or one of the other categories of a system. The particular analysis consists in extracting a sequence of temporal features from the input and compute discrete Fourier transform (DFT) coefficients. It is known that the Fourier transform represents the signal in the frequency domain. It is not unreasonable to believe that by clustering these frequency coefficients, the different input categories will be recovered.

The rest of the article is structured as follows. In Section 2 the motivation for employing spectral features is introduced. Particular details of the methodology follow in Section 3. Experiments are reported in Section 4 and conclusions are drawn in Section 5.

2   Categorization Based on Frequential Patterns

The main idea in this work is that the different input elements (text, graphics, etc.) are produced in different ways by users moving their hand. For instance, writing a cursive word involves many loops, many angular changes, a global left-to right motion, etc. In contrast, most graphic elements such as symbols usually consist of regular patterns with long straight segments and few dominant directions.

There have been some works in the handwriting recognition domain (see [11]) that use what are called motor models. This kind of models take into account how the writing is physically produced. Usually, this is considered to be output of a process of coupled oscillations. In the present work we do not use such a model, but we are inspired by the fact that the different kinds of "behaviour" needed to draw text or symbols must be reflected in the form of these oscillations, and therefore in their frequencies. Thus the analysis of digital ink signals in the frequency domain can provide some useful information e.g. for distinguishing textual from non-textual elements.

3   Methodology

As mentioned in the introductory section of this work, the employed methodology obeys the following steps. First, a sequence of temporal features (in this case the direction feature) is determined from the input signal. Then, a number of DFT coefficients are computed from the previously obtained signal to build a fixed-length representation of the signal. Finally, this feature vector is classified into one of the possible categories.

3.1   Direction Feature

It is common to represent a digital ink element X as a sequence of features $X = X_1 X_2 \ldots X_t \ldots X_T$, where each X_t denotes the features observed at time t. These features are typically the horizontal and vertical coordinates (x_t and y_t) but this choice is not restrictive and other features can be included, such as pen pressure, height or even features computed from other features, as velocity, angles or angle differences. In any case, each of the features can be seen as a time signal.

In our work, after a preliminary study, the use of the direction feature alone showed best performance for our particular problem. We follow the proposal by Yu [12] to compute this feature. First, we determine the stroke segments between all pairs of consecutive points. Under these conditions, the direction feature is computed as:

$$d_t = \frac{\sum_{t-l}^{t+l} \theta(i, i+1)}{2l+1},$$
(1)

where $\theta(i, i+1)$ denotes the angle between segment i and segment $i+1$. The parameter l determines how many points are selected for computing each local direction and will produce a smoothing effect if its value is high enough. A value of $l = 1$ was employed in the present study.

3.2   DFT Coefficients

Instead of representing the difference feature using the raw values, discrete Fourier transform coefficients (DFT) [13] are computed. This has a number of advantages. On the one hand, an equivalent fixed-length feature vector is obtained. On the other hand, the analysis of the frequencies of the direction signal can provide some information on the way elements are sketched, following the ideas presented in the introduction.

Given the direction features $d = d_0 d_1 \ldots d_{n-1}$, its DFT coefficient of order k is computed using the common expression:

$$D_k = \sum_{m=0}^{n-1} d_m e^{-2\pi imk/n},$$
(2)

where $i = \sqrt{-1}$. For illustration purposes, in Fig. 2 we represent the distribution of text and non-text elements extracted from the database of [7] using 100 components. A PCA [14] is performed in order to reduce the dimensionality to 2D

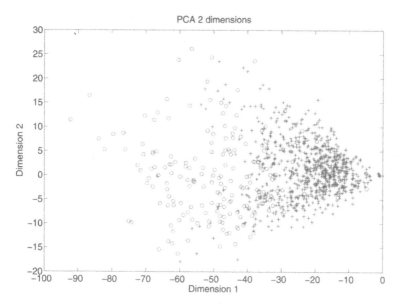

Fig. 2. Text (o) and non-text (+) categories in the space spanned by the two first principal components

and being able to visualize these results. Note that, even in the reduced space of 2 dimensions, text and non-text elements have distinct distributions using the proposed approach.

3.3 Category Detection

In the experimental section two experiments will be presented. The first one performs a clustering without using label information to find out the natural categories in which elements group themselves, which will be assessed visually and numerically. This part will be carried out using a fuzzy K-means [14] clustering. The second experiment will assume there are two categories (text and symbols) and we will formulate the problem as a binary classification using a Fisher linear discriminant.

As suggested in the previous subsection, a PCA can be used to reduce the dimensionality of the feature vectors, which might have a positive impact on performance.

4 Experiments

This section provides a description of the performed experiments as well as their results and some discussion. It should be mentioned that although the results presented here are promising, this is a work in progress.

The following experiments are carried out on a database of more than 900 symbols extracted from the application presented in [7]. The kind of elements one can encounter are similar to the ones that were shown in Fig. 1 and the ones that will be presented later in Fig. 3. These elements have been manually labelled as text and non-text. Indeed, it was considered that an element can be called text if it is visually identified as a text string of at least two characters. This choice is motivated by the fact that some symbols can unavoidably be confused with characters, such as crosses and "X", circles and "O", etc.

4.1   Unsupervised Clustering of Input Categories

The first experiment is designed as follows: given a set of input elements, they are clustered to determine which natural groups appear. As an unsupervised case, these labels will not be used other than for evaluation of the resulting clusters. As mentioned above, the clustering method is fuzzy K-means.

Several parameter combinations were tried. In the following, 100 DFT components are computed, and the two most informative dimensions are kept after a PCA. This unusually high number of harmonics and the posterior dimensionality reduction suggest to employ a descriptor presenting better energy compaction as future work, such as one based on discrete cosine transform (DCT).

The result using K=2 centroids leads to a cluster with majority of words (62.5% purity) and another with majority of non-words (97.2% purity). As usual, the purity of a cluster is defined as the fraction of most popular elements. For reference, the proportion of words / non-words in the dataset is about 20/80%.

It is interesting to see what the clusters will look like when more centroids are considered, for instance K=3. A priori, one could think that we can see a new category emerge. But in this dataset, we have not observed that the third cluster contains elements that can be considered distinct. Instead, it contains a mixture of text and symbols that have low degree of membership in the other two classes. This is positive, since it leads to an increase in purity for the other two clusters: 86.4% for the text cluster and 98.5% for the symbol cluster. Thus, the "intermediate" cluster could be considered a rejection cluster for holding the most unconfident elements. In particular, its purity is 78.1% (of symbols), thus not very significant with respect to the actual rates of text and symbols in the whole dataset.

Although the clusters, especially the word cluster, contain many samples of the "wrong" category, it is worth sorting the samples by the value of the membership function. In Fig. 3, the first 24 ranked samples in the text and symbol clusters are displayed, for the case of K=3. One can appreciate how by selecting the samples with highest degree of membership one is confident about the categorization decision.

Further increasing the number of clusters does not lead to any additional information about categories.

Although the clustering seems to work well in practice, the results show that there are only two relevant categories in this dataset. Under these circumstances, it is more reasonable to apply directly a discriminative classifier, which is done in the next section.

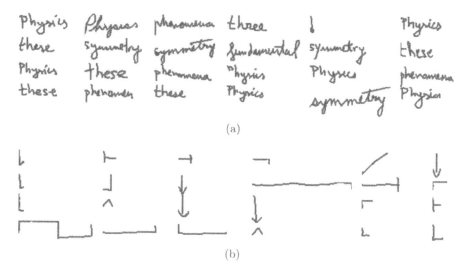

(a)

(b)

Fig. 3. First ranked 24 elements in class 1 (a) and class 2 (b)

4.2   Text Detection

The second experiment emulates a more practical framework in which a new ink element is presented to the input and the system must flag it with a text/non-text label. Due to the moderate number of samples, we use 2-fold cross validation for training and testing a Fisher linear discriminant. The distance of an element to the linear boundary (negative for elements at the other side of it) is used as confidence score.

The performance of the confidence scores can be assessed via tradeoff curves such as precision-recall or false rejection (FR) versus false acceptance (FA). At each point of such a curve, the system assumes a decision threshold that leads to a point of e.g. FR and FA. By varying the threshold, the curves can be obtained. One possible way of displaying such a curve is in the so-called Detection Error-Tradeoff (DET) plot [15]. In Fig. 4, the DET plot obtained using the DFT coefficients of the direction feature is shown as a solid line.

In our attempt to compare the proposed approach to an alternative system, we have built a wrapper around a commercial handwriting recognition (HWR) engine also to be able to distinguish between text and not text. The idea is very simple but it works in practice. First, we pass any input element to the HWR engine. If it is recognized as a word, the confidence given by the engine (a value between 0 and 1) is taken as the confidence score of being a word. In case that the output is not a word, or that is a word with just one character (for keeping the above criterion), the confidence is set to 0 since the chances of confusion are high and we would prefer to reject it. It is to be expected that most text is recognized as text with a high confidence score and that symbols are either not recognized as text or recognized as text with very low confidence.

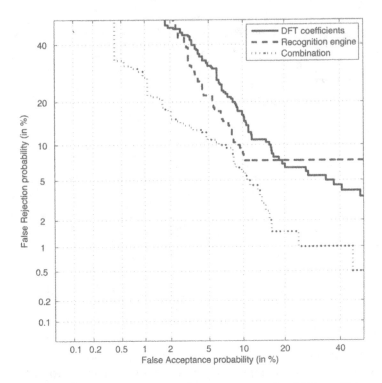

Fig. 4. DET plot comparing the curve obtained using the DFT coefficients of the direction feature (solid line), the confidence score given by a commercial recognition engine (dashed line), and a normalized combination of both scores (dotted line)

Under these conditions we have plotted the DET curve of this system in Fig. 4 as a dashed line. Its performance is a couple of points better than the DFT coefficients in terms of FA percentages for a fixed FR. However, in favour of the DFT coefficients, it should be mentioned that the cost of computing these features is less than twice the cost of running the HWR engine. Therefore, even if the HWR engine based system works better in practice, when one is concerned about the speed, the DFT features can be used if a loss of a couple of points is acceptable. It is often the case in many digital ink applications that they have to fulfill a real time requirement, so that the proposed assumption is not unreasonable.

But if one is really concerned about obtaining the best performance regardless of the speed, there is one more interesting experiment that can be run, namely the combination of both scores. If they provide different information, their combination can be beneficial. For each sample, the average of both scores is taken, where the scores have been previously standardized (i.e. set to mean 0 and standard deviation 1). The obtained DET plot is shown as a dotted line in Fig. 4. It can be appreciated how the combination of the HWR engine baseline system with the proposed approach produces a reduction of FA and FR which is better than any of the two individual systems.

5   Conclusions and Perspectives

This work is concerned with the detection of textual patterns in a digital ink application where the input can be either text or a symbol. Although this work is still in progress, we report interesting results using DFT coefficients of the direction feature, that we summarize below.

First, an unsupervised clustering of elements leads to separated clusters for text and graphics with high purities. These purities can be increased by introducing an "auxiliary" cluster that captures the doubtful samples and can be employed as rejection class. The purity is especially high if one considers samples with a degree of membership above a sufficiently high threshold.

Second, in a text/non-text classification experiment, the proposed features perform a couple of points worse (in terms of FA percentages) compared to a baseline system based on a HWR engine. If this loss can be tolerated, it will be a reasonable alternative when one is more concerned about the speed. But more interestingly, combining the HWR engine based system with the proposed one even leads to better performance.

As future work, three main objectives can be identified. Some of them are related to the current limitations of the system. The first one is continue the research on features for improving the performance of the proposed system at least to reach the performance of the HWR engine based system. Such a system would present the significant advantage of the reduced computational cost. Second, the generalization of this approach to the more general case of free-form annotations. In this work, the dataset has only allowed categorization of two types of elements, but in general the method could be applied to a richer input. At this point it could be also interesting to compare to a HMM-based approach such as the one proposed in [10]. Also, this method will be integrated in the proofreading system [7] with the objective of improving its interpretation performance.

Acknowledgements

This work has been partially supported by the Spanish projects TIN2006-15694-C02-02, and CONSOLIDER-INGENIO 2010 (CSD2007-00018).

References

1. Saund, E.: Games: the killer app for pen computing? In: Sketch-based interfaces and modeling 2006 (2006) (Opening Keynote)
2. Juchmes, R., Leclercq, P., Azar, S.: A freehand-sketch environment for architectural design supported by a multi-agent system. Computers & Graphics 29(6), 905–915 (2005)

3. Sánchez, G., Valveny, E., Lladós, J., Mas, J., Lozano, N.: A platform to extract knowledge from graphic documents. Application to an architectural sketch understanding scenario. In: Marinai, S., Dengel, A.R. (eds.) DAS 2004. LNCS, vol. 3163, pp. 389–400. Springer, Heidelberg (2004)
4. LaViola, J.J., Zeleznik, R.C.: Mathpad2: a system for the creation and exploration of mathematical sketches. ACM Transactions on Graphics 23(3), 432–440 (2004)
5. Wang, X., Shilman, M., Raghuparthy, S.: Parsing ink annotations on heterogeneous documents. In: Stahovich, T., Sousa, M.C., Jorge, J.A. (eds.) Sketch-based interfaces and modeling 2006, pp. 43–50 (2006)
6. Golovchinsky, G., Denoue, L.: Moving markup: repositioning freeform annotations. In: UIST 2002: Proceedings of the 15th annual ACM symposium on User interface software and technology, pp. 21–30. ACM Press, New York (2002)
7. Rodríguez, J.A., Sánchez, G., Lladós, J.: Automatic interpretation of proofreading sketches. In: Stahovich, T., Sousa, M.C. (eds.) Eurographics Workshop on Sketch-Based Interfaces and Modeling, Eurographics Association, pp. 35–42 (2006)
8. Rodríguez, J.A., Sánchez, G., Lladós, J.: A pen-based interface for real-time document edition. In: Ninth International Conference on Document Analysis and Recognition (ICDAR 2007) (2007)
9. Shilman, M., Wei, Z.: Recognizing freeform digital ink annotations. In: Document Analysis Systems, pp. 322–331 (2004)
10. Bishop, C.M., Svensen, M., Hinton, G.E.: Distinguishing text from graphics in on-line handwritten ink. In: IWFHR 2004: Proceedings of the Ninth International Workshop on Frontiers in Handwriting Recognition (IWFHR 2004), Washington, DC, USA, pp. 142–147. IEEE Computer Society, Los Alamitos (2004)
11. Plamondon, R., Srihari, S.N.: On-line and off-line handwriting recognition: a comprehensive survey. IEEE Transactions on Pattern Analysis and Machine Intelligence 22, 63–82 (2000)
12. Yu, B.: Recognition of freehand sketches using mean shift. In: IUI 2003: Proceedings of the 8th international conference on Intelligent user interfaces, pp. 204–210. ACM Press, New York (2003)
13. Cooley, J.W., Tukey, J.W.: An algorithm for the machine calculation of complex fourier series. Math. Comput. 19, 297–301 (1965)
14. Duda, R.O., Hart, P.E., Stork, D.G.: 4. Pattern Classification. In: Nonparametric techniques. Wiley-Interscience Publication, Chichester (2000)
15. Martin, A., Doddington, G., Kamm, T., Ordowski, M., Przybocki, M.: The DET curve in assessment of detection task performance. In: Proceedings of EuroSpeech 1997, pp. 1895–1898 (1997)

A Figure Image Processing System

Linlin Li[1], Shijian Lu[2], and Chew Lim Tan[1]

[1] School of Computing, National University of Singapore
{lilinlin,tancl}@comp.nus.edu.sg
[2] Institute for Infocomm Research, A*STAR, Singapore
slu@i2r.a-star.edu.sg

Abstract. Patent document images maintained by the U.S. patent database have a specific format, in which figures and descriptions are separated into different pages. This makes it difficult for users to refer to a figure while reading the description or vice versa. The system introduced in this paper is to prepare these patent documents for a friendly browsing interface. The system is able to segment an imaged page with several figures into individual figures and extract caption and label information from the figure. After obtaining captions and labels, figures and the relevant description are linked together, and thus users could easily refer from a description to the figure or vice versa.

Keywords: Graphics Recognition, Graphics Segmentation, User Interface.

1 Introduction

The U.S. patent database, run by the United States Patent and Trademark Office, maintains both patent text and images. The Web Patent Full-Text Database (PatFT) contains the full-text of over 3,000,000 patents, while the Web Patent Full-Page Images Database (PatImg) contains over 70,000,000 images, including every page of over 7,000,000 patents from 1790 to the most recent issue week. Most patents are presented in both full-text and image format. The retrieval requirement is huge in this database. The Web Patent Databases now serve over 25,000,000 pages of text (over 150,000,000 hits) per month to over 350,000 customers each month [1].

Unfortunately, although serving such a huge amount of customers, the layout of patent documents is not optimal for reading purpose. Firstly, a patent document is divided into five sections, including abstract, figure, description (text), claim and reference sections, each of which occupies a few consecutive pages. All figures of a patent document appear together in the figure section before the description section. Therefore, when a user is reading a description paragraph and wants to refer to some figures, he/she has to scroll back to the figure section. Secondly, as shown in figure 1, a typical figure, which illustrates an invention, generally has many labels, each of which represents a particular part of the invention. In order to locate the exact sentences about a label, a user may have

W. Liu, J. Lladós, and J.-M. Ogier (Eds.): GREC 2007, LNCS 5046, pp. 191–201, 2008.

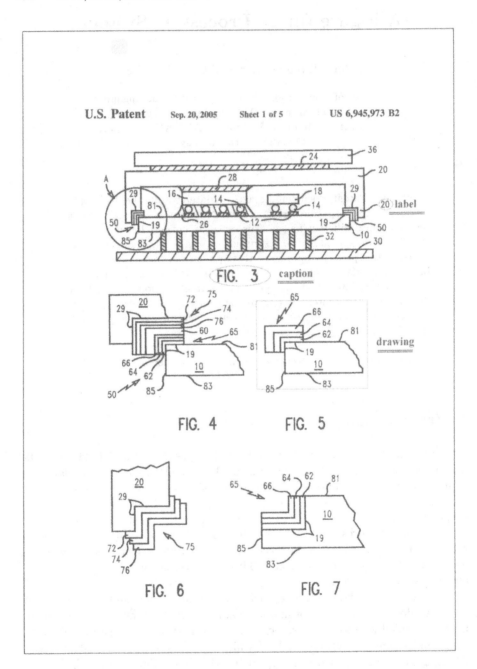

Fig. 1. A figure image of a patent document with several figures. A typical figure has a caption, drawings and several labels.

to go through the whole description section, which definitely will slow down the reading speed.

We have designed and implemented a system to link captions and labels in the description section and those in the figure section together, in order to provide users with a friendlier interface. In particular, two problems mentioned in the last paragraph are addressed: individual figures are cut out form a whole figure page and captions and labels were extracted. Hence, users are able to efficiently jump to the relevant description by clicking the captions or labels in a figure, or vice versa.

This paper is organized as follows. In sections 2, processing steps and techniques employed in the system are introduced in detail. In section 3, the testing results based on a self-prepared dataset are presented and discussed. Section 4 shows the interface in which descriptions and figures are paralleled. Section 5 and section 6 are the conclusion and acknowledgement section respectively.

2   System Description

The workflow of the system is illustrated in figure 2. Firstly, a page is checked whether it is a figure page or text page, by a black pixel density threshold. If it is a figure page, it is sent for further processing, otherwise thrown away. Secondly, the image is segmented into individual figures. Thirdly, bounding boxes of the caption and labels are located for each figure, regardless of whether the page is a rotated page or not. A rotated page is a page whose caption and label text is parallel to the vertical edge, an example of which is shown in figure 3. Because OCR will not accept rotated characters, the next step is to identify and rectify rotated pages. Then, content of the target bounding boxes are recognized by OCR software. After that, a post processing step is employed to filter out recognition errors and words out of our interest. Finally, captions and labels in figures are linked to those in the description by html functions. Users are able to swiftly search a caption or label in the description by clicking the one in the figure. A preliminary browsing interface of the system is shown in section 4.

2.1   Image Segmentation

As shown in figure 1, a page may have several figures. Several algorithms, including XYcut [2], Docstrum [3], and Voronoi Diagram [4], are presented in previous studies to segment a document image into zones. All these algorithms assume that distances among components within a zone are always smaller than distances between components from different zones. However, this is not true in these figure images. In figure 1, for example, the distance between drawings of FIG 4 and FIG 5 is much smaller than the distance between the drawing and the caption of either figure. In order to tackle this problem, a bottom-up segmentation method is proposed.

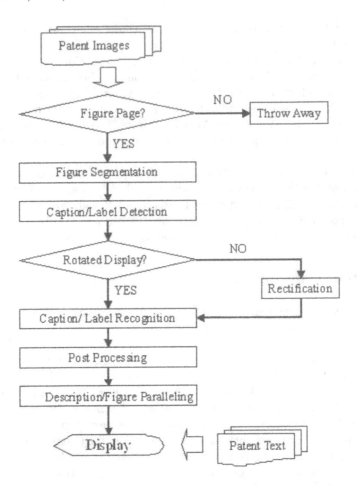

Fig. 2. The workflow of processing patent images

The assumption of a figure is that, there is only one main drawing in each figure (such as the drawing shown in figure 1, although a few figures do have several drawings, these drawings are quite near to each other and can be connected by smearing) and other small components scatter near around the main drawing. The segmentation method is as follows:

(1) A smearing algorithm [5] is used to merge separated parts of a figure. Although we assume there is only one drawing for each figure. The drawing may fall into several parts due to dashed lines, bad image quality or some other reasons. Also, a few patent pages have more than one drawing. Therefore, smearing is necessary for connecting parts of the drawing.

(2) A connected component analysis [6] is employed to detect all connected components in the image.

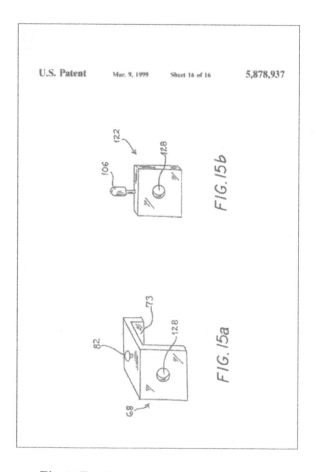

U.S. Patent   Mar. 9, 1999   Sheet 16 of 16   5,878,937

FIG.15b

FIG.15a

Fig. 3. Text in some pages is vertically posed

(3) Components are classified as either "seed" or "fragment" using heuristic rules based on width and height constraints. A seed is the drawing in a figure, while a fragment is a small component like a label.

(4) Each fragment F_i is merged into a seed S_j, which has the minimum distance to the fragment among all seeds. It is easy to understand that, if F_i and S_j are near to each other, they are very likely to be in the same figure. The distance between a fragment and a seed is defined as:

$$dis(F_i, S_j) = abs(C_x(F_i) - C_x(S_j)) + abs(C_y(F_i) - C_y(S_j)) \qquad (1)$$

where $abs(.)$ is the absolute value function, and $(C_x(A), C_y(A))$ is the centroid of component A, which is defined as:

$$C_x(A) = \frac{\sum_{i=1}^{n} p_{ix}}{n} \qquad (2)$$

$$C_x(A) = \frac{\sum_{i=1}^{n} p_{iy}}{n} \tag{3}$$

$p_i \in A$ is the black pixel index of A, and p_{ix} and p_{iy} are the xy coordinates of p_i.

(5) After merging, each seed is considered as a figure.

2.2 Caption/Label Detection

In this step, the caption and label detection method will be introduced. The suggested method is able to locate text in a figure regardless of the orientation of the text. Namely, the text can be detected whether it is vertically posed (as shown in figure 3), horizontally posed, or even upside-down. In addition, text can be successfully detected when skew is present in the page.

After all connected components are extracted, each component is classified as either a text component (character) or a graphics component according to five criteria: the width, height, width/height ratio, number of black pixels, and black pixel density (the number of black pixels over the area of the bounding box). The decision heuristic rule is shown in table 1. If a parameter of a component is within the suggested range, it is classified as a text component, otherwise a graphics component. The thresholds are based on observation of hundreds of patent images.

Table 1. Component classification criteria

	Width (pixels)	Height (pixels)	Width/Height Ratio	# of Black Pixels	Black Pixel Density
Upper	100	100	10	1500	0.9
Lower	5	5	1	900	0.2

Text components are grouped into words or phrases by the grouping function below [7]:

$$f(s_1, s_2) = \sqrt{\frac{k s_1 s_2}{s_1 + s_2}} \tag{4}$$

where s_1 and s_2 are the areas of two components; the coefficient k is a constant value, which can be adjusted according to the batch of samples in use ($k = 20$ in the experiment). If the Euclidian Distance between s_1 and s_2 are smaller than $f(s_1, s_2)$, they are considered as in the same group (a word).

2.3 Rotated Page Rectification

The rotated page detection is based on the text components (characters) obtained by the Caption/Label Detection method. The method employed in our system is similar to Docstrum algorithm [3]. The nearest neighbor character, denoted by N, of a character C is found, and the central-line defined by the centroids of both N and C is computed. Angles of all central-lines are collected, and the dominant angle decides whether the page is rotated or not.

2.4 Post Processing

This step is to pick up valid captions and labels from OCR output. Many undesirable contents other than captions and labels appear in the OCR output: the bounding box detection step may generate false boxes without text information, which lead to non-sense strings; a figure may contain text other than labels or captions; OCR process introduces recognition errors, too. Hence, the patterns of captions and labels are strictly defined, and only strings of these patterns are kept, otherwise thrown away. A valid label comprises of n consecutive digits ($n < 3$), such as '231'. A valid figure caption may be of either of two different patterns: "Fig *" and "Figure *". For example, both "Fig.1" and "Figure.1_(a)" are valid captions.

3 Experimental Results and Discussion

The system was tested on 50 patent images. To prepare the ground truth, these images were manually segmented. Captions and labels were manually located. The segmentation, detection and recognition output were compared with the prepared ground truth respectively. The results are shown in table 2. Our system segmented 84 out of 95 figures, 95% of which were correct. The system detected and recognized 80 out of 95 captions, all of which are correct. The system also detected and recognized 523 out of 571 labels, 95% of which were correct. The experiment result showed an encouraging performance.

Table 2. Figure segmentation, label and caption extraction results

	Ground truth	System output	Precision%	Recall%
Figures	95	84	95	84.00
Captions	95	80	100	84.21
Labels	571	532	95	88.51

It was found that, in the figure segmentation step, errors occurred when two figures are very near each other, such as FIG. 5 and FIG. 6 shown in figure 4. In this case, two figures were merged together by the smearing algorithm, thus were considered as a whole figure.

In the caption detection step, 100% precision is archived, namely, all captions outputted by the system were correct. However, some captions were missing due to extremely large character size. Characters of these captions were classified as graphics components in the component classification.

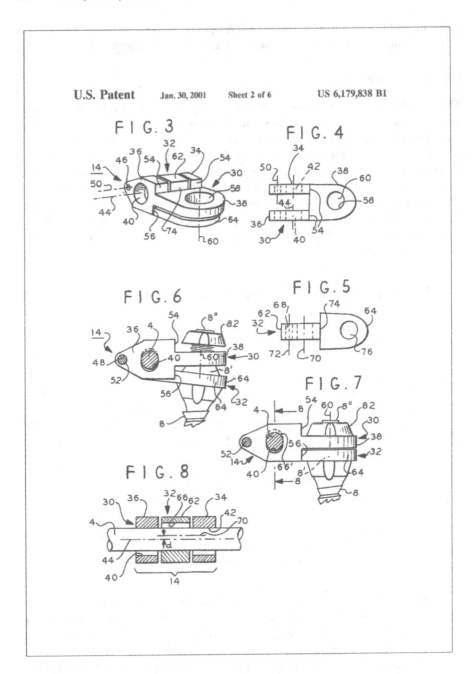

Fig. 4. Figures in a page are quite near to each other, which lead to errors in the figure segmentation

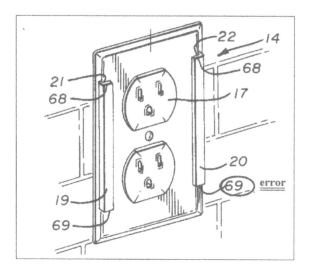

Fig. 5. Label 69 is connected to the drawing, thus it is classified as a graphics component together with the drawing

There were tens of labels in a figure usually. The main reason which caused errors in the label detection step was that, a few labels were connected to the drawing, such as label 69 shown in figure 5. Those labels could not be detected and then were missing in the final output. Further more, small graphics components might be classified as text-components, which also degraded the precision.

After checking with the results, we find that the bottle-neck of the performance is the text/graphics components classification step. In our current method, the classification depends on a few measures of the component, and many thresholds are set manually based on an observation of a small amount of patents. In the future, a more sophisticated classification will be included, which is able to automatically learn the thresholds from a large amount of training data.

4   User Interface Demo

Currently, the system provides Boolean retrieval of captions and labels in the description section, if a user clicks the corresponding areas of the figure. A snapshot of the preliminary user interface of our system is shown in figure 6. The left part of the interface is a text display window, and the right part is a figure display window. Individual figures are cut out and shown in order. When a label is clicked, the corresponding label occurrences in the text window are located and highlight. Figure 6 shows an example when label 23 was clicked.

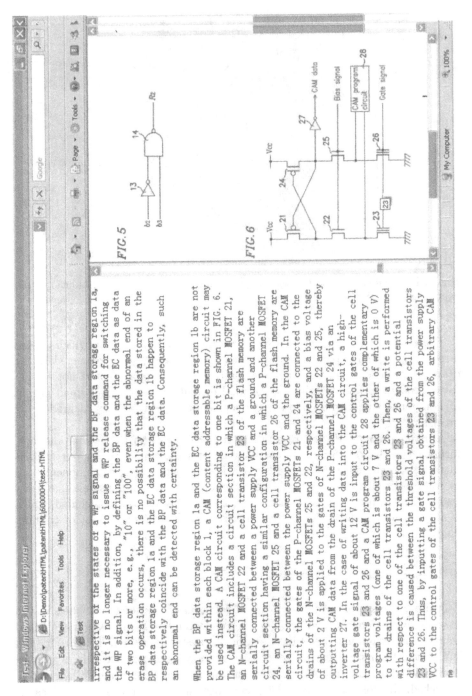

Fig. 6. A snapshot of the system interface, when label 23 had been clicked, and occurrences in the text were highlighted

5 Conclusion and Future Work

This paper introduces a figure image processing system, which is able to segment individual figures out of an image and extract their captions and labels, regardless of the orientation of the figure. This information is useful when preparing a better browsing interface of patent documents for users. The experiment results show an encouraging performance. After analyzing the performance, we find that the bottle-neck of the performance is text/graphics component classification. Our future work will focus on developing a more sophisticated classification method.

Acknowledgement. This project is supported by the NUS/MDA grant R252-000-325-279. The permission to use the data and domain knowledge provided by Patsnap Inc is gratefully acknowledged.

References

1. The U.S. Patent Database, http://www.uspto.gov/patft/help/contents.htm
2. Nagy, G., Seth, S., Viswanathan, M.: A Prototype Document Image Analysis System for Technical Journals. Computer 25, 10–22 (1992)
3. Gorman, L.: The Document Spectrum for Page Layout Analysis. IEEE Trans. Pattern Analysis and Machine Intelligence 15, 1162–1173 (1993)
4. Kise, K., Sato, A., Iwata, M.: Segmentation of Page Images Using the Area Voronoi Diagram. Computer Vision and Image Understanding 70, 370–382 (1998)
5. Wong, K.Y., Casey, R.G., Wahl, F.M.: Document analysis system. IBM Journal of Research and Development, 647–656 (1982)
6. Gonzalez, R., Woods, R.: Digital Image Processing, ch. 2. Addison-Wesley Publishing Company, Reading (1992)
7. Yuan, B., Kwoh, L.K., Tan, C.L.: Finding the best-fit bounding-boxes. In: 7th IAPR Workshop on Document Analysis Systems, New Zealand (February 13-15, 2006)

A Segmentation Scheme
Based on a Multi-graph Representation:
Application to Colour Cadastral Maps

Romain Raveaux, Jean-Christophe Burie, and Jean-Marc Ogier

L3I Laboratory – University of La Rochelle, France
{Romain.Raveaux01,Jean-Christophe.Burie,
Jean-Marc.Ogier}@Univ-lr.fr

Abstract. In this paper, a colour segmentation process is proposed. The novelty relies on an efficient way to introduce a priori knowledge to isolate pertinent regions. Basically, from a pixel classification stage a suitable colour model is set up. A hybrid colour space is built by choosing meaningful components from several standard colour representations. Thereafter, a segmentation algorithm is performed. The region extraction is executed by a vectorial gradient dealing with hybrid colour space. From this point, a merging mechanism is carried out. It is based on a multi-graphs data structure where each graph represents a different point of view of the region layout. Hence, merging decisions can be taken considering graph information and according to a set of applicative rules. The whole system is assessed on ancient cadastral maps and experiments tend to reveal a reliable behaviour in term of information retrieval.

Keywords: Colour Segmentation, Colour Space, Graphics Recognition, Document Understanding.

1 Introduction

Technical documents have a strategic role in numerous organisations, composing somehow a graphic representation of their heritage. In the context of a project called "ALPAGE", a closer look is given to ancient French cadastral maps related to the Parisian urban space during the 19th century. Hence, the data collection is made up of 1100 images issued from the digitalization of Atlas books and where each image contains a vast number of domain-objects, ie. Parcels, water collection points, stairs, windows/doors... From a computer science point of view, the challenge consists in the extraction of information from colour documents in the objective of providing a vector layer to be inserted in a GIS (Geographical Information System). Despite the large number of proposed document interpretation methods [1], only a handful of them focus in colour document analysis. In particularly, we state the case of Falco X.A [2] who proposed an object-line extraction method building a regular-mesh graph from a bitmap image where every pixel is treated as a node and putting an edge between every pair of adjacent nodes. Then the extraction problem can be considered

W. Liu, J. Lladós, and J.-M. Ogier (Eds.): GREC 2007, LNCS 5046, pp. 202–212, 2008.

as finding an optimal path between a start node and an end node. The limitation lies in the local aspect of this approach since the mesh graph construction is not achievable on the whole image; the start node is then user -defined to reduce the graph complexity. Among this reduced set of paradigms working with colour documents, the intention of Poh Kok Loo et al [3] is not much to deal with object extraction but more likely to distinguish text and graphic information. From this statement, a real place does exist for our system which aims to extract high level objects (parcel, water well ...) in a noisy environment because of the presence strong time due degradations: colour degradation, yellowing of the paper, pigment fading... Our paradigm is structured around three main ideas. The first, one investigates the colour image restoration in the objective to power up ancient colours. Secondly, a comparative study explores the several colour spaces in order to choose the best model for the segmentation process. Finally, a knowledge-based segmentation method is proposed where a priori information is introduced through the use of a multi-graphs data structure.

2 Pre-processing Steps: Colour Restoration and Colour Spaces

2.1 Colour Restoration

In introduction, we expressed the difficulties to analyse ancient documents which were deprecated due to the time, usage condition or storage environment. So clearly, a real need for image restoration has come up. A pre-process, a faded colour correction has been executed to bring colours back to original or at least to unleash colour significance. It works automatically by increasing non-uniformly the colour saturation of washed-out pigments without affected the dominant colour.

Let X be the colour vector for a given pixel: $X = \begin{bmatrix} R \\ G \\ B \end{bmatrix}$

Let Y be the data in an independent system axis:

$$Y = T(X - \mu)$$

Where:

T are the singular vectors of the covariance matrix.
μ is the mean vector.

Let Y' be the data extended according the direction the main factorial axis:

$$Y' = KY$$

$$K = \begin{bmatrix} k1 & 0 & 0 \\ 0 & k2 & 0 \\ 0 & 0 & k3 \end{bmatrix}$$ Coefficients are chosen experimentally.

Fig. 1. Original image

Fig. 2. Restored image

The restoration matrix is given as follow:

$$M = T^{-1}KT$$

Let X' be the vector containing the restored values: $X' = T^{-1}KT(X - \mu) + \mu$

2.2 Colour Space Selection

The choice of a relevant colour space is a crucial step when dealing with image processing tasks (segmentation, graphic recognition...). In this paper, a colour space selection system is proposed [Fig 3]. This step aims to maximize the distinction between colours while being robust to variations inside a given colour cluster. Each pixel is projected into nine standard colour spaces in order to build a vector composed of 25 colour components. Let C be a set of colour components. $C = \{Ci\}_{i=1}^{N} =$ {R,G,B, I1,I2,I3, L*, u*,v*,...} with Card(C) = 25. From this point, pixels represent a raw database, an Expectation Maximization (EM) clutering algorithm is performed on those raw data in order to label them. Each feature vector is tagged with a label representing the colour cluster it belongs to. Feature vectors are then reduced to a Hybrid Colour Space made up of the three most significant colour components. Hence, the framework can be split up in two parts: on one hand, the selection feature methods to decrease the dimension space and on the other, the evaluation of the suitability of a representation model. The quality of a colour space is evaluated according to its ability to make colour cluster homogenous and consequently to improve the data separability. This criterion is directly linked to the colour classification rate. The colour representation choice is done on-line after a pixel classification stage. Eleven colour spaces are evaluated according to their recognition rates [Table 2]. Hybrid spaces are built thanks to feature selection methods [4], [5], [6], [Table 1]. On all colour spaces, a 1-NN classifier using a Euclidian metric is performed in order to obtain the corresponding colour recognition rates.

Table 1. Selection feature methods in use

Name	Type	Evaluation	Searching algorithm
CFS [4]	Filter	CFS	Greedy stepwise
DHCS [6]	Filter	Principal Component Analysis(PCA)	Ranker
GACS [5]	Wrapper	Classification	Genetic Algorithm
OneRS [4]	Wrapper	Classification	Ranker

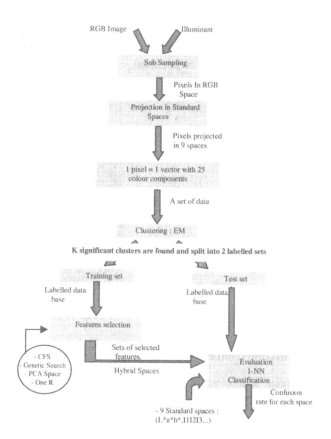

Fig. 3. A framework for colour space selection

Table 2. Pixel data bases description and Colour classification rate

Image	Type	# of clusters	$\left\|X_{training}\right\|$ pixels	$\left\|X_{test}\right\|$ pixels
Image of document	Ancient Cadastral Map	14	110424	110424

Cadastral Map			
Colour Spaces	Rate	Colour Spaces	Rate
RGB	0.4556	HIS	0.6334
I1I2I3	0.7778	La*b*	0.7334
XYZ	0.4223	L*u*v*	0.6667
YIQ	0.6889	**DHCS**	**0.64**
YUV	0.6223	**CFS**	**0.9667**
AC1C2	0.7	**GACS**	**0.8112**
PCA	0.7556	**OnRS**	**0.5889**

3   Indice Extractions

3.1   Black Layer and Colour Layer Separation

Basically, the black layer is extracted by using an Otsu binarization [9] on the luminance channel (Y channel of the YIQ colour space). Figure 4, 5 show the two extracted layers.

Fig. 4. Black layer

Fig. 5. Colour layer

3.2   Colour Segmentation from Hybrid Colour Space

Once the source image is transferred into a suitable hybrid colour space, an edge detection algorithm is processed. This contour image is generated thanks to a vectorial gradient according to the following formalism. The gradient or multi-component gradient takes into account the vectorial nature of a given image considering its representation space (RGB for example or in our case hybrid colour space). The vectorial gradient is calculated from all components seeking direction for which variations are the highest. This is done through maximization of a distance criterion according to the L2 metric, characterizing the vectorial difference in a given colour space. The approaches proposed by DiZenzo[7] first, and then by Lee and Cok under a different formalism are methods that determine multi-components contours by calculating a colour gradient from the marginal gradients.

Given 2 neighbour pixels P and Q characterizing by their colour attribute A, the colour variation is given by the following equation:

$$\Delta A(P,Q) = A(Q) - A(P)$$

The pixels P and Q are neighbours, the variation ΔA can be calculated for the infinitesimal gap: dp = (dx, dy)

$$dA = \frac{\partial A}{\partial x}dx + \frac{\partial A}{\partial y}dy$$

This differential is a distance between pixels P and Q. The square of the distance is given by the expression below:

$$dA^2 = \left(\frac{\partial A}{\partial x}\right)^2 dx^2 + 2\frac{\partial A}{\partial x}\frac{\partial A}{\partial y} dxdy + \left(\frac{\partial A}{\partial y}\right)^2 dy^2$$

$$= adx^2 + 2bdxdy + cdy^2$$

$$a = \left(G_x^{e1}\right)^2 + \left(G_x^{e2}\right)^2 + \left(G_x^{e3}\right)^2$$

$$b = G_x^{e1}G_y^{e1} + G_x^{e2}G_y^{e2} + G_x^{e3}G_y^{e3}$$

$$c = \left(G_y^{e1}\right)^2 + \left(G_y^{e2}\right)^2 + \left(G_y^{e3}\right)^2$$

Where, E can be seen as a set of colour components representing the three primaries of the hybrid colour model. And where G_n^m can be expressed as the marginal gradient in the direction n for the mth colour components of the set E.

The calculation of gradient vector requires the computation at each site (x, y): the slope direction of A and the norm of the vectorial gradient. This is done by searching the extrema of the quadratic form above that coincide with the eigen values of the matrix M.

$$M = \begin{pmatrix} a & b \\ b & c \end{pmatrix}$$

The eigen values of M are:

$$\lambda_\pm = 0.5\left(a + b \pm \sqrt{(a-c)^2 + 4b^2}\right)$$

Finally the contour force for each pixel (x,y) is given by the following relation:

$$Edge(x, y) = \sqrt{\lambda_+ - \lambda_-}$$

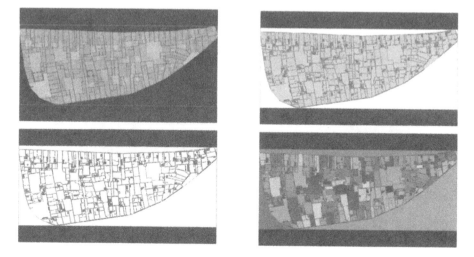

Fig. 6. Top Left: the source image; Top right: the gradient values; Bottom left: the binary image from edge values, Bottom right: the white connected components analysis

These edge values are filtered using a two class classifier based on an entropy principle in order to get rid off low gradient values. At the end of this clustering stage a binary image is generated. This image will be called as contour image through the rest of this paper. Finally, regions are extracted by finding the white areas outlined by black edges [Fig 6].

4 Syntaxic Level: A Multi Graph Data Structure

Each graph represents a point of view of the region layout. The three graphs in use rely on the same basement where one node corresponds to one region [Fig 7]. However, edges and edge attributes change from one graph to another. The graph definitions are explains to the next paragraph.

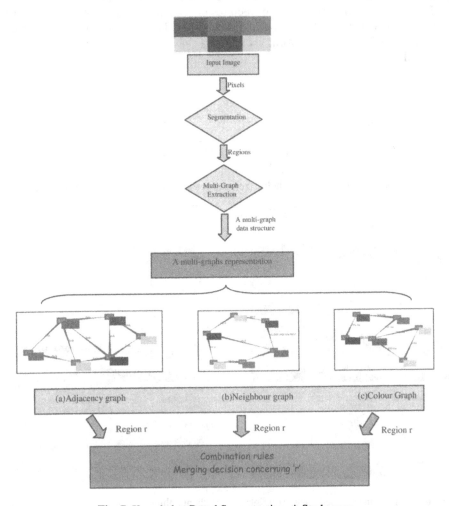

Fig. 7. Knowledge-Based Segmentation. A Study case.

4.1 Graph Definitions

4.1.1 Adjacency Graph Definition

Each region represents a vertex in this graph. Then, edges are built using the following rule: two vertices are linked with an undirected and an unlabelled edge if one of the nodes is connected to the other node in the corresponding image.

4.1.2 Neighbour Graph Definition

Each region represents a vertex in this graph. Then, edges are built using the following rule: two vertices are linked with an undirected edge if one of the nodes is one of the h nearest neighbours of the other node in the corresponding image. The h value, concerning the number of significant neighbours, is issued from a comparative study. This graph is a representation of the spatial layout of the regions. Edges are labelled with the spatial distance between the two region centres.

4.1.3 Colour Graph Definition

Each region represents a vertex in this graph. Then, edges are built using the following rule: two vertices are linked with an undirected edge if one of the nodes is one of the k closest neighbours of the other node in a colorimetric point of view. The colour graph expresses information concerning the colour distance between regions. This graph is an interpretation from the colour point of view of the region organisation. Edges are labelled with the colour distance between the two regions.

5 Semantic Level: Merging Rules

The segmentation process [7] is not enough to reconstruct high level information just because the black layer or an important colour difference can obstruct the spatial progression of the edge detection. From this fact, graphs are used to guide the region merging system. Each graph provides a context for a given node n. and this later is merged if it fulfils four rules [Fig 7], these conditions are a *priori* knowledge representation. The merging mechanism is described through the following algorithm:

Algorithm: Merging scheme for multi-graphs data structure

Require: the colour similarity threshold: T_{color}

Require: the spatial distance threshold: $T_{spatial}$

Ensure: A list of M Regions.

```
1: Start
2: MergingFlag=true
3: While MergingFlag == true do
4:         MergingFlag=false
5:             for i=1 to Number of Nodes do
6:               CurNode = GetCurrentNode(i)
```

```
7:              MNode=CombinationRules(CurNode, T_color, T_spatial)
8:              if MNode exist then
9:                 MergeNodesInMultiGraphs(CurNode,MNode)
10:                MergingFlag=true
11:                break
12:             end if
13:          end for
14: end while
15: return the remaining nodes.
16: End
```

When a node $n1$ is merged with another one $n2$, the whole structure has to be updated, the three graphs have to be coherent, hence, the merged node n1 is deleted in the three graphs and its edges are linked to $n2$.

Elements:
Regions
Black layer

Syntax:
[element1] operator [element2]

Operators:
$\subset$: A region is inside another one
$\otimes$: Close Spatially: Threshold (Tspatial)
$\oplus$: Close Colour: Threshold (Tcolour)
$\bullet$: Connectedness
$\bar{x}$: Not.

Function:
Merge (R1,R2) $\equiv$ R1 $\cup$ R2

Merging rules:
[Rule 1] (R1 $\subset$ R2) $\wedge$ (R2 $\neq$ BlackLayer) $\rightarrow$ R1 $\cup$ R2
[Rule 2] (R1 $\bullet$ R2) $\wedge$ (R1 $\oplus$ R2) $\rightarrow$ R1 $\cup$ R2
[Rule 3] (R1 $\otimes$ R2) $\wedge$ (R1 $\oplus$ R2) $\rightarrow$ R1 $\cup$ R2
[Rule 4] (R1 $\bullet$ R2) $\wedge$ (R1 $\overline{\bullet}$ BlackLayer) $\rightarrow$ R1 $\cup$ R2

Fig. 8. Merging rules

6 Experimental Results

In order to compare the segmentation defined by an expert and the results generated by our segmentation algorithm, the Vinet [8] criterion is chosen.

The Vinet's measure is calculated by counting common pixels between the user defined image and the result computed by an image processing task. This can be expressed by the given formula:

$$Vinet = 1 - \frac{R_i - \left(R_i \cap V_j\right)}{R_i} \quad \text{between } [0, 1]$$

Where R_i, V_j are respectively a region from the ground truth image and a region from the image segmented by the computer. The higher is the measure the better is the segmentation. Figure 8 illustrates the segmentation results while in Table 3 Vinet measures are reported. Experiments highlight the profit to include different point of view of a single segmentation to guide a merging method.

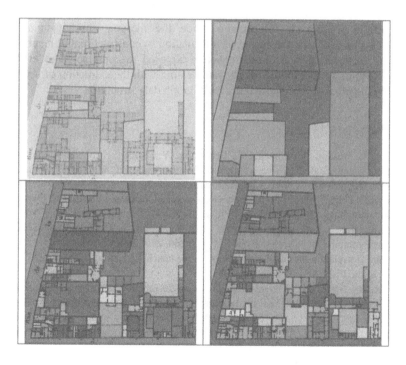

Fig. 9. Top left: Source Image; Top right: Ground truth composed of 17 regions; Bottom left: Segmentation without merging; Bottom right: Segmentation after the application of merging rules

Table 3. Segmentation results

Segmentation processes	# of regions	Vinet measure
Region growing[5]	1301	0.253
Our approach without merging	516	0.51794
Our approach with merging	74	0.64423

7 Conclusion

In this paper, we have been interested in an original problem, the colour graphic document analysis and an application to ancient cadastral maps. The aim was to identify graphic elements, which are defined by their colour homogeneity. Our contribution relies on a document oriented segmentation scheme. Firstly, a vectorial gradient working in a hybrid colour space is applied in order to achieve the partition in regions. Thereafter, a higher point view is given by a multi-graph representation, these multiple sources of information guide the merging mechanism. Finally, applicative rules condition the fusion of information to construct higher level objects. In addition, research perspectives are being explored to combine more efficiently black and colour layers. Ongoing works are dealing with the extraction of a visibility graph from the black layer to enrich the multi graph data structure.

References

1. Ogier, J.M., Mullot, R., Labiche, J., Lecourtier, Y.: Semantic coherency: the basis of an image interpretation device application to the cadastral map interpretation. IEEE Transactions on Systems, Man, and Cybernetics, Part B 30(2), 322–338 (2000)
2. Falco, X.A., Udapa, J.K., et al.: User-steered image segmentation paradigms: live wire and live lane. Graphical Models and Image Processing 60(4), 233–260 (1998)
3. Loo, P.K., Tan, C.L.: Adaptive Region Growing Color Segmentation for Text Using Irregular Pyramid. Document Analysis Systems, 264–275 (2004)
4. Hall, M.: Correlation-based feature selection for machine learning. Thesis in Computer Science at the University of Waikato (1998)
5. Raveaux, R., Burie, J.C., Ogier, J.M.: A colour document interpretation: Application to ancient cadastral maps. In: The 9th International Conference On Document Analysis, pp. 1128–1132 (2007)
6. Rugna, J.D., Colantoni, P., Boukala, N.: Hybrid color spaces applied to image database. Electronic Imaging, SPIE, 254–264 (2004)
7. Di Zenzo, S.: A note on the gradient of a multi-image. Computer Vision, Graphics, and Image Processing 33 (1986)
8. Rosenberger, C.: Adaptative evaluation of image segmentation results. In: 18th International Conference on Pattern Recognition, pp. 399–402 (2006)
9. Otsu, N.: A Threshold Selection Method from Graylevel Histogram. IEEE Transactions on System, Man, Cybernetics 19(1), 62–66 (1978)

Smoothing a Network of Planar
Polygonal Lines Obtained with Vectorization

Alexander Gribov and Eugene Bodansky

Environmental Systems Research Institute (ESRI)
380 New York St., Redlands, CA 92373-8100, USA
agribov@esri.com

Abstract. A new method of smoothing polygonal lines obtained as the result of vectorization and creating the network is suggested. This method performs not only smoothing but also filtering of vectorization errors taking into account that these errors appear not only as the errors of vertices but as errors of node coordinates as well. An important part of this algorithm is a technique of building piecewise polynomial base functions for local approximation of the polylines of the network. The suggested algorithm has a linear computational complexity for exponential weight functions. The necessity of using finite weight functions is shown. Algorithms of calculating tangents and curvatures are derived. Shrinking errors and errors of parameters are analyzed. A method of compensation of the shrinking errors is suggested and how to do smoothing with variable intensity is shown.

Keywords: smoothing, error filtering, local approximation, polygonal lines, network of polylines, weight functions, vectorization, line drawings.

1 Introduction

The result of vectorization of line drawings (maps, engineering drawings, electrical schematics, etc.) is usually an ordered set of points located close to the centerlines of linear objects or borders of solids. The set of points is used to build *source polygonal lines* that approximate the planar curves of the line drawing.

Curves depicted on the line drawings are called *ground-truth curves*. The differences between the source polygonal line and the appropriate ground-truth curve are *vectorization errors*. The sources of these errors are discretization of line drawings and scanning noise, errors of raw vectorization and post-processing, etc.

Vectorization errors cause not only approximation errors but also corrupt properties of ground-truth curves, for example, smoothness.

A smoothing problem is resolved if there is an algorithm for the calculation of the coordinates of any point of the smooth curve that approximates a ground-truth curve with some required accuracy. Smoothing source polygonal lines can be required not only for smoothing, when it is known *a priori* that *ground-truth curves* are smooth, but also for resolving such problems as building of a tangent at a given point on the curve or calculation of the local curvature of the curve.

W. Liu, J. Lladós, and J.-M. Ogier (Eds.): GREC 2007, LNCS 5046, pp. 213–234, 2008.

The input of these algorithms is the vertices of the source polygonal lines, some control parameters, and thresholds.

There are many different types of smoothing algorithms. We use a smoothing algorithm based on the convolution with exponential kernel [8] that possesses many good properties. These algorithms can be used to smooth isolated curves. Note that these methods can shift extremities of smoothing curves relative to the extremities of source polygonal lines. If curves create a network, the extremities of the curves after such smoothing can disperse and fail to coincide with appropriate nodes. This means that the topology of the network will be changed (see Fig. 1a).

Usually, curves of the network are smoothed as stand-alone curves. In this case, *a priori* information that extremities of the appropriate curves belong to the same nodes of the ground-truth lines is not taken into account. Pay attention to the fact that while solving problems of smoothing and vectorization error filtering of polygonal lines, it is necessary to try to use all *a priori* information about properties of ground-truth curves (shape, continuousness, smoothness, etc.). Disregarding of this information degrades accuracy.

Usually, to save the network topology, the positions of the ends of the curves should be fixed (Fig. 1b). To do this, approximating curves approach the source polygonal lines as they get nearer to their extremities. This solution to the problem has essential shortcomings. In proximity to the extremities of the curves, errors of approximation are not reduced but remain identical to the errors of the source polygonal lines. Another shortcoming is the necessity of the special algorithm usage, so this method of smoothing a network of curves will require not one but two algorithms.

In the article, we suggest a smoothing method that saves the topology of the network and performs smoothing and vectorization error filtering of all lines in the network (Fig. 1c). This method permits moving network nodes together with the extremities of the smoothed lines. In this case, smoothed approximating curves depend not only on the single appropriate polygonal line but on the other polygonal lines of the network as well. A similar task is resolved in [4], [5] but nevertheless there is an essential difference between this task and our problem.

In this article, a new method of building piecewise polynomial base functions is suggested. Usage of these functions gives us a possibility to upgrade our algorithm of

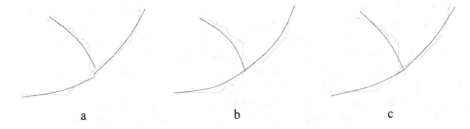

a   b   c

Fig. 1. Smoothing a network of curves. a) Each curve is smoothed as a stand-alone curve. b) The positions of the curves' ends are fixed before smoothing. c) The network of curves is smoothed with the suggested method.

smoothing isolated lines to the algorithm of smoothing a network of polygonal lines. It is shown that in this case weight functions have to be finite, functions may differ from zero only at the finite interval. Usage of the exponential weight functions gives a possibility to suggest an algorithm with a linear computational complexity, even for smoothing a network of polygonal lines.

The article describes the derivation of the algorithms for the calculation of tangents at given points on the curve and the calculation of the local curvatures of the curve. It analyzes shrinking errors inherent to smoothing algorithms that use convolution and suggests a new method for reduction of these errors. It also suggests the technique for changing smoothing intensity and shows that this technique can be used for compensation of errors generated when using a parametric description of curves.

It is important that the suggested method can be used for the smoothing of any polygonal lines, not only those obtained by vectorization.

In Appendix A, the technique of building a piecewise polynomial basis for a network of polygonal lines can be found. Appendix B contains an algorithm of calculation of integrals used to smooth polygonal lines with an exponential weight function and piecewise polynomial basis. This algorithm reduces the computational complexity of the suggested method of smoothing to $O(n)$.

2   Statement of the Problem

Given the network of source polygonal lines $L_i(x, y) = 0$ obtained by vectorization of line drawings. These lines form a graph G_z without loops. Lines are edges of the graph; nodes and pseudo-nodes (corners) are vertices.

Given prototypes of the source polygonal lines (ground-truth curves) are smooth.

It is necessary to build smooth curves $\hat{L}_i(x, y) = 0$ that approximate the source polygonal lines with minimizing of accumulated weighted least-squares residual. These curves have to create graph G_a with the same topological properties as graph G_z. The weighted function depends on statistical information about ground-truth curves and vectorization errors.

Fig. 2 shows an example of the network of the source polygonal lines, which consists of N lines. p_i is the number of the line that precedes the i-th line on the

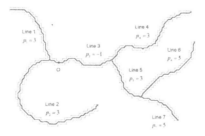

Fig. 2. The network of the polygonal lines

trajectory from point O (an origin) to the i-th line. If the origin O belongs to the i-th line, $p_i = -1$.

Curves are described with parametric equations. The parameter l is the distance from the origin (point O) of the curve to the current point P measured along the curve. Because the distance grows monotonically it is possible to substitute l for t, which is the time required to move along the curve from the origin to P with a given velocity V. So the parametric equations of curves are $x = x(t)$ and $y = y(t)$. For definiteness assume that $V = 1$ unless stipulated otherwise.

Each of these functions is defined on the segment $[S, E]$. Each function can be smoothed separately. For simplicity, a designation $u = u(t)$ is used, where u is in turn x and y. Note that the speed of growing u (the derivative of u with respect to t) is restricted and cannot be more than V.

When a network of polygonal lines is analyzed, the location of any point is defined not only with parameter t but with the number of the line i as well, and the parametric equations of the line become $u = u_z(i, t)$.

In the simplest case (an approximation of 0-th order), coordinates $u_a(i_P, t_P)$ of the smooth curve corresponding to i_P-th polygonal line of the network can be estimated with a convolution

$$u_a(i_P, t_P) = \frac{\sum_{i=1}^{N} \int_{S_i}^{E_i} w(t - t_P) \cdot u_z(i, t) \cdot dt}{\sum_{i=1}^{N} \int_{S_i}^{E_i} w(t - t_P) \cdot dt}. \tag{1}$$

Here $[S_i, E_i]$ is the domain of definition of $u(i, t)$ of i-th polygonal line, $w(t)$ is a weight function, and the i_P is the number of the line to which an origin (point O) belongs. Fig. 3 shows behavior of the weight function in the vicinity of a node.

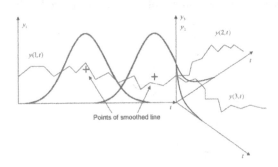

Points of smoothed line

Fig. 3. Behavior of the weight function in the vicinity of the node

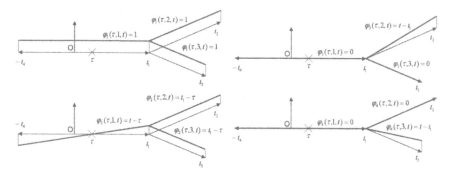

Fig. 4. Piecewise polynomial basis functions ($i^* = 1, N = 3, K = 1$)

There is another way for construction of the weight function such as applying diffusion process over the network of polygonal lines [13], [14]. While this approach constructs kernels with physical properties it is computationally expensive.

The better result can be obtained using a local polynomial approximation. The behavior of a smooth curve in the vicinity of any point P belonging to the same curve as the origin O can be described with a linear combination of piecewise polynomial basis functions $\varphi_j(\tau,i,t)$:

$$u_a(\tau,i,t) = \sum_{j=1}^{M} \alpha_j^u(\tau) \cdot \varphi_j(\tau,i,t) . \tag{2}$$

Here the parameter τ defines the location of the point P; t defines the location of the arbitrary point of the line L_i; j is the number of the basis function, $j = \overline{1,M}$; M is the number of the basis functions, $M = N \cdot K + 1$, and K is the order of the polynomial approximation.

The technique of building basis piecewise polynomial functions for a network without loops is described in Appendix A. Fig. 4 shows an example of such basis polynomial functions.

It is necessary to find the coordinates of the point $P(i^*,\tau)$ of the smooth approximation of the network of polygonal lines that minimize the functional

$$F^u = \sum_{i=1}^{N} \int_{S_i}^{E_i} w(t-\tau) \cdot \left(u_a(\tau,i,t) - u_z(i,t)\right)^2 \cdot dt . \tag{3}$$

Because generally free curves cannot be described by a polynomial with constant coefficients (due to some reasons it is not recommended to use approximating polynomials with an order more than 3), the absolute value of residuals $|u_a(\tau,i,t) - u_z(i,t)|$ becomes greater when $|t - \tau|$ grows. So usually the weight function with the maximum value in $t = \tau$ and damping with increasing $|t - \tau|$ is used. The weight function gives possibility to perform a local approximation. The more $u(i,t)$ (coordinates of ground-truth curves) differs from a polynomial with constant coefficients, the more the speed of damping of the weight function has to change.

To calculate coordinates of the point P of the smooth approximating curve, it is necessary to find the values of the coefficients $\alpha_j^u(\tau)$ that guarantee the minimum of F^u. Designate these coefficients $\hat{\alpha}_j^u(\tau)$. The best approximation of the source polygonal line with a smooth curve in point P is

$$\hat{u}(i^*,\tau) = \hat{u}_a(\tau,i^*,\tau) = \sum_{j=1}^{M} \hat{\alpha}_j^u(\tau) \cdot \varphi_j(\tau,i^*,\tau) \ . \tag{4}$$

So our task was reduced to solution of the regression equations for each point of polygonal lines. The solution has to be a network of continuous smooth curves minimizing the accumulated weighted squares of residuals for each point.

3 The Optimal Solution

Equations defining optimal values of coefficients $\hat{\alpha}_j^u(\tau)$ can be derived from

$$\frac{\partial F^u}{\partial \alpha_j^u} = 0 \ , \tag{5}$$

where $j = 1,...,M$ and M is the number of basis functions. In the matrix form, derived equations can be written as

$$A(\tau) \cdot \alpha^u(\tau) = b^u(\tau) \ , \tag{6}$$

where $A(\tau)$ is a square matrix $(M \times M)$ with elements equal

$$a_{i,j}(\tau) = \sum_{k=1}^{N} \int_{S_k}^{E_k} w(t-\tau) \cdot \varphi_i(\tau,k,t) \cdot \varphi_j(\tau,k,t) \cdot dt \tag{7}$$

and $b^u(\tau)$ is a matrix $(M \times 1)$ with elements equal

$$b_i^u(\tau) = \sum_{k=1}^{N} \int_{S_k}^{E_k} w(t-\tau) \cdot \varphi_i(\tau,k,t) \cdot u_z(k,t) \cdot dt \ , \tag{8}$$

therefore,

$$\alpha^u(\tau) = A^{-1}(\tau) \cdot b^u(\tau) \ . \tag{9}$$

Here $A^{-1}(\tau)$ is an inverse matrix. Some difficulties can appear when the system (6) is ill conditioned. It happens when the weight function is defined on a finite interval (about finite weight functions see below) and the end of this interval is very close to some node. This could be the cause of changing the dimension of the matrix $A(\tau)$.

To eliminate these difficulties, pseudo-inverse matrix $A^{+}(\tau)$ can be used instead of the corresponding inverse matrix [6].

$$\alpha^{u}(\tau) = A^{+}(\tau) \cdot b^{u}(\tau) . \qquad (10)$$

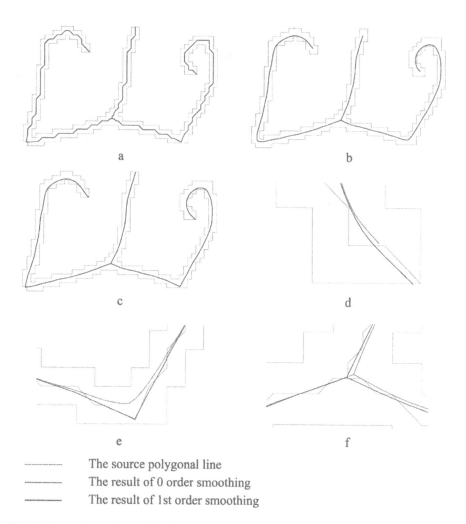

a   b

c   d

e   f

--------   The source polygonal line
--------   The result of 0 order smoothing
--------   The result of 1st order smoothing

Fig. 5. Comparison of approximating curves obtained with 0 and 1st order smoothing: a) the source polygonal lines, b) 0 order smoothing, c) 1st order smoothing, d) extremity, e) corner, f) intersection

There are papers dedicated to smoothing isolated curves and polygonal lines. A convolution with a Gaussian kernel is used in [1]-[3]. This is a special case of our task for $N=1$ (an isolated curve) and $K=0$ (a polynomial basis function is a constant). The method of approximation of isolated polygonal lines with smooth curves with the minimum of the accumulated squares residuals was described in [7]. It uses the polynomial basis functions of arbitrary order, but the functional does not use a weight function, and it is a global (not local) approximation method. It is more appropriate for approximating regular (not arbitrary) functions.

The task, resolved in [8], is also a special case of the task resolved in this paper but in a more common case than the task analyzed in [1]-[3], [7]. The suggested method is intended for smoothing of isolated polygonal line where $N=1$, but, in contrast to [7], a weight function is used (i.e., it is a local smoothing), and, in contrast to [1]-[3], there are used polynomial basis functions of an arbitrary order. Fig. 5 shows examples of influence of the approximation order on the result. Close to nodes and extremities, the first order approximation ($K=1$) gives a better result than the zero order approximation ($K=0$). Besides, the smoothing with $K=0$ corrupts corners (pseudo nodes), because the basis consists only of the constant function. The smoothing with $K=1$ significantly improves the result in corners. Fig. 5d, 5e, and 5f show improved result of smoothing in extremities, corners, and intersections.

In [4], [5] the task of the approximation of functions defined at the network of the curves is analyzed. However, a method was not developed for building basis functions for the network of the curves. Other differences between this task and our problems are parameter t is known without errors, the values of the functions may not be equal in nodes, the network is described with an oriented graph, and weight functions are one-sided functions.

4 Weight Functions

Computational complexity of the smoothing algorithms from [1]-[3] based on convolution with the Gaussian kernel (weight function) $g(t) = \dfrac{1}{\sqrt{2\pi}\sigma} e^{-\frac{t^2}{2\sigma^2}}$ equals $O(n \cdot \log n)$ if the fast Fourier transform is used [9].

In [8] an exponential weight function $w(t) = e^{\frac{-|t-\tau|}{\sigma}}$ was used, where σ is the radius of smoothing. For simplification of equations, suppose that $\sigma = 1$. To save the intensity of smoothing, change the scale of parameter t. To do this it is enough to change the speed V of moving along the line, assigning $V = \sigma$ because

$$\int_{S}^{E} w\left(\frac{t-\tau}{\sigma}\right) \cdot f(t) \cdot dt = \sigma \cdot \int_{S/\sigma}^{E/\sigma} w(x-\tau') \cdot f(x \cdot \sigma) \cdot dx, \quad \text{where} \quad x = \frac{t}{\sigma}, \text{ and}$$

$$\tau' = \frac{\tau}{\sigma}.$$

In case of an exponential weight function and polynomial basis functions, coefficients $a_{i,j}(\tau)$ and $b_i^u(\tau)$ defined with (7) and (8) can be presented with the sum of the integrals $\int\limits_S^E e^{-|t-\tau|}\cdot(t-\tau)^k\cdot dt$ and $\int\limits_S^E e^{-|t-\tau|}\cdot(t-\tau)^k\cdot f(t)\cdot dt$.

Computational complexity of the smoothing algorithm in this case is defined with calculation of these integrals or the integral

$$I = \int\limits_S^E e^{-|t-\tau|}\cdot(t-\tau)^k\cdot f(t)\cdot dt \ , \tag{11}$$

because another integral is a special case of the integral (11).

In [8] an iterative algorithm of calculation of integrals I was suggested. The computational complexity of this algorithm is $O(n)$. This algorithm is described in Appendix B.

For any network of polygonal lines the weight function should not produce singular matrix $A(\tau)$ from (6); otherwise, the solution of (6) becomes unstable. For example, the Butterworth filter does not possess this property and, therefore, cannot be used as a weight function for the smoothing of a network of polygonal lines. Another example, the weight function constructed by diffusion process [14] can be used only for zero order smoothing.

Both weight functions, exponential and Gaussian, are infinite functions. They are defined and differed from 0 on the infinite line. In smoothing, calculation of derivatives and curvature, and solving other applied problems, these infinite functions are used on a finite interval and they are usually simply truncated to fit the interval. In [10] it was shown that this is a source of major errors in finding second and higher order derivatives because truncation introduces discontinuities that cannot be differentiated, and a very simple method to avoid this error was described. The method consists of using weight functions that are defined on the finite intervals. If the weight functions are piecewise smooth, finite, and equal to 0 on the ends of the intervals where these functions are defined, approximating smooth lines have derivatives of the first and second order.

One of the weight functions that satisfy these conditions on the interval $[1,-1]$ is the function $w(t) = (1-|t|)\cdot e^{-3|t|}$, $t\in[-1;1]$ (see Fig. 6a). Function $w(t) = (1+|t|-5\cdot t^2+3\cdot|t|^3)\cdot e^{-t}$, $t\in[-1;1]$ shown in Fig. 6b has the same properties and, in addition, derivatives of the function in $t=0$ and on the ends of the interval equal 0. The same properties have function $w(t) = \dfrac{1}{e^4-6\cdot e^2+1}\times$

$\times[(1-5\cdot e^2+(3\cdot e^2-1)\cdot|t|)\cdot e^{|t|}+(e^4-e^2+(e^4+e^2)\cdot|t|)\cdot e^{-|t|}]$ as shown in Fig. 6c.

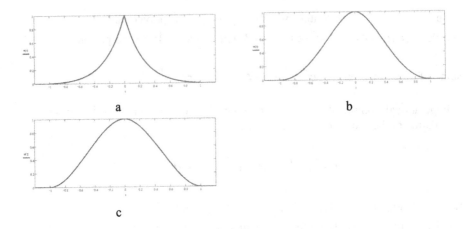

a

b

c

Fig. 6. Examples of finite weighting functions

All these weight functions can be described as

$$w(t) = \sum_{i=0}^{n} a_i \cdot |t|^i \cdot e^{-|t|} + \sum_{i=0}^{m} b_i \cdot |t|^i \cdot e^{|t|} . \qquad (12)$$

Smoothing with such weight functions can be calculated with complexity $O(n)$ because an iterative technique of calculation of integrals (11) suggested in Appendix B can be used not only for infinite but for finite exponential functions as well. Note that because it is possible to approximate Gaussian function with functions (12) it is possible to do convolution with the Gaussian kernel with linear computational complexity.

Using finite weight functions allow you to smooth fragments of a network of polygonal lines without an edge effect and to smooth a network of polygonal lines with loops.

Smoothing fragments without edge effects could be required, for example, for joining adjacent fragments. For protection from an edge effect it is necessary to process the fragment that is more than the source one and to clip the result. The problem is how to evaluate the size of the processing fragment. If the weight function is finite and its radius is r, the increment of the source fragment dimension has to be not less than r plus maximum shift approximating line from the source polygonal line. For a finite weight function the maximum shift for zero order approximation is not more than r, and for the first order approximation is not more than $2r$. It is impossible to evaluate the shift of approximated curve if the order of approximation is more than 1. So for smoothing a fragment of polygonal lines using the first order approximation is recommended. The results of zero and first order approximations differ when the interval, where the weight function is defined, overlaps extremities of polylines, corners, or intersections.

Using finite weight functions with piecewise polynomial basis functions permit smoothing a network of polygonal lines with loops. It is impossible to do this if weight functions are infinite because, in this case, the number of the piecewise polynomial basis functions can be infinite.

Sometimes, it is necessary to smooth a network of lines at the same time insuring that some points of the lines remain fixed. If the weight function is symmetric it is enough to do a skew-symmetric prolongation of functions $x(t)$ and $y(t)$ with respect to $t = \tau$, where τ is parameter values corresponding to fixed points of the lines [11]. Such prolongation and smoothing with infinite weight functions can produce the new network with an infinite number of lines. Usage of finite weight functions guarantees that the number of lines of the new network will be finite and, therefore, the basis will consist of the finite number of functions.

5 Estimation of Derivatives

There are two very important applied problems: building of a tangent $\vec{q}$ to the curve and calculating a local curvature C. If a line is defined with functions $x(t)$ and $y(t)$, then

$$\vec{q} = (x', y') , \tag{13}$$

$$C = \frac{\left| x' \cdot y'' - y' \cdot x'' \right|}{(x'^2 + y'^2)^{\frac{3}{2}}} . \tag{14}$$

Here x', y', x'' and y'' are first and second derivatives at t (see, for example, [12]).

Using the suggested method of smoothing, first and second derivatives at t can be calculated directly without previous smoothing of source polygonal lines.

It follows from (4) that

$$\frac{d}{d\tau}\hat{u}(i^*, \tau) = \frac{d}{d\tau}\left(\sum_{j=1}^{M} \hat{\alpha}_j^u(\tau) \cdot \varphi_j(\tau, i^*, \tau) \right) =$$
$$= \sum_{j=1}^{M}\left[\frac{d}{d\tau}\hat{\alpha}_j^u(\tau) \cdot \varphi_j(\tau, i^*, \tau) + \hat{\alpha}_j^u(\tau) \cdot \frac{d}{d\tau}\varphi_j(\tau, i^*, \tau) \right]. \tag{15}$$

Because of using the technique of building basis functions described in Appendix A,

$$\hat{u}(i^*, \tau) = \alpha_1^u(\tau) . \tag{16}$$

So (15) may be simplified as following

$$\frac{d}{d\tau}\hat{u}(i^*, \tau) = \frac{d}{d\tau}\alpha_1^u(\tau) . \tag{17}$$

It means that it is necessary to find $\dfrac{d}{d\tau}\alpha^u(\tau)$. Using (6) obtain

$$A(\tau)\cdot\frac{d}{d\tau}\alpha^u(\tau) = \frac{d}{d\tau}b^u(\tau) - \frac{d}{d\tau}A(\tau)\cdot\alpha^u(\tau) . \tag{18}$$

It follows from (7) and (8) that $A(\tau)$ and $b^u(\tau)$ can be written as linear combinations of

$$\psi(\tau) = \int_S^E w(t-\tau)\cdot\tau^k\cdot f(t)\cdot dt . \tag{19}$$

It is possible to use such technique of building basis functions that $\varphi_j(\tau,i,t)$ doesn't depend on τ. Then it is possible to simplify differentiation of $\psi(\tau)$ at τ, because in this case $\psi(\tau) = \int_S^E w(t-\tau)\cdot f(t)\cdot dt$.

To calculate $\dfrac{d}{d\tau}\hat{u}(i^*,\tau)$, it is necessary to find derivatives of $\psi(\tau)$ at τ, calculate the right part of (18), and resolve this equation (calculate $\dfrac{d}{d\tau}\alpha_1^u(\tau)$).

To find second derivatives, differentiate (15)

$$\frac{d^2}{d\tau^2}\hat{u}(i^*,\tau) = \frac{d^2}{d\tau^2}\left(\sum_{j=1}^M\hat{\alpha}_j^u(\tau)\cdot\varphi_j(\tau,i^*,\tau)\right) =$$

$$= \sum_{j=1}^M\left[\frac{d^2}{d\tau^2}\hat{\alpha}_j^u(\tau)\cdot\varphi_j(\tau,i^*,\tau) + 2\cdot\frac{d}{d\tau}\hat{\alpha}_j^u(\tau)\cdot\frac{d}{d\tau}\varphi_j(\tau,i^*,\tau) + \right.$$

$$\left. + \hat{\alpha}_j^u(\tau)\cdot\frac{d^2}{d\tau^2}\varphi_j(\tau,i^*,\tau)\right]. \tag{20}$$

Taking into account (17), obtain

$$\frac{d^2}{d\tau^2}\hat{u}(i^*,\tau) = \frac{d^2}{d\tau^2}\alpha_1^u(\tau) . \tag{21}$$

The second derivative of $\alpha_j^u(\tau)$ can be obtained by differentiation of (18):

$$A(\tau)\cdot\frac{d^2}{d\tau^2}\alpha^u(\tau) =$$

$$= \frac{d^2}{d\tau^2}b^u(\tau) - 2 \cdot \frac{d}{d\tau}A(\tau) \cdot \frac{d}{d\tau}\alpha^u(\tau) - \frac{d^2}{d\tau^2}A(\tau) \cdot \alpha^u(\tau). \qquad (22)$$

To find second derivatives x'' and y'' it is necessary to find the second derivatives of $\psi(\tau)$ (19), calculate the right part of (22), find $\frac{d^2}{d\tau^2}\alpha^u(\tau)$ resolving this equation, and to use (20).

Higher order derivatives can be found similarly.

Fig. 7 illustrates polygonal lines, approximating smooth curves, and tangent lines calculated with the described algorithm.

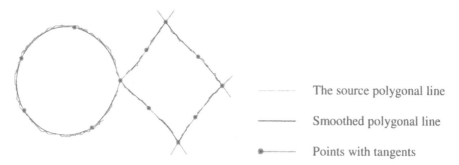

-------	The source polygonal line
———	Smoothed polygonal line
•———	Points with tangents

Fig. 7. Tangents to smooth curves

6 Correction of Shrinking Error

The shrinking error is inherent for any smoothing method based on averaging.[1] This problem of smoothing methods was analyzed in [3]. "Convolution with any averaging filter will cause each point to migrate toward the center as a monotonic function of curvature and degree of smoothing . . . For any application in which it is important to know the location of a curve in the image . . . this variable migration would be a critical defect." Fig. 8 illustrates a shrinking error and its correction.

The author of [3] suggested a method using the assumption that a smooth curve in the vicinity of any point can be approximated with good precision with circular arcs. The next assumption is the radius of the curvature of the ground-truth curve is close to the radius of the curvature R of the resulting smooth line. In [3] smoothing is performed with a Gaussian convolution. The equation has been derived of the relation between the shrinking error (shift of the current point), the curvature radius R, and

---

[1] Low-pass filters address the shrinkage problem for stand-alone polygonal lines. These filters are not nonnegative functions and cannot be applied directly to smoothing of the network of polygonal lines (see Section 4). The solution offered in [14] is computationally expensive and, therefore, is not considered in this paper.

The source polygonal line

The result of smoothing

The result of smoothing
with shrinking correction

Fig. 8. Examples of shrinking errors and compensation of these errors

the parameter of the weight function σ. The solution of this equation is used for correction of the shrinking error. The shortcoming of this method is that an estimation of the curvature radius can be unstable because equation used for evaluation of the curvature radius has two solutions and estimation depends on the σ. Besides, the first assumption (good local approximation of a smooth curve with circular arcs) is not always valid.

In this paper, a new method for correction shrinking errors is suggested. The location of the given point on the polygonal line relative to the appropriate point on the smoothed curve is defined by the vector of a residual $\vec{V}(t)$. Let (ξ, η) be a rotating right-hand orthogonal coordinate frame with an origin coincident with a current point on the smoothed curve. An angle of rotation of the axis ξ relative to the tangent of the current point is constant. In Fig. 9a this angle equals 0 (i.e., the axis ξ coincides with the local tangent). Designate projections of $\vec{V}(t)$ at the axes of the rotating coordinate frame $s_\xi(t) = \vec{V}(t) \cdot \vec{i}_\xi$, $s_\eta(t) = \vec{V}(t) \cdot \vec{i}_\eta$, where $\vec{i}_\xi$ and $\vec{i}_\eta$ are appropriate unit vectors.

The idea behind this suggested method for compensation of shrinking errors is very simple. The difference between the source polygonal line and the smooth approximating curve $\vec{V}(t)$ can be measured. Projections $s_\xi(t)$ and $s_\eta(t)$ consist of high- and low-frequency components. Shrinking error is a low-frequency component. The period of other components is less than the radius of the weight function σ. Smoothing $s_\xi(t)$ and $s_\eta(t)$ with the method suggested in this article, obtain a shrinking error $\hat{s}_\xi(t)$ and $\hat{s}_\eta(t)$ and use it for compensation.

$$\hat{\vec{u}}(t) = \vec{\hat{u}}(t) + \hat{s}_\xi(t) \cdot \vec{i}_\xi + \hat{s}_\eta(t) \cdot \vec{i}_\eta . \tag{23}$$

Fig. 8 illustrates the effect of the compensation of shrinking errors.

This method can be used for a network of curves as well. If a network of curves does not have loops, and the first curve is chosen, the curves of the network can be ordered; therefore, for each i-th curve the previous curve (p_i-th curve) is known.

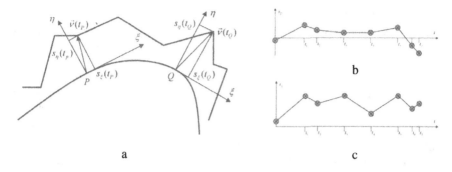

Fig. 9. The difference between the source polygonal line and the smooth approximating curve. a) Vectors of residuals $\vec{V}(t)$, b) and c) projections $s_\xi(t)$ and $s_\eta(t)$ of $\vec{V}(t)$ at the axes of the rotating coordinate frame.

Let the number of the first curve be i^*. Let the rotating coordinate frame of i-th curve $(\xi,\eta)_{i^*,i}$ at the common node for curves i-th and p_i-th coincides with the rotating coordinate frame of p_i-th curve $(\xi,\eta)_{i^*,p_i}$. Calculate projections $s_\xi(i,t)$ and $s_\eta(i,t)$ of the residual vectors $\vec{V}(i,t)$. Using the method of smoothing suggested in this paper with $s_\xi(i,t)$ and $s_\eta(i,t)$ instead of $x(i,t)$ and $y(i,t)$, obtain estimations of shrinking errors $\hat{s}_\xi(i,t)$ and $\hat{s}_\eta(i,t)$. Selection of another first curve $i^{**} \neq i^*$ causes rotating frames $(\xi,\eta)_{i^*,i}$ and $(\xi,\eta)_{i^{**},i}$ at a constant angle for any i. Therefore, this technique of building rotating coordinate frames guarantees that the result will not be depended on the choice of the first curve, and topology of the network will be saved.

In section 2 was noted that signals $x(i,t)$ and $y(i,t)$ have limited derivatives. In contrast to these signals, derivatives of projections $s_\xi(i,t)$ and $s_\eta(i,t)$ can be infinite. Because of that, estimations $\hat{s}_\xi(i,t)$ and $\hat{s}_\eta(i,t)$, obtained with smoothing of first order and higher, could be unlimited near nodes. So it is impossible to guarantee the absence of edge effects using a first order smoothing algorithm for processing fragments of curves. For a zero order smoothing, a maximum value of shrinking error estimation is not more than the radius of the weight function. Therefore, for correction of shrinking errors, using a zero order smoothing is recommended.

7 Smoothing with Variable Intensity

In some cases, the frequency of the vectorization error can change as a function of t, thickness of line, etc. It can require modifying an intensity of smoothing that can be done with changing a smoothing radius $\sigma(t)$ of the weight function. But changing σ can break some properties of the resulting approximating curve (smoothness, existing of second derivatives, etc.). Because of variability of σ, it is impossible to represent coefficients $a_{i,j}(\tau)$ and $b_i^u(\tau)$ as linear combinations of integrals I defined with expression (11). So calculation of new integrals can be required for evaluation of coordinates of each point of the smoothed curve, and computational complexity of the algorithm can be increased and become $O(n^2)$. Algorithms for calculation of derivatives and curvatures of the resulting curves become more complicated as well because, in this case, it is necessary to take into account an influence of σ changing.

This paper suggests another method based on changing the scale of parameter t. As was shown in section 4, a scale of parameter t depends on the velocity V. So it is possible to change the velocity V inversely proportional to changing noise frequency. In this case, the intensity of smoothing will be changed in spite of the constant radius of the weight function: the intensity will increase with increasing V and decrease with decreasing V. After appropriately changing the scale of the parameter, it is possible to use the algorithm of smoothing and calculation of derivatives described in this paper without any modifications. The properties of the resulting smoothed curve and the computational complexity will not be changed.

8   Errors of a Parametric Representation of Curves

For smoothing plain curves they have to be described with two parametric equations $x = x(t)$ and $y = y(t)$. Each of these functions is processed separately. Until now, it was supposed that the true value of the parameter t is known. In reality it is not true and this is the main difference between processing curves and time signals. The parameter t is a scaled length of the segment of the processing curve (a source polygonal line) between the initial and current points. Therefore, the parameter t depends on vectorization errors. Fig. 10 shows a ground-truth line, source polygonal line, and function $y = y(t)$.

The ground-truth line is a horizontal line; the source polygonal line has a saw-tooth noise. If parameter t is measured as a distance along the ground-truth line, the function $y(t)$ and the source polygonal line are coincided, but if parameter t is measured along the source polygonal line, $y(t)$ will be again a saw-tooth function, but frequency of this function will be decreased. The error of parameter t changes a scale of t, and, consequently, this error can be compensated with intensity of smoothing (see a previous section), but it can be done only if there is enough information about the vectorization error. The error of parameter t can also be compensated by performing several iterations if to use the resulting smoothing curve for measuring values of parameter t after each iteration.

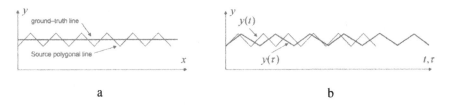

Fig. 10. An error of calculation of parameter t (t was calculated along the ground-truth line; τ was calculated along the source polygonal line)

9 Conclusion

The problem of smoothing and approximating of the network of source polygonal lines was considered. The result has to be the curves that are the best approximation of the source polygonal lines creating the network with the same topology as the network of appropriate ground-truth curves and possessing some of their properties (for example, continuity, smoothness, and existence of second derivatives). Under "the best approximation", we understand the minimum of accumulated weighted squares of the residual. Usually, to receive such results, locations of line ends and nodes are fixed and each line is smoothed independently of others. One of the results of smoothing is filtering vectorization errors. Therefore, it is possible to reckon that the smooth line approximates the ground-truth curve better than the source polygonal line. Fixing line ends worsens precision of approximation of ground-truth lines in the vicinity of line ends and nodes. The suggested method can be used for smoothing a network of lines with free line ends and nodes saving a topology of the source network of lines.

The suggested method is based on smoothing separate lines with a convolution with an exponential weight function and local approximation of functions $x = x(t)$ and $y = y(t)$ – parametric descriptions of the curve – with a linear combination of polynomial basis functions [8]. The weighting function takes into account the impossibility of good approximation of a source polygonal line with the polynomial of limited order with constant coefficients and restricted order. A smoothing algorithm using a Gaussian weight function has a computational complexity $O(n \log n)$. Using exponential weighting functions, it is possible to receive an iteration algorithm of smoothing with a linear computational complexity. If necessary, a Gaussian weight function can be approximated with exponential functions.

The construction of the optimum weight functions require further research and will be addressed in the future works.

For smoothing a network of lines, the basis functions have to be piecewise smooth. Using such basis, it could be possible to smooth curves without corruption of corners. A technique of building piecewise polynomial basis functions is described in Appendix A. The number of basis functions depends on the order of the polynomial and the number of lines of the network. When a smoothing algorithm is used, cycles of a network of lines are treated as infinite periodical lines. Due to this, the number of basis functions becomes infinite. So the suggested method of smoothing with an infinite weight function can be used only for the network of lines without loops (a tree graph). If the weight function is not equal to zero only at finite segment of t, the cycles are opened and basis functions could be built for arbitrary networks of lines.

Using finite weight functions simplifies resolving some important tasks. One of them is smoothing a fragment of the network of lines without an edge effect. It is easy to understand how important this problem is taking into account that usually universal vectorization systems have to have not only automatic vectorization but also (very popular among users) interactive modes of vectorization. Such modes as tracing and raster snapping are very useful for vectorization of complicated line drawings (with several thematic layers and complex line types) and raster images of poor quality. Using these modes, it is possible to obtain the result in real time independently on the size of the source document. It can be produced by vectorizing and post-processing not the full image but only such fragment that is of interest. However, it is necessary to do it without an edge effect. Then it could be saved and matched later with adjacent fragments. Using finite weight functions and approximation of 1st order simplifies solution of these tasks.

The suggested method of smoothing can be used for evaluation of derivatives of $x(t)$ and $y(t)$, which are necessary to build tangents and calculate a local curvature of the smoothed curve.

If smoothing has to have variable intensity, it is possible to change not the radius of smoothing but the scale of t. After an appropriate change to the scale of t, the suggested smoothing method can be used without any modification.

The result of smoothing depends on the error of evaluation of parameter t. This property is inherent to all methods that use parametric description of curves – calculating t as a length of segments of source lines – and smooth functions $x = x(t)$ and $y = y(t)$ independently. This error can be decreased with recurrent smoothing and recalculation of t for each iteration.

Smoothing methods based on averaging, like the method suggested in this paper, have shrinking errors. The suggested method for correction of these errors gives a good result if vectorization errors and ground-truth curves have different frequency characteristics. The method can be used for compensation of shrinking errors obtained after smoothing a network of lines. An algorithm of shrinking correction has to be based on the finite weight function and zero order approximation.

The suggested smoothing method is invariant to rotation of a network of lines and can be used to the dimensions that are more than 2.

All examples illustrating the suggested method and algorithms were obtained with universal vectorization system ArcScan for ArcGIS.

References

1. Asada, H., Brady, M.: The curvature primal sketch. IEEE Transactions on Pattern Analysis and Machine Intelligence 8(1), 2–14 (1986)
2. Mokhtarian, F., Mackworth, A.: Scale-Based Description and Recognition of Planar Curves and Two-dimensional Shapes. IEEE Transactions on Pattern Analysis and Machine Intelligence 8(1), 34–43 (1986)

3. Lowe, D.G.: Organization of smooth image curves at multiple scales. International Journal of Computer Vision 3(2), 119–130 (1989), http://www.cs.ubc.ca/spider/lowe/pubs.html, http://www.cs.ubc.ca/~lowe/papers/iccv88.pdf

4. Ver Hoef, J.M., Peterson, E., Theobald, D.: Spatial statistical models that use flow and stream distance. Environmental and Ecological Statistics 13(4), 449–464 (2006), http://www.springerlink.com/content/a36r0kt6570537q0/fulltext.pdf

5. Cressie, N., Frey, J., Harch, B., Smith, M.: Spatial Prediction on a River Network. Journal of Agricultural, Biological & Environmental Statistics, American Statistical Association 11(2), 127–150 (2006)

6. Harville, D.A.: Matrix Algebra From a Statistician's Perspective. Springer, Heidelberg (1997)

7. Pavlidis, T.: Algorithms for Graphics and Image Processing. Computer Science Press (1982)

8. Bodansky, E., Gribov, A., Pilouk, M.: Smoothing and Compression of Lines Obtained by Raster-to-Vector Conversion. In: Blostein, D., Kwon, Y.-B. (eds.) GREC 2001. LNCS, vol. 2390, pp. 256–265. Springer, Heidelberg (2002), http://agribov.blogspot.com

9. Knuth, D.E.: The Art of Computer Programming. Addison-Wesley, Reading (September 1998)

10. Weiss, I.: High-Order Differentiation Filters That Work. IEEE Transactions on Pattern Analysis and Machine Intelligence 16(7), 734–739 (1994)

11. Burghardt, D.: Controlled Line Smoothing by Snakes. GeoInformatica 9(3), 237–252 (2005), http://www.springerlink.com/openurl.asp?genre=article&eissn=1573-7624&volume=9&issue=3&spage=237

12. Korn, G.A., Korn, T.M.: Mathematical Handbook for Scientists and Engineers: Definitions, Theorems, and Formulas for Reference and Review. McGraw-Hill Book Company, New York (1968)

13. Kondor, R.I., Lafferty, J.: Diffusion Kernels on Graphs and Other Discrete Structures. In: Machine Learning, Proceedings of the Nineteenth International Conference, pp. 315–322 (July 2002), http://www1.cs.columbia.edu/~risi/papers/diffusion-kernels.pdf

14. Taubin, G.: Curve and Surface Smoothing without Shrinkage. In: Proceeding, Fifth International Conference on Computer Vision, pp. 852–857 (June 1995)

Appendix

A   A Technique of Building Piecewise Polynomial Basis Functions

Given a network of lines L_i, $i = \overline{1, N}$, creating a graph without loops. There is an origin point O that belongs to i^*-th line. Any point P of the graph is defined by the number of the line i and by the length t of the trajectory between O and P. Coordinates of the current point P are $x(i,t)$ and $y(i,t)$. Sign of the parameter t that defines a location of the point P is defined by the mutual location of O and P (from what side of O the point P is located). p_i is an index of the line that precedes i-th line at the trajectory from point O to i-th line. $p_{i^*} = -1$. Fig. 2 shows an example of a network of lines and values of characteristics of each line.

The number of piecewise polynomial basis functions for the network of smooth lines is $M = N \cdot K + 1$, where K is an order of polynomial basis (an order of approximation).

A suggested technique of building basis functions takes into account a location of the point in which smoothing will be done. Coordinates of this point are $x_z(i^*, \tau)$ and $y_z(i^*, \tau)$.

Define basis functions as

$$\varphi_j(\tau, i, t) = \begin{cases} 1, j = 1 \\ v_i^{r(j),k(j)}(\tau, t), 1 < j \leq M \end{cases}, \quad j = \overline{1, M} .$$

Here j is a number of the basis function; i is an index of the line.

$r(j) = \lfloor (j-2)/K \rfloor + 1$; $\lfloor x \rfloor$ is the greatest integer that is not more than x,

$k(j) = (j-2) \bmod K + 1$; $x \bmod y$ is a remainder of the integer division x by y,

$$v_i^{r,k}(\tau, t) = \begin{cases} (t-\tau)^k \cdot \delta_{i,r}, i = i^* \vee t = 0, \\ v_{p_i}^{r,k}(S_i) + (t-S_i)^k \cdot \delta_{i,r}, i \neq i^* \wedge t > 0, \\ v_{p_i}^{r,k}(E_i) + (t-E_i)^k \cdot \delta_{i,r}, i \neq i^* \wedge t < 0, \end{cases} \qquad \begin{array}{l} i = \overline{1, N}, \\ r = \overline{1, N}, \\ k = \overline{1, K}. \end{array}$$

$$\delta_{i,j} = \begin{cases} 1, i = j \\ 0, i \neq j \end{cases} - \text{Kronecker delta.}$$

Fig. 4 shows an example of the basis functions for $N = 3$, $K = 1$.

B Iterative Algorithm for Calculating Integrals

$$I = \int_S^E e^{-|t-\tau|} \cdot (t-\tau)^k \cdot f(t) \cdot dt$$

Computational complexity of the algorithm of smoothing with an exponential weight function and polynomial basis functions is defined by the complexity of calculation of integrals I. In [8] an iterative method of calculation of integrals was suggested that has a linear complexity. There is a short description of this method.

Analyze three cases.

1. $S \leq \tau \leq E$

Let τ_i be parameters appropriated to vertices of the source polygonal line.

$$\int_S^E e^{-|t-\tau|} \cdot (t-\tau)^k \cdot f(t) \cdot dt =$$

$$= \int_S^\tau e^{t-\tau} \cdot (t-\tau)^k \cdot f(t) \cdot dt + \int_\tau^E e^{\tau-t} \cdot (t-\tau)^k \cdot f(t) \cdot dt .$$

Designate the first term $J_k^L(\tau)$ and the second term $J_k^R(\tau)$. Obviously $J_k^L(S) = 0$.

Given $J_k^L(\tau_j)$. Then

$$J_k^L(\tau_{j+1}) = \int_S^{\tau_{j+1}} e^{t-\tau_{j+1}} \cdot (t-\tau_{j+1})^k \cdot f(t) \cdot dt =$$

$$= \int_S^{\tau_j} e^{t-\tau_{j+1}} \cdot (t-\tau_{j+1})^k \cdot f(t) \cdot dt + \int_{\tau_j}^{\tau_{j+1}} e^{t-\tau_{j+1}} \cdot (t-\tau_{j+1})^k \cdot f(t) \cdot dt .$$

Obviously,

$$\int_S^{\tau_j} e^{t-\tau_{j+1}} \cdot (t-\tau_{j+1})^k \cdot f(t) \cdot dt =$$

$$= e^{\tau_j-\tau_{j+1}} \cdot \int_S^{\tau_j} e^{t-\tau_j} \cdot (t-\tau_j+\tau_j-\tau_{j+1})^k \cdot f(t) \cdot dt =$$

$$= e^{\tau_j-\tau_{j+1}} \times \int_S^{\tau_j} e^{t-\tau_j} \cdot \sum_{i=0}^k \left\{ C_k^i \cdot (t-\tau_j)^i \cdot (\tau_j-\tau_{j+1})^{k-i} \right\} \cdot f(t) \cdot dt =$$

$$= e^{\tau_j-\tau_{j+1}} \times \sum_{i=0}^k \left\{ C_k^i \cdot (\tau_j-\tau_{j+1})^{k-i} \cdot \int_S^{\tau_j} e^{t-\tau_j} \cdot (t-\tau_j)^i \cdot f(t) \cdot dt \right\} .$$

where C_k^i are binomial coefficients.

Now obtain an iterative formula

$$J_k^L(\tau_{j+1}) = e^{\tau_j-\tau_{j+1}} \cdot \sum_{i=0}^k \left\{ C_k^i \cdot (\tau_j-\tau_{j+1})^{k-i} \cdot J_i^L(\tau_j) \right\} +$$

$$+ \int_{\tau_j}^{\tau_{j+1}} e^{t-\tau_{j+1}} \cdot (t-\tau_{j+1})^k \cdot f(t) \cdot dt .$$

(B.1)

It is possible similarly to obtain the formula

$$J_k^R(\tau_{j-1}) = e^{\tau_{j-1}-\tau_j} \cdot \sum_{i=0}^{k} \left\{ C_k^i \cdot (\tau_j - \tau_{j-1})^{k-i} \cdot J_i^R(\tau_j) \right\} +$$

$$+ \int_{\tau_{j-1}}^{\tau_j} e^{\tau_{j-1}-t} \cdot (t - \tau_{j-1})^k \cdot f(t) \cdot dt .$$

(B.2)

If it is necessary to calculate values of integrals for some τ, where $\tau_j < \tau < \tau_{j+1}$, it is enough to substitute τ instead of τ_{j+1} and τ instead of τ_{j-1} in formulas (B.1) and (B.2) respectively, and the complexity of such calculations is $O(1)$. So it is possible to calculate integrals I with a computational complexity $O(n)$.

2. $\tau < S$

$$\int_S^E e^{-|t-\tau|} \cdot (t-\tau)^k \cdot f(t) \cdot dt = \int_S^E e^{\tau-t} \cdot (t-\tau)^k \cdot f(t) \cdot dt =$$

$$= e^{\tau-S} \times \int_S^E e^{S-t} \cdot (t-S+S-\tau)^k \cdot f(t) \cdot dt =$$

$$= e^{\tau-S} \times \int_S^E e^{S-t} \cdot \sum_{i=0}^{k} \left\{ C_k^i \cdot (t-S)^i \cdot (S-\tau)^{k-i} \right\} f(t) \cdot dt =$$

$$= e^{\tau-S} \times \sum_{i=0}^{k} \left\{ C_k^i \cdot (S-\tau)^{k-i} \cdot \int_S^E e^{S-t} \cdot (t-S)^i \cdot f(t) \cdot dt \right\} .$$

It means that the complexity of calculation of $\int_S^E e^{-|t-\tau|} \cdot (t-\tau)^k \cdot f(t) \cdot dt$, when

integrals $\int_S^E e^{S-t} \cdot (t-S)^k \cdot f(t) \cdot dt$, $k = 1, ..., K$ are known, is $O(1)$. So

calculation of integrals $\int_S^E e^{-|t-\tau|} \cdot (t-\tau)^k \cdot f(t) \cdot dt$ has complexity $O(n)$.

3. Similar results can be obtained for the case $\tau > E$.

So calculation of elements $a_{i,j}(\tau)$ and $b_i^u(\tau)$ with iterative formulas (B.1) and (B.2) has complexity $O(n)$.

Verification of the Document Components from Dual Extraction of MRTD Information

Young-Bin Kwon and Jeong-Hoon Kim

Departmen Computer Science and Engineering, Chung-Ang University,
221 Heuksukdong, Dongjakku, Seoul, Korea
ybkwon@cau.ac.kr, jhkim@cvlab.cau.ac.kr

Abstract. This paper proposes a method for character region extraction and picture separation in a passport by adopting a preprocessing phase for the passport recognition system. Character regions required for the recognition make black pixel and remainder of the passport regions makes white pixel in the detected character spaces. This method uses the MRZ sub-region in order to automatically calculate the threshold value of the binary image which is also applied to the other character regions. This method also applies horizontal/ vertical histogram projection in order to remove the picture region from the binary image. After region detection of picture area, the image part of the passport is stored in the database of face images. The remainder of the passport is composed of character part. Recognition on the extracted character is performed on the various different passports. From the obtained information, auto-correlation of extracted characters within a given passport are accomplished after character recognition. A cross-check process of MRZ information and field information of passport on similarity is implemented. For this purpose, this paper uses the auto-correlation between the binarization method based on the color information and an extracted image from a passport to propose a characteristic extraction method to prevent passport forgery.

Keywords: passport, verification, MRTD(Machine Readable Travel Document), cross-check, auto-correlation.

1 Introduction

Keys used for the personal identification method, which is generally used in the airport, is the passport number and personal records including picture and character information. The number of passport issued by each country is increasing. With the increase in the number of people travelling, the number of passports processed by the immigration is also increased. Immigration control is done to control immigration, find forged passports, emigration and immigration forbidder, wanted criminal, and emigration and immigration ineligibility persons such as alien. An efficient and accurate passport recognition system is required for the immigration control judgment. Most passport recognition system only processes the MRZ (Machine Readable Zone) code and the picture on the passport. MRZ code refers to the recognition code used to express the information on the passport using 44 OCR-B

W. Liu, J. Lladós, and J.-M. Ogier (Eds.): GREC 2007, LNCS 5046, pp. 235–244, 2008.
© Springer-Verlag Berlin Heidelberg 2008

font characters per line for passport recognition. By comparing the data on the passport with the information on the MRZ code, we can improve the forged passport detection rate through the recognition and verification method.

The existing methods only extracted the MRZ code and picture from the passport for the recognition [1, 2]. These methods are not able to utilize the other information on the passport and are vulnerable to passport forgery if the MRZ code is altered with. In this paper, we propose a method that extracts the printed personal information as well as the MRZ code to cross-check the information for a more thorough recognition. The use of the MRZ proves effective in passport processing at the immigration. However, the simple structure of the MRZ is vulnerable to forgery and the auto-recognition process makes it even easier to pass-by with forged passports. Checking for the validity of a passport is limited with the MRZ auto-recognition method. In this paper, the redundant information appearing on both the MRZ area and the printed personal information area are extracted for comparison to check for the validity of the passport.

Extent recognition system does not require any special binarization method [5] because it is only used for the MRZ code recognition. However, a binarization method that separates the characters in the passport from the background is required to for the recognition of the data on the passport. This paper proposes a binarization method which uses the character's RGB property and histogram of the MRZ region. Extraction of picture region is proposed by a method which executes horizontal/vertical histogram projection and analyzes the result value. After extraction of each character, a recognition method based on template matching and feature comparison is implemented. A self-checking of the traced characters samples is performed in order to compare the similarity of extracted characters. This method is simple and easy to compare all recognized characters within a same passport. If the same characters are not found in the same passport, we utilize the standard OCR-B font to compare the similarity. A cross-check for the same field information from MRZ and upper personal information area is also performed.

2   Passport Data Extraction

2.1   Passport Data Extraction Process

In order to extract the data part from passport, it is necessary to divide the regions into the picture area, the MRZ area, and the upper printed personal information area. The MRZ area is easily detected through the horizontal projection method because they are composed of two lines at the bottom of the passport and each line contains 44 characters. During the projection, threshold value of 180 on each RGB component is used to detect the gray scale value of printed characters. The separation of the picture area and the upper printed personal information area is accomplished by detecting the boundary line between two regions. As shown in Fig. 2, a vertical projection of the passport except the MRZ area can detect the separation line. There is no passport which uses more than half of its area for the picture area. Thus, only the left half part of the passport is used for vertical projection. To decide the line boundary, the right

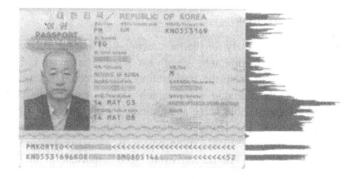

Fig. 1. Detection of MRZ area

Fig. 2. Detection of Picture & Personal Information areas

most point of the zero values is selected. If no zero value is found after projection, we may adjust the threshold value to create the zero values for the boundary.

After detection of each area, the picture area is saved as an image. Binarization is only performed on the MRZ area and the upper personal information area. Passports with hologram backgrounds are more difficult to process. However, the knowledge of the printed characters gray value information which has same value when they are issued helps us to detect the printed value correctly. Character recognition based on template matching using OCR-B font and feature comparison is performed after binarization. The location information on each character filed is already standardized by ICAO helps us to improve the recognition ratio for the numbers and alphabets [6, 7]. After recognition of printed characters, cross-check verification on each region is also performed in order to detect the possible fraud passport. But some countries use different fonts between two areas. Thus, a development of a self-correlation method which calculates a correlation between extracted characters is needed. We call it extracted auto-correlation character analysis.

2.2   Colour Image Binarization

The photo area is stored separately for the face recognition and the information from the MRZ and personal information area is each extracted. The image is binarized to distinguish the character part from the background. Most documents are black and white or have a simple background color which makes it easier to perform the binarization using a fixed threshold value and a method such as the Trier, Text [4] or iterative binarization method [5]. However, passports usually use various colors for the background with various features. Some passports even use holograms for the background which makes the binarization even more important in the process of accurately extracting the passport information.

The MRZ area of the passport has to be checked by binarization in order to extract the passport information. In previous studies [1], the average value of the background and character pixels was calculated from the initial threshold value for the binarization. In other words, the initial threshold value is set by dividing the sum of all the RGB values by the number of pixels, before comparing it against the current threshold value. If this number is bigger than the threshold value, the RGB value is added to the object. If the value is smaller, the process of adding the object and black and dividing it by 2 is repeatedly applied to get the value of R, G and B before the binarization. The threshold value for the Red, Green and Blue were acquired for binarization. The value is put through the median filter to get the final result. The background of the personal information area is more complex than the MRZ area which makes it difficult to get good results by simply applying binarization. The RGB elements of most printed letters on the MRZ and personal information area are similar in most cases. In the personal information area, an improved binarization method of laying over is used which consists of two stages. In the first stage, the RGB value of the original image acquired using the above method is extracted and stored as the threshold value. In the second stage, the RGB element, which is the RGB value, is acquired. The RGB element of the pixels in the personal information area is compared against the RGB value of the threshold value. After acquiring the R:G:B ration of the personal information area, this value is compared with the R:G:B ratio of the MRZ area to add or subtract the range value. After this process, the same binarization method applied to the MRZ area is applied repeatedly. This method has an advantage of more clearly expressing the contrast for a more thorough binarization result.

2.3   Extracted Character Analysis by Automata

The MRZ information stores the information based on the information of the location of each field. MRZ has a certain pattern with the '<' as the identifier. The result of template matching using the '<' identifier is used to design the automata for the segmentation in each field. The starting point of each field is decided by the analysis of each MRZ line and the information is extracted using the automata at each location. The extracted information is stored along with its number of characters for cross-checking with the personal information, as shown in Fig. 3.

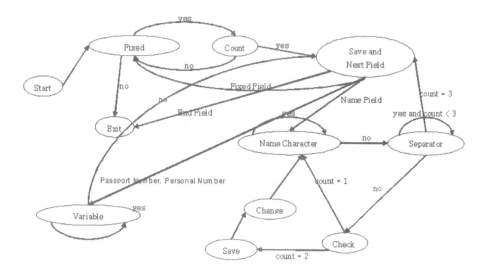

Fig. 3. Automata for the Segmentation of MRZ

2.4 Character Extraction and Auto-Correlation Analysis by Alphabet

The field can easily be divided in the MRZ area based on the regular order and the identifier, '<', but there is a few more steps to process it in the personal information area. Each extracted character has to be expressed as a set of characters for the information extraction, as shown in Fig. 4 and characters either too small or too big are filtered.

Fig. 4. Character set

The personal information area is regrouped into a few sets after this process. Each set is recognized using the method shown in Figure 4 and this result is string-matched against the field extracted form the MRZ to recognize the set most similar to the MRZ. A different algorithm has to be applied for the field matching of the date of birth and date of expiry because of the different format used to express dates by personal information area and the MRZ area. The date on the MRZ area is expressed using 6 digits whereas the expression used in the personal information area is different in various formats. Fig. 5 shows a few examples.

Although there are many different formats used to express the date in the personal information area, the order of appearance, from day to month to year, is the same as in the MRZ area. The matching algorithm can be produced using this pattern. The year is first located before locating the day field after which the month field can easily be

Fig. 5. Examples of Data Expression

Fig. 6. '/' Identifier Example

found by finding the number in-between. This algorithm is applied differently by the expression of the year. The year data is first extracted from the MRZ area, and if the first digit is a 0 or 1, the digit 20 is added on to express the years after the year 2000, and 19 is added otherwise. To express the years after 1900, the same set of digits is looked up for in the personal information area, and the same digits used to express the day in the MRZ area is looked up for to find the day. Finally, if the height of the day and year area is the same, the set between the two is set as the month. The same algorithm is applied to the personal information area where the year is expressed with only two digits. Finally, a few exceptions handling process is required to divide the field within the personal information area. Chinese passports divide the English and Chinese name using the '/' identifier as shown in Fig. 6. The '/' identifier is used to divide the Chinese and English area into two different sets.

A different algorithm is required to divide the information in the personal information area. If the value of each set is smaller than the height of each character multiplied by 1.5, the two characters are recognized to be a single set of characters. The extracted MRZ and personal information can be used for cross-checking since they contain the same information.

A single character from 0~9 and 'A' to 'Z', extracted from the passport is used for the comparison. For example, if you chose 'A', both 'A' from the MRZ and upper personal information area, is printed on the screen. The result of match comparing the 'A' in the MRZ and the upper personal information area is then printed. The average shape of the 'A' in the MRZ and the upper personal information area are compared for the difference in their font shape. The shape that is most irrelevant from the others can be displayed. The matching score for each character is acquired through dividing the compared score by n. The comparison with the extracted characters becomes possible using the auto-correlation. The type and number of characters extracted from the information on each passport is different due to the different names and nationalities. The information on the MRZ and the personal information area share at least two same alphabets since they contain the same information. Therefore, it is possible to look for the cross-characteristic based on this fact. The relation between the two areas is very high since the information on a passport is printed at the same time. Various auto-correlation can be found from the result for result analysis.

3   Results

Character extraction experiment is performed using 50 different passport images. The passports used for the experiment are from 13 Asian/Oceania countries, 5 European countries, 2 American countries and 2 African countries. Fig. 7 shows the improved result of the binarization method applied in this paper.

The image on the left shows the result of applying the iterative binarization method on the image and the image on the right shows the result of applying the improved binarization method explained in section 2.2.. The E and K on the name field and the E on the date of expiry are expressed beyond recognition on the left side image whereas they are expressed properly on the right side image. Field extraction result demonstrates perfect extraction results. The recognition of characters using template matching and feature comparison is implemented. The recognition rate of character is 99.9% and personal information extraction is 99.8% with an average speed of 2.3 second required for the recognition of the whole page. Fig. 8 shows the result of the cross-checking.

Fig. 9 and Fig. 10 show the result of single character analysis. Each figure illustrates the extracted data from each region and the rate of auto-correlation between extracted characters.

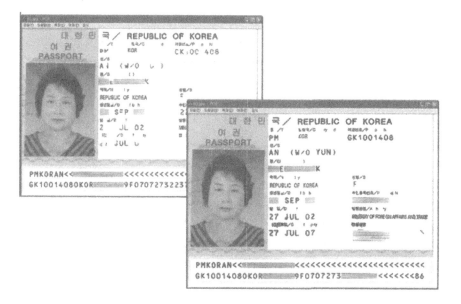

Fig. 7. Iterative Binarization (left) and Proposed Colour Binarization

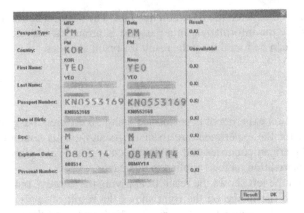

Fig. 8. Result of Cross Check

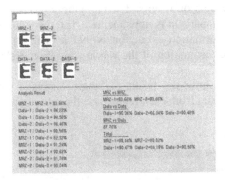

Fig. 9. Analysis Result of 'E'

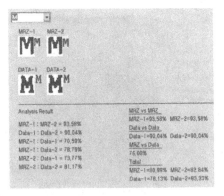

Fig. 10. Analysis Result of 'M'

In the result of Fig. 9, we can see that the similarity is higher when comparing the same area type than comparing the MRZ from the Personal Information area. The second 'E' on the Personal Information area, DATA-2, shows 84.19% similarity level which means that it is the most heterogeneous one.

Fig. 10 shows the result of analyzing the character 'M' on a European passport. The similarity level is at a low level of 76% between the MRZ and Personal Information area. This means that the British passport uses a different font for the MRZ and Personal Information area. We can construct a knowledge database of the passports for the date of issue and the country based on the detected information. Forgery detection may be performed using this knowledge base.

4 Conclusion

In this paper, we have proposed a method for extracting and utilizing the information on the MRZ area and personal information area through the structure analysis which showed a high recognition rate through experiments. This method also applies an automata for the sorting of the MRZ information into different fields for a more effective classification. The concept of cross-checking can be used to check for fraud information using redundant information on the passport for more accuracy. This will contribute to the identification of forged passports by comparing the information on the passport image itself by using the characteristic of each image. Experiments have been performed by using 50 different passports from 22 different countries. The characters on the MRZ area of these passports consist of the same font type and have a simple background image for easier recognition. However, the background image is complex in the personal information area which made the recognition of only 22 passports possible through this method. Therefore, a method for removing the hologram from the passport and a method for extracting the characters from a background with the same color is required for a more precise recognition.

In this paper, we have implemented and experimented with the concept of passport structure analysis and character recognition module. A method for removing the hologram and a module for extracting the characters from the background has to be developed for a more precise passport recognition system.

Acknowledgments. This work is supported by Seoul R&BD program (10544).

References

1. Kim, T.J., Kwon, Y.B.: Crosscheck of Passport Information for Personal Identification. In: GREC 2005, pp. 162–172 (2005)
2. Kim, K.B., Kim, Y.-J., Oh, A.-s.: An Intelligent System for Passport Recognition Using Enhanced RBF Network. In: Zhang, J., He, J.-H., Fu, Y. (eds.) CIS 2004. LNCS, vol. 3314, pp. 762–767. Springer, Heidelberg (2004)
3. Trier, P.D., Taxt, T.: Evaluation of Binarization Methods for Document Images. IEEE Trans. On PAMI 17(3), 312–315 (1995)

4. Dawoud, A., Kamel, M.S.: Iterative Multimodel Subimage Binarization for Handwritten Character Segmentation. IEEE Trans. On IP 13(9), 1223–1230 (2004)
5. Gonzalez, R.C., Woods, R.E.: Digital Image Processing. Addison Wesley Longman (1992)
6. Tan, H.L.: Hybrid feature-based and template matching optical character recognition system, United States Patent 5077805 (1991)
7. ICAO, Document 9303, http://mrtd.icao.int/content/view/33/202/

A System to Segment Text and Symbols from Color Maps

Partha Pratim Roy[1], Eduard Vazquez[1], Josep Lladós[1], Ramon Baldrich[1],
and Umapada Pal[2]

[1] Computer Vision Center, Universitat Autònoma de Barcelona, 08193,
Bellaterra (Barcelona), Spain
[2] Computer Vision and Pattern Recognition Unit, Indian Statistical Institute,
Kolkata - 108, India

Abstract. Automatic separation of text and symbols from graphics in
document image is one of the fundamental aims in graphics recognition.
In maps, separation of text and symbols from graphics involves many
challenges because the text and symbols frequently touch/overlap with
graphical components. Sometimes the colors in a single character are
gradually distributed which adds extra difficulty in text and symbol separation from color maps. In this paper we proposed a system to retrieve
text and symbol from color map. Here, at first, we separate the map
into different foreground layers according to color features and then in
each layer, connected component features and skeleton information are
used to identify text and symbol from graphics on the basis of their geometrical features. Lastly, segmentation results of the individual layers
are combined to get final segmentation results. From the experiment we
obtained encouraging results.

1 Introduction

Automatic discrimination of text/graphics in document image is one of the fundamental aims in graphics recognition [2],[3],[6]. Here, the aim is to segment the
document into two layers: a layer assumed to contain text and symbols and the
other one containing the rest of graphical objects such as street, river, border of
the regions etc. The problem has received a great deal of attention in the literature because of the different processing approach of text and graphics. At the
component level the problem is not too intensed. The spatial distribution of the
components and their sizes, can be measured in a number of ways, and fairly reliable classification can be obtained. Difficulties arise however, when either there
is text and symbol embedded in the graphics components, or text and symbol
touched with graphics. It includes, frequent intersection of text and symbols with
graphical lines and curves and segmentation of such documents is very difficult.
The separation problem of text and graphics intersections has not yet been dealt
with successfully although there exist many pieces of published paper.

Fletcher and Kasturi [6] proposed an algorithm to separate text-string from
mixed graphics. They used simple heuristics based on the characteristics of text
characters. The method is insensitive in text font style, size and orientation. One

W. Liu, J. Lladós, and J.-M. Ogier (Eds.): GREC 2007, LNCS 5046, pp. 245–256, 2008.
© Springer-Verlag Berlin Heidelberg 2008

of the assumptions was that the text character should not touch with graphics or other characters and each text character forms an isolated component. Another assumption of the method is that, the character components of a string are aligned straight. Luo et al. [1]uses the directional mathematical morphology approach for separation of character strings from maps. The idea is to separate all long linear segments by directional morphology and histogram analysis of these segments. Long segments are considered as part of graphics; effectively leaving small text character segments. Tan et al. [7] illustrates a system to extract text strings from a mixed text/graphics image using *Pyramid* structure. Multi-resolution representations of such a pyramid structure help to select different regions for segmentation. Cao and Tan [3] proposed a method of detecting and extracting text characters that are touched to graphics. It is based on the observation that the constituent strokes of characters are usually short segments in comparison with those of graphics. They looked for the lines in the overlapped region on the vectorized image to interpret intersection of text and graphics. More consolidated method is proposed by Tombre et al. [2]. This method is based on the analysis of the connected components. The algorithm has covered a number of improvements to make it more stable for graphics-rich documents.

In color maps the color carries a lot of information which is useful for separation. Using color features, complexity of text/graphics separation can be reduced drastically. Because, in color maps, different colors are used to represent text and symbol from graphics. By color properties, one can separate touching/overlapped text symbols from graphics without exploring much graphics recognition methodologies. There exist color segmentation methods for separating colors used in a document. Recent work includes a variety of techniques: for example, morphological watershed based region growing [12], JSEG segmentation [13], Mean Shift algorithm [15] etc. If an image contains only homogeneous color regions, clustering methods in color space [14,16] are sufficient to handle the problem. However, in color text documents sometimes even, it appears at first appearance, that the character seems to be printed in a single color, but actual measurements reveals that the colors in a single character are gradually distributed. This degradation effect of color in text layer sometimes makes it more difficult to segment. It causes over/under -segmentation and such over/under -segmentation creates problem in separating foreground layer (text graphics layers of different colors) efficiently. Here it is quite challenging to remove the noise and to extract the intended characters. Among color segmentation methods, clustering is one of the simplest, and has been widely used. The method proposed by Coleman and Andrews [16] is a clustering algorithm based on k-means which operates in an "unsupervised" mode and does not require training prototypes. Since, k-means algorithm requires the number of cluster to initiate, in order to have this information, the knowledge about the image data is necessary. This method iterates on a number of clusters, and evaluate the quality of the clusters by within-cluster and between-cluster scatter matrices. Moreover, the influence of resolution of the input image is an important factor in color document analysis. When a color document is scanned at high resolution, mesh noise occurs in the

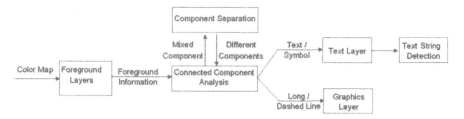

Fig. 1. (a)Block diagram of our approach

digital color image resulting over-segmentation, when a segmentation algorithm is applied. Hase et al. [17] studied offset printing color documents such as book covers, posters. They discussed a method to absorb the variation of color distribution of color segmentation. Their algorithm tries to prevent over-segmentation and fusion with the background while maintaining real-time usability. They have described a selected local color averaging technique to remove the problem of mesh noise. Dhar and Chanda [10] presented a method for extraction and recognition of text and symbol from topographic maps. Here, color segmentation is done by a supervised clustering algorithm and in each color layer, symbols are recognized on the basis of symbol-specific geometrical features.

There are various types of color maps, and they contain thousands of text and symbol in different shapes. So, we need a sophisticated process, which can be used in general without many heuristic measures. The problems are mainly due to presence of graphical components with texts and symbols in different color layers. To handle such situations, the objective of this paper is to combine different features like color information, connected component analysis, skeleton information etc. to get better segmentation results. Our proposed technique is language and font independent. Also, our method does not depend on number of colors present in the document.

In our proposed method, the color map is first analyzed using color information and separated into different layers. Text/graphics components are extracted from these layers. A new algorithm, combining connected component and skeleton analysis, has been proposed to identify the isolated character, joined character, dash and long line components from each layer. The components in which both character and long line present due to overlapping are considered as mixed components. Using Hough transform and skeleton analysis, these mixed components are analyzed for their segmentation into character and line. Extracted character components are grouped into string according to their color and proximity features. Block diagram of our approach has been shown in Fig.1.

2   Color Segmentation and Foreground Layer Selection

The problem of foreground detection in color maps could be defined as that of detecting layers containing text, symbols and graphical objects. If the text layer is assumed to be of dark color in a light background, then the problem can be solved by converting the RGB color-space to YIQ color-space, and applying a threshold in Intensity (Y channel) image. In color degraded image, this method is not efficient to separate foreground layer. For our map handling, due to image degradation, we performed the color segmentation to get different color layers and this is done by a combination of color feature and spatial information. This is followed by selection of foreground layer considering the features of text/graphics information. It is done by applying a heuristic measures on color volume and edge information of each color layers. The detail of color segmentation is discussed as follows.

First, we apply the method of Vazquez et al. [4] to find dominant colors in a d-dimensional histogram Ω_d. The method proposed is a two-steps operator. The creaseness operator, MLSEC-ST[5] is introduced in order to spurn non-representative data as well as to enhance meaningful information. This process assigns a high creaseness value at the center of elongated objects by means of divergence calculation of the Structural Tensor Field.

Formally, given a symmetric neighborhood of size σ_i centered at point x, say, $N(x, \sigma_i)$ the Structure Tensor is defined as:

$$S(x, \sigma_i) = N(x, \sigma_i) * (w(x) \cdot w^t(x)) \tag{1}$$

where a gaussian of standard deviation σ_d is used to make the calculation of ω. Afterward, if $\omega'(x, \sigma_i)$ is the eigenvector corresponding to the largest eigenvalue of $S(x, \sigma_i)$ then, the dominant gradient vector in the neighborhood of size proportional to σ_i centered at x is:

$$\overline{w}(x, \sigma_i) = sign(\omega'^t(x, \sigma_i) \cdot \omega(x))\omega'(x, \sigma_i) \tag{2}$$

Next, the creaseness value is associated to a point x_k, $\forall k \in \Omega$ as follows:

$$k(p) = -Div(\overline{w}_p) = -\frac{d}{r} \sum_{k=1}^{r} \overline{w}_k^t(\sigma_i) \cdot n_k \tag{3}$$

As a result, we have Ω_d', a representation of Ω_d where each point is represented by its creaseness value. Next, we find the dominant structures of the histogram by applying a ridges extraction algorithm [11]. The ridges extraction procedure will join several points with a high creaseness under one unique ridge. Therefore, we will have as many ridges as dominant structures. This is done as follows:

If a given point $x \in \Omega_d$ is a local maxima, then we visit the neighbor of x which becomes a local maxima when we remove x from Ω_d. The process stops when it reaches a flat region.

Formally, let $\tau(\Omega_d')$ be the set containing all local maxima in Ω_d', and $neigh(x)$ be the r-connected neighborhood of a point x with a zero-crossing in its gradient for a given direction, i.e. we calculate the gradient values not for all points but for ridge points and its closer neighbors. We also define $\eta(x, n_j)$ as the common neighbors between x and n_j, where $x \in \Omega_d'$ and $n_j \in neigh(x)$:

$$\eta(x, n_j) = \{neigh(x) \bigcap neigh(n_j)\} \tag{4}$$

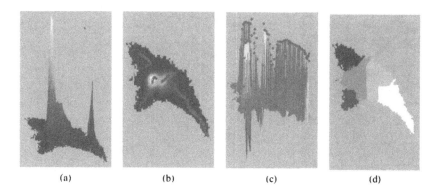

Fig. 2. (a)Ω_3, original 3-dimensional distribution. (b)Ω_3' Creaseness representation of a) and ridges found are represented with dots. (c)Ridges found in d)fitted on original distribution. 2-Dimensional view. (e)Clustering of the original distribution.

It is worth to note that neither x nor n_j are included in $\eta(x, n_j)$. Then, we define the ridge points in a creaseness representation image Ω_d', as:

$$\tau_z(\Omega_d') = \tau_{z-1}(\Omega_d') \bigcup \{n \in neigh(l) \mid l \in \tau_{z-1}(\Omega_d'), \mu(l, n) = 0\} \qquad (5)$$

$$\mu(x, n_j) = \sharp \{y \in \tau(x, n_j) \mid \Omega_d'(y) \geq \Omega_d'(n_j)\} \qquad (6)$$

Next, we perform a clustering to this distribution. For this, the points in $\tau_z(\Omega_d')$ are used as marks in a watershed procedure. To avoid the drawbacks of the watershed algorithm [11], the method is combined with a Voronoi spatial partitioning. This will perform a clustering of the original histogram Ω_d in as many regions as dominant colors are found. Fig. 2 shows an example of the whole process. Concretely, Fig. 2(a) shows a synthetic 3-dimensional distribution, i.e., Ω_3. Creaseness representation of it, Ω_3', is depicted in Fig. 2(b) where dots are the ridges found. The same ridges obtained in Ω_3' are fitted in the original distribution in 2(c). Finally, the clustering of Ω_3 is showed in Fig. 2(d) and there are 5 different cluster in this figure. Based on the number of cluster, color volume of each cluster and edge information of each color layer, the number of foreground layers is decided.

3   Text and Symbol Separation

After selecting different foreground layers we separate text and symbol from graphics in each of the layers. Different steps used in each foreground layers are discussed as follows.

3.1   Connected Component Analysis with Skeleton Information

Text, symbol and graphical lines are present in foreground layer. We considered the connected component analysis developed by Tombre et al. [2] for initial

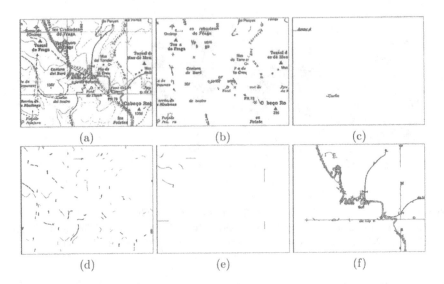

Fig. 3. (a)Example of a foreground Layer (b)Isolated text and symbol (c)Joined characters (d)Small elongated components (e)Long line (f)Mixed components

segmentation. A few criteria based on geometrical features of the connected component are good enough to group a component into one between text or graphics layer. But, there are some constraints. For example, some characters cannot be split due to touching. We will call them as "joined characters". If joined characters touch with long lines, they will not be separated by simple rules, because their features will be different from isolated character. We integrate skeleton information along with geometrical features to detect the long segments and to analyze them accordingly. We separate the components into 5 groups namely, Isolated characters, Joined characters, Dash components, Long components and Mixed components. In skeleton image of each component, if there exists no long segment, then the component is included into one of the isolated character/symbol or joined character or dash component group. Otherwise, it is considered as mixed or long component. The description of each of them is given below. The Fig. 3 shows different components of a foreground layer.

In *Isolated Character* group, normal text alphabets and small symbols are included. These are selected by the connected component size histogram analysis [2]. *Joined Characters* consists of the components where more than one isolated character touch each other. These connected components have larger aspect ratio than isolated characters and do not contain long skeleton segments. *Dash Characters* are mainly small elongated components. These include the dash segments from the dash line along with some isolated characters, such as "1","l" etc. These characters are combined into dash character group, because, at the pixel level analysis, they hold the same property as dash segments. The *Long Line* is the graphics layer of our algorithm. The segments obtained from the skeleton of this component are all larger compared to the size of text characters. Straight and curve both types of line can be possible. *Mixed Component* consists of the components where both long line and isolated/joined characters are present. This happens due to overlapping with each other and we can not separate such mixed component during component labelling. Segmentation of the mixed components is done in two stages. In the first stage long straight lines are removed from the mixed components. Next,long curve parts are detected and removed from the remaining part of the image. These two stages are discussed as follows.

3.2 Long Straight Line Removal

We perform Hough Transform to detect the straight lines present in the binarized image. In Hough space, all the collinear pixels of a straight line will be found intersecting at the same point (ρ, θ), where ρ and θ identify the line equation. Depending on accumulation of pixels, the straight lines are sorted out. Some characters may touch with this straight line portion(see Fig. 4(a)). Hence our objective is to remove the non-character part from the straight line. We used stroke-width information in our approach to get the line width (L_w). Stroke width of a straight line segment is computed using the statistical mode of the black run-lengths, obtained by scanning the segment in horizontal, vertical and two diagonal directions. The portions of the line where the width is more than L_w, are separated from straight line. Fig. 4(b) shows the remaining part of the image after straight line removal in Fig. 4(a) by Hough Transform analysis.

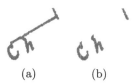

(a)   (b)

Fig. 4. (a)Characters are joined with a long line in a Mixed component (b)Isolated characters after removal of straight long line

3.3 Long Curve Line Removal

According to text and graphics feature, it is assumed that the length of segments of the characters are smaller compared to that of graphics. The mixed-component segmentation method proposed by Cao and Tan [3] which is based on the continuation of the strokes in the skeleton works well for documents, where the text and lines are of more or less thin in nature. But, there are some limitations. When a line touches a symbol or text of blob like shape (dense pixels), the thinned image is always not perfect for the arrangement of segments. It needs post-processing, which is a difficult job. To overcome the drawback we compute the skeleton and all the segments are decomposed at the intersection point of

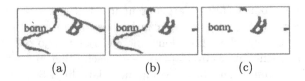

(a)   (b)   (c)

Fig. 5. (a)A Mixed component (b)Long straight part is removed (c)Extracted part after removing the long curve lines

the skeleton. Based on the bounding box (BB) information of a segment, the major axis (L_s) is calculated as:

$$L_s = Max(Height_{BB}, Width_{BB}) \tag{7}$$

The segments having L_s larger than average character height (here, average height is obtained by averaging the heights of isolated characters) are chosen for elimination. The remaining portion after removal of long straight and curve line are considered as either isolated characters or joined characters according to their feature. For example, Fig. 5(a) demonstrates an initial mixed component with touching characters. The component after removing long straight line is shown in Fig. 5(b). After doing curve line removal, the remaining portions are shown in Fig. 5(c).

3.4   Character String Extraction

After passing through different separation methods, the mixed components will get separated. The long lines of mixed component will be in the graphics layer. Text and symbol will be in isolated or joined character layer. These isolated and joined characters are combined to get all the text and symbol components. Now we cluster the isolated characters into individual words. In general, the gap T_w between two words is larger than the gap T_c between two characters in a word and the grouping is formed by the characters of similar colors. Using this positional information and color information of different isolated character/symbol we cluster different words. For example see Fig. 6(b), where text and symbols are marked by rectangular box.

(a)   (b)

Fig. 6. (a)Original Image (b)Segmented text and symbol layer. Here, segmented text and symbol parts are marked by rectangular box.

4   Experimental Results and Discussion

We have taken maps from different scripts to test our method. 26 maps are selected from "Spanish","English","Russian", and "Bengali" and the average size of the test maps are 350x450 pixels. We considered a large varieties of data for our experiment and some examples of such data are discussed as follows. The background can be of single color (Russian) or multi-color. Foreground contains numerous colors to represent text and symbol or graphical lines. The text characters are connected in Bengali maps. In other maps these are generally isolated, but sometimes joined text strings are also found due to printing or noise issue. In graphics part, lines can be dashed or continuous. There exist both straight and curve lines. The long continuous lines are touched/overlapped with text in many places. The overlapped text and lines sometimes are in similar color or in different color. In text string, the arrangement of characters are of both linear and curvilinear. The maps are not noise free always. In Bengali maps the noise/dithering effect is prominent.

Our proposed methodology combines color information, connected component analysis and skeleton information for segmentation. The thinning was done by the algorithm proposed by Ahmed and Ward [9] which works in rotation invariant nature. The length of a segment is used to classify it either as long lines or as small segments. String construction and joining missing characters are done using morphological operation. Scale invariance is also incorporated by computing histogram analysis of the components' size, considering aspect ratio of components.

From the experiment of color separation, we noted that our system shows very good results when different colors are distinct. For example, we obtained 100% color segmentation result in Russian map, because Russian maps contain distinct color layers. But from Bengali and Spanish map, we obtained 75% (on an average) results because of the color degradation. We computed this percentage on pixel label. In our dataset, after color separation, we found total 55 mixed components where long lines exist. Among them 22 were straight and rest were curve. The component label accuracy of removal of long straight lines are 100% and for long curve line it is 85%.

In Fig. 7c(2), it is interesting to see that we have recovered the number "79" from a curve line. This extracted number is shown by dotted box in Fig. 7c(2) and this number was touched with the curve line. There are 1400 isolated character and 60 joined characters in our dataset. From the experiment, we noted that more than 98% cases our method segment text and symbol of a map into isolated and joined character group. We also noted that, if a long line is not fully straight and it contains a sufficient straight part, this part will be detected by Hough Transform and will be removed leaving the other small parts. These small segments will fall into isolated and joined character group. This is same for long curve line. In skeleton analysis, a curve line may not be fully removed, if it visits many junctions in the travel path. This will result false alarms. For example, see near right-bottom part of Fig. 6(b), where we can see some false alarms, which have been generated because of this problem. In our present work, grouping

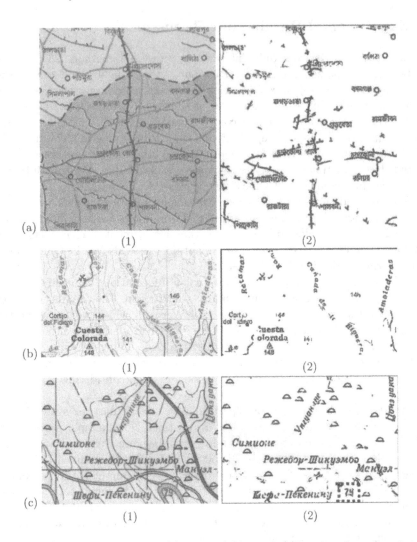

Fig. 7. Images in different scripts (a)Bengali (b)Spanish (c)Russian. In each script, (2) shows the extracted text and symbols of the corresponding color images (1).

of selected dash-like components into text/symbol layer is not considered. Due to this, missing of dash components like text characters may be obtained (see Fig. 6(b)). In future, we plan to use context information to solve these problem as follows. The isolated dash components are likely to be of dash lines, if they are arranged in a linear fashion. The other dash-shaped characters may be included for text part. For the false alarms generated by skeleton analysis, we should combine the neighborhood text region information to include them in text layer.

Almost all the previous approaches in text/graphics separation either used binarized image or converted the color image to binarized image for this purpose. The separation of text/graphics layer in degraded color image using color analysis is not an easy task. There exists no methodology to evaluate the correctness of color segmentation result. The validity of the results vary according to human perception and thus focus for measuring in terms of qualitative rather than quantitative. Color separation analysis using creaseness operator reduces the amount of variability of color information effectively. In our test maps, where the color degradation is less, it outperforms. But in maps of "Spanish" and "Bengali" the noise is very prominent and we got some over segmentation. Here, we selected the foreground layers manually and used these layers for our text/graphics separation purpose. To get the idea of segmentation results of different scripts, see Fig. 7, where text and symbols are extracted from the color images.

5   Conclusion

In this paper we proposed a language and font insensitive system to retrieve text and symbol from color maps. Here, at first, we separated the maps into different layers according to color features and then in each layer, connected component features, skeleton information, geometrical features are used to identify text and symbol from graphics. We tested our method on documents of different languages like English, Spanish, Russian, Indian etc. and from the experiment we obtained encouraging results. In future we plan to test our system on more documents of different languages. Also, we plan to use the contextual information to remove small non-text part that are included into isolated and joined character group, mistaken by our approach.

Acknowledgement

This work has been partially supported by the Spanish projects TIN2006-15694-C02-02 and CONSOLIDER-INGENIO 2010 (CSD2007-00018).

References

1. Luo, H.Z., Agam, G., Dinstein, I.: Directional mathematical morphology approach for line thinning and extraction of character strings from maps and line drawings. In: ICDAR 1995, Washington, DC, USA, vol. 1, p. 257 (1995)
2. Tombre, K., Tabbone, S., Peissier, L., Lamiroy, B., Dosch, P.: Text /graphics separation revisited. In: Lopresti, D.P., Hu, J., Kashi, R.S. (eds.) DAS 2002. LNCS, vol. 2423, pp. 200–211. Springer, Heidelberg (2002)
3. Cao, R., Tan, C.L.: Text/graphics separation in maps. In: Proceedings of 4th IAPR International Workshop on Graphics Recognition, Kingston, Ontario, Canada, September 2001, pp. 44–48 (2001)
4. Vazquez, E., Baldrich, R., Vazquez, J., Vanrell, M.: Topological histogram reduction towards colour segmentation. In: Lecture Notes in Computer Science - Pattern Recognition and Image Analysis, pp. 55–62 (2007)
5. López, A.M., Lloret, D., Serrat, J., Villanueva, J.J.: Multilocal Creaseness Based on the Level-Set Extrinsic Curvature. Computer Vision and Image Understanding: CVIU 77(2), 111–144 (2000)

6. Fletcher, L.A., Kasturi, R.: A Robust Algorithm for Text String Separation from Mixed Text/Graphics Images. IEEE Transactions on PAMI 10(6), 910–918 (1998)
7. Tan, C.L., Ng, P.O.: Text extraction using pyramid. Pattern Recognition 31(1), 63–72 (1998)
8. Roy, P.P.: An Approach to Text / Graphics Separation from Color Maps. M.S. thesis, CVC, UAB, Barcelona (February 2007)
9. Ahmed, M., Ward, R.: A Rotation Invariant Rule-Based Thinning Algorithm for Character Recognition. IEEE Transactions on PAMI 24(12), 1672–1678 (2002)
10. Dhar, D.B., Chanda, B.: Extraction and recognition of geographical features from paper maps. IJDA 8(4), 232–245 (2006)
11. Lopez, A.M., Villanueva, J.J., Lumbreras, F., Serrat, J.: Evaluation of methods for ridge and valley detection. IEEE Transactions on PAMI 21(4), 327–334 (1999)
12. Shafarenko, L., Petrou, M., Kittler, J.: Automatic watershed segmentation of randomly textured color images. IEEE Transactions on Image Processing 6(11), 1530–1544 (1997)
13. Deng, Y., Manjunath, B.S., Shin, H.: Color image segmentation. In: IEEE Computer Society Conference on Computer Vision and Pattern Recognition CVPR 1999, vol. 2, pp. 446–451 (1999)
14. Comaniciu, D., Meer, P.: Robust Analysis of Feature Spaces: Color Image Segmentation. In: IEEE Conf. on CVPR, pp. 750–755 (1997)
15. Comaniciu, D., Meer, P.: Mean Shift Analysis and Applications. In: ICCV 1999: Proceedings of the International Conference on Computer Vision, Washington, DC, USA, vol. 2, p. 1197 (1999)
16. Coleman, G.B., Andrews, H.C.: Image segmentation by clustering. Proceedings of IEEE 67, 773–785 (1979)
17. Hase, H., Yoneda, M., Tokai, S., Kato, J., Suen, Y.: Color segmentation for text extraction. IJDAR 6(4), 271–284 (2003)

A Non-symmetrical Method of Image Local-Difference Comparison for Ancient Impressions Dating

Étienne Baudrier[1], Nathalie Girard[2], and Jean-Marc Ogier[2]

[1] Laboratoire Signal Images Communication
Université de Poitiers, Bvd Marie et Pierre Curie, BP 30179,
86962 FUTUROSCOPE CHASSENEUIL CEDEX, France
etienne.baudrier@sic.univ-poitiers.fr
[2] Laboratoire d'Informatique, Image et Interactions
Université de La Rochelle, Avenue Crépeau, 17042 LA ROCHELLE CEDEX 1, France
nathalie.girard@etudiant.univ-lr.fr, jean-marc.ogier@univ-lr.fr

Abstract. In this article, we focus on the dating of images (impressions, ornamental letters) printed starting from the same stamp. This difficult task needs a good observation of the differences between the compared images. We present a method, based on a local adaptation of the Hausdorff distance, that evaluates locally the image differences. It allows the user to visualize these differences. A description of the pertinent differences for the dating allows us to evaluate our method visualization ability. Then our method is successfully compared to the existing method. Finally, a framework for a future automatic dating method is presented.

Keywords: Image comparison, binary images, Hausdorff distance, local dissimilarity measure, visualization, ancient images, dating.

1 Introduction

As ancient documents are being digitized, systems for retrieving documents or images can now be found in Digital Libraries [1]. With regard to illustrations, the content-based image retrieval is difficult and the user often needs to check visually the similarity of the retrieved images. In this article, we focus on the dating of images (impressions, ornamental letters) printed starting from the same stamp. This issue is difficult even for the expert eyes. In a perfect case where printings are perfectly conserved, digitized and registered, the sign of the map of pixel-to-pixel gray level difference (PPDMap) would be sufficient to conclude: if the difference is positive, it means that the second printing has been printed with the stamp in a degraded state, and so the second printing is less old. Nevertheless, in a real case, printings starting from the same stamp can differ for various reasons:

1. The printing degradation state,
2. The digitalization which can cause variations in the gray level, the resolution, etc,

W. Liu, J. Lladós, and J.-M. Ogier (Eds.): GREC 2007, LNCS 5046, pp. 257–265, 2008.

3. If there is a binarization step, the method used can cause differences in the binarized images,
4. The registration can cause a slight shift or/and rotation resulting in differences in the visualization,
5. The differences due to the wood stamp ageing.

The only interesting differences for the user are the ones due to the stamp degradation. We call the other ones *perturbations*.

In this frame, we present a method that can help the expert's dating. This method, based on a local adaptation of the Hausdorff distance, evaluates locally the differences. It minimizes the perturbation impact and allows a better visualization of the interesting differences. No other visualization method exists than the PPDMap. Our method performance has been successfully compared to the PPDMap in [2]. Finally, the frame of an automatic dating method is detailed.

2   Dissimilarity Measure Based on the Hausdorff Distance

Among dissimilarity measures over binary images, the Hausdorff distance (HD) has often been used in the content-based retrieval domain and is known to have successful applications in object matching [3] or in face recognition [4]. For finite sets of points, the HD can be defined as [3]:

Definition 1 (Hausdorff distance). *Given two non-empty finite sets of points $F = (f_1, \ldots, f_n)$ and $G = (g_1, \ldots, g_m)$ of $\mathbb{R}^2$, and an underlying distance d, the HD is given by*

$$HD(F, G) = \max\left(h(F, G), h(G, F)\right) \tag{1}$$

where

$$h(F, G) = \max_{f \in F} \left(\min_{g \in G} d(f, g) \right). \tag{2}$$

$h(F, G)$ is the so-called *directed Hausdorff distance*.

The classical HD presents interesting properties but measures the most mismatched points between F and G, and presents as main drawback its sensitivity to noise [5]. Indeed, considering two images containing the same pattern and one point added to the first image, far from the pattern, then the HD will measure the distance between the pattern and the point.

Several modifications of the HD have been proposed to improve it such as: the partial HD [3], the modified HD (MHD) [6], the censored HD [5], the "doubly" modified Hausdorff distance [4], the least trimmed squared HD [7] and the weighted Hausdorff distance [8]. Those improved HD are detailed in [9]. These measures stay global and do not take into account local dissimilarities. Indeed, if $D_H(F, G) = \alpha$, it means that there are $f \in F, g \in G$ that realize the maximum and the minimum in Eq. (1): $d(f,g) = \alpha$. But the measure $D_H(F,G) = \alpha$ does not allow one to say if the couple (f,g) is unique or if there are several couples of points realizing the distance α and, in this case, if the couples are gathered in a part of the images or distributed everywhere in the image, which corresponds to different degrees of dissimilarity. These observations motivated us to define a local and parameter-free HD which in the next paragraph.

2.1 Definition of the Windowed Hausdorff Distance

The main reasons of the modification is that the DH is not defined for empty sets and this case is possible in a window. Moreover, the obtained measures when the window is sliding or growing must be consistent. A solution is to introduce the distance to the window side as it follows:

Definition 2 (Windowed Hausdorff distance). *Let F, G be two bounded sets of $\mathbb{R}^2$.*

$$HD_W(F,G) = \max\left(h_W(F,G), h_W(G,F)\right)$$

where there are three cases

1. If $F \cap W \neq \emptyset$ and $G \cap W \neq \emptyset$,

$$h_W(F,G) = \max_{f \in F \cap W}\left[\min\left(\min_{g \in G \cap W} d(f,g), \min_{w \in Fr(W)} d(f,w)\right)\right],$$

2. if $F \cap W \neq \emptyset$ and $G \cap W = \emptyset$,

$$h_W(F,G) = \max_{f \in F \cap W}\left[\min_{w \in Fr(W)} d(f,w)\right],$$

3. if $F \cap W = \emptyset$,

$$h_W(F,G) = 0.$$

Remark 1

- *In case both of the sets are non-empty, the only difference with the classical definition is the term $\min_{w \in Fr(W)} d(f,w)$ which is the distance from the point f to the edge.*
- *In case there is exactly one set without point in W, one of the two directed distances is equal to 0 and the expression of the other one takes into account the distance to the edge.*
- *In case there is no point of F or of G in W, both of the directed distances are equal to 0 and therefore the global distance too. This is consistent with the fact that the two extracted parts are equal.*

The definition of the windowed HD enables to make a local distance but it introduces a parameter which is the window size. It can be chosen by the user, or automatically and globally, or locally according to the local surrounding. The following properties of the windowed HD allow to fix locally the window size and then to evaluate the local dissimilarity.

Property 1 (Identity). *Let F, G be two bounded sets of points of $\mathbb{R}^2$, and W a convex closed subset of $\mathbb{R}^2$.*

$$HD_W(F,G) = 0 \iff F \cap W = B \cap W \tag{3}$$

The following properties need the window W to be a ball. Prop. 2 ensures that the new pieces of information that are taken into account when the window is enlarged do not reduce the former dissimilarity-measure value. Prop. 3 gives a maximum to the windowed HD.

Property 2 (growth). *Let $V = B(x_v, r_v)$ and $W = B(x_w, r_w)$ be two close discs such as $V \subset W$ then $HD_V(F,G) \leq HD_W(F,G)$.*

Property 3 (Boundary). *Let $x \in \mathbb{R}^2$ and $r > 0$, and let define $W = B(x, r)$ then $HD_W(F,G) \leq HD(F,G)$.*

An algorithm for the computation of the local HD map is proposed below (alg. 1). It consists of a sliding window whose radius is locally adapted to find the local optimal radius.

Algorithm 1. Computation of LDMap

---

compute $D_H(F,G)$
for all pixel x **do**
   $n := 1$ {initialization of the window-size}
   while $HD_{B(x,n)}(F,G) = n$ and $n \leq HD(F,G)$ **do**
      $n := n + 1$
   end while
   $LDMap(x) = HD_{B(x,n-1)}(F,G) = n - 1$
end for

It shows the way to adapt the window to the local dissimilarity. This step is done in the *while* loop.

Nevertheless, this algorithm is time consuming. Indeed, the computation complexity is in $O(m^4)$ for two $m \times m$ pixel images. The next section presents a formula for the measure that saves most of the time computation and gives the same result (detail can be found in [10]). The computation is faster but the interpretation –in terms of local dissimilarity measure– comes from Alg. 1.

2.2   Local Dissimilarity Map

Theorem 1 (LDMap mathematical formula)

$$\forall x \in \mathbb{R}^2, \; LDMap(x) = |G(x) - F(x)| \max(d(x, F), d(x, G))$$

The formula gives for each pixel x a value that depends on the distance transformation from the sets F and G. Fast algorithms have been developed for distance

Fig. 1. Asymmetry illustration : Two images and the sign of their SILDMap

transformation. Their computation complexity are $O(m^2)$ for $m \times m$ images. So the LDMap complexity with the formula is a $O(m^2)$, which is linear in the pixel number.

The LDMap provides symmetric differences measures:

$$LDMap(F, G)(x) = LDMap(G, F)(x).$$

In order to date a printing against another, an asymmetry is introduced in the LDMap, in the following way: if a pixel group is present (resp absent) in the former printing and not in the latter, it is negatively (resp positively) measured (see Fig. 1 where the sign in SiLDMap is represented).

The Signed Local Dissimilarity Map (SiLDMap) gathers all the signed measures in a map:

Theorem 2 (SiLDMap)

> *For x, a pixel of the images,*
>
> $$SiLDMap(x) = (G(x) - F(x)) \max(d(x, F), d(x, G)).$$

3   Experiment and Perspectives

3.1   Experiment

To assess the efficiency of LDMap, let's evaluate first its ability to minimize the impact of perturbations and its ability to render the relevant ones. We have used a database coming from the digital library BVH [11] which includes 168 images of ornamental letters. The database contains four versions of each ornamental letter stamp, coming from four distinct books (so the database contains impressions of 168/4=42 distinct stamps). The four versions of the same stamp that are available provide some perturbations in the visualization: perturbations of ageing, digitization and registration (see Fig. 2). The tested methods are used to produce maps (see Fig. 3) that are classified by a support vector machine (SVM). The experiment protocol is as follows: the comparison of the 168 images gives 14028 visualization maps that are separated in two classes, one gathering the 252 maps comparing images from the same stamp C_{sim} and one including the 13776 maps comparing images from distinct stamps C_{dissim}. A SVM learning stage is

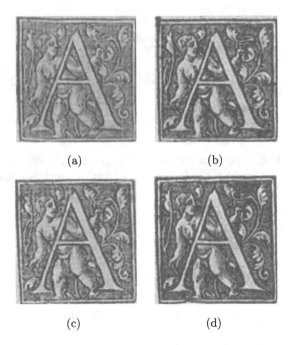

(a)   (b)

(c)   (d)

Fig. 2. A group of four distinct printings coming from the same stamp

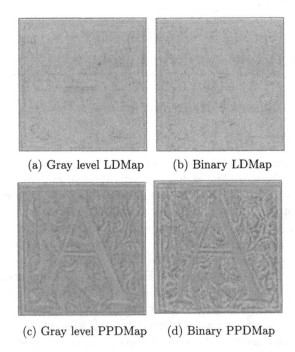

(a) Gray level LDMap   (b) Binary LDMap

(c) Gray level PPDMap   (d) Binary PPDMap

Fig. 3. Visualization Maps between the the ornamental letters 2(a) and 2(c). PPDMaps contain more high values than the LDMap: they are more sensitive to perturbations.

Table 1. Results of the classification for the gray level and binary LDMap and PPDMap

Successful retrieval	found in C_{sim}	found in C_{dissim}
Gray level LDMap	**95%**	**97%**
Binary LDMap	93%	95%
Gray level PPDMap	70%	75%
Binary PPDMap	70%	69%

done on a part of the two classes and a test is realized on the other part. The classification results are compared with those obtained manually. The results, reflecting the average found on 100 tests, are gathered in Tab. 1. Precision and recall measures do not bring more information because they are up to 96% since the first item retrieval rate. Results show that the LDMap allows the SVM to make a better classification than the PPDMap. So the perturbations are less represented in the LDMap than in the PPDMap. The LDMap visualization is therefore better than the PPDMap one. One reason is that a PPDMap does not enable the user to distinguish between a simple translation and a real difference. The result is a successful retrieval rate of 96%, which proves the LDMap robustness against perturbations. A study of the LDMap robustness to ink-stain and erasing can be found in [10]. The robustness is really better for a stain than for an erasing. One reason is that treated information is the one of black pixels. As a consequence, the stain does not change so much the LDmap values whereas the erasing produces a great increase of the LDMap values. Nevertheless, for stains and erasing with a surface smaller than 20% of the total image surface, the robustness is really good.

3.2   Visualisation and Discussion

Thus, the SiLDMap, the signed version of the LDMap, can help in the printing dating by quantifying and localizing relevant differences. The easiest way is to use the SILDMap to visualize the printing differences. Fig. 4 gives an example from two printings of an ornamental letter, and their SILDMap. It contains positive and negative values so the dating is not trivial. Four significant differences have been surrounded in blue. The one at the top of the letter "L" is the only negative and is due to a inking difference. The one at the bottom of the "L" can be interpreted as a missing piece of wood in the first printing, which leads to think that the wood stamp was older when it has been used for the first printing than for the second printing.

The SILDMap measures the differences locally so it is sensible to registration. Fig. 5 shows an example of registration issue: the values in the centre of the SILDMap are low which proves that the affine registration is good, nevertheless high values increase toward the bottom right corner and the top left corner. As both of the printing come from the same stamp, it means that one of their digitized images is deformed (which is an interesting piece of information). The deformation brings high values in the SILDMap that hide the pertinent differences.

Fig. 4. Two impressions and their SiLDMap

Fig. 5. Two impressions and their SiLDMap illustrating the perturbation impact

So a non-linear registration is necessary (and the SILDMap may contain information for the registration) to exploit the SILDMap efficiently.

3.3   Perspectives

The next step is to date automatically the printings thanks to their SILDMap. As Fig. 4 shows it, the dating is not only based on the difference values, but also on elements like local connectedness, inking pressure... An automatic diagnosis should then associate high level information with the SiLDMap to be efficient.

References

1. Baird, H.S.: Digital library and document image analysis. In: Proc. of the 7th Int. Conf. on Document Analysis and Recognition (ICDAR), IAPR, pp. 1–13 (2003)
2. Baudrier, E., Riffaud, A.: A method for image local-difference visualization. In: Proc. of the 9th Int. Conf. on Document Analysis and Recognition (ICDAR), Brazil, IAPR (2007)
3. Huttenlocher, D.P., Klanderman, D., Rucklidge, W.J.: Comparing images using the Hausdorff distance. Trans. on Pattern Analysis and Machine Intel. 15(9), 850–863 (1993)
4. Takàcs, B.: Comparing faces using the modified Hausdorff distance. Pattern Recognition 31(12), 1873–1881 (1998)

5. Paumard, J.: Robust comparison of binary images. Pattern Recognition Letters 18(10), 1057–1063 (1997)
6. Dubuisson, M.P., Jain, A.K.: A modified Hausdorff distance for object matching. In: Proc. of the Int. Conf. on Pattern Recognition (ICPR), IAPR, pp. 566–568 (1994)
7. Sim, D.G., Kwon, O.K., Park, R.H.: Object matching algorithms using robust Hausdorff distance measures. IEEE Trans. on Image Processing 8(3), 425–429 (1999)
8. Lu, Y., Tan, C., Huang, W., Fan, L.: An approach to word image matching based on weighted Hausdorff distance. In: Proc. 6th Internat. Conf. on Document Anal. Recogn., pp. 921–925 (2001)
9. Zhao, C., Shi, W., Deng, Y.: A new Hausdorff distance for image matching. Pattern Recognition Letters (2004)
10. Baudrier, E., Millon, G., Nicolier, F., Ruan, S.: Binary-image comparison with local-dissimilarity quantification. Pattern Recognition 41(5), 1461–1478 (2008)
11. Ramel, J., Busson, S., Demonet, M.: Agora: the interactive document image analysis tool of the BVH project. In: Conf. on Document Image Analysis for Library, pp. 145–155 (2006)

Generating Ground Truthed Dataset of Chart Images: Automatic or Semi-automatic?

Weihua Huang[1], Chew Lim Tan[1], and Jiuzhou Zhao[1]

[1] School of Computing, National University of Singapore
3 Science Drive 2, Singapore 117543
{huangwh,tancl,zhaojiuz}@comp.nus.edu.sg

Abstract. Ground truthing tools mainly fall into two categories: automatic and semi-automatic. In this paper, we first discuss the pros and cons of the two approaches. We then report our own work on designing and implementing systems for generating a chart image dataset and multi-level ground truth data. Both semi-automatic and automatic approaches were adopted, resulting in two independent systems. The dataset as well as the ground truth data are publicly available so that other researchers can access them for evaluating and comparing performances of different systems.

Keywords: Ground truth generation, Maps and Charts Interpretation.

1 Introduction

Ground truthing and performance evaluation has been recognized as an important factor in advancing research in various fields. In the document analysis field, George Nagy addressed the importance of "application-oriented benchmarking" in each research area in document image recognition [1]. Ground truthed datasets that are both well established and publicly accessible are needed to evaluate and compare the performance of different image recognition and analysis systems.

As research on scientific chart recognition and understanding is a relatively young topic, there is no well established public dataset with ground truth that is specifically established for evaluating chart recognition systems. We believe that by making such a public ground truthed dataset, more attention can be drawn from other researchers that might be interested in this relatively new area. The desired dataset should have the following features:

1. The dataset should contain a sufficient number of chart images, to test the efficiency of a system working on a large scale of images.
2. The dataset should include both synthetic images and real-life images. Synthetic images are easier to generate to a large scale, while real-life images are used to present real-life effects.
3. The chart images in the dataset should cover most commonly used chart types to maintain good variety in the test images.

W. Liu, J. Lladós, and J.-M. Ogier (Eds.): GREC 2007, LNCS 5046, pp. 266–277, 2008.

4. The ground truth data should contain details in multiple aspects, so that the dataset can be used to evaluate recognition systems in various ways or at various levels.

Traditional ground truthing tools mainly fall into two categories: automatic and semi-automatic. Our work here adopts both approaches, using the automatic approach to generate synthetic images with ground truth, and the semi-automatic approach for getting ground truth from real-life images. In this paper, we are going to summarize the two systems developed by us for creating chart image datasets with ground truth. In a previous paper [2], the ground truthing system based on the semi-automatic approach was already reported. So in this paper, more emphasis will be put on the second system built based on the automatic approach.

The remaining sections of the paper cover our work in details. Section 2 surveys ground truthing works in both approaches and discusses their pros and cons. Section 3 revisits the semi-automatic system reported previously. Section 4 presents the automatic approach. Section 5 describes the final ground truthed dataset. Section 6 gives a conclusion to this paper.

2 Ground Truthing: Automatic vs. Semi-automatic

Most ground truthing systems reported in the literature are semi-automatic. A semi-automatic ground truthing system may involve human correction following automatic processing steps [3,4,5], or it can consist of a mixture of auto-processing steps and human inputs [8,9]. The semi-automatic approach has certain advantages. First of all, a semi-automatic system can extract ground truth data from a wide range of images with complex layout and varying types, as long as the basic processing functions are available to handle them. Secondly, a semi-automatic system is good to extract ground truth from real-life images, as human inputs or corrections can minimize the error raised from noise and distortions. Thus the resulting ground truthed dataset can reflect real-life noise and distortions. On the other hand, there are also drawbacks of the semi-automatic approach. Firstly, the process is not very efficient as it involves human effort during the process. As a result, it will be either very time consuming or very labour intensive to form a large data collection. Secondly, human verification and correction at low-level still leave certain chance to introduce inaccurate ground truth data. For example, the start point and end point of the vectorized lines may be a few pixels from the true end-points. Although the error is insignificant for most of the time, it is undesired as what we are looking for is ground "truth".

On the other hand, there are also fully automatic ground truthing systems, such as [6,7]. An automatic ground truthing system usually makes use of existing document/graphics generation packages to create datasets and captures intermediate results as the ground truth. Through literature review, we found out that automatic ground truthing is used when the targeted ground truth data only require high level details, such as the number of cells in a table or the font type of the text string. If low-level details are to be included, such as the boundary lines of a cell in a table or the bounding box of a character, then semi-automatic approach seems to be a better choice unless such details are directly available. A typical automatic ground truthing system is computationally efficient and thus is good for the generation of a dataset with a large scale. Furthermore, the ground truth data obtained through automatic process are highly accurate. But the automatic approach also has some drawbacks. Firstly, the amount of ground truth data that can be automatically obtained is restricted, and the low-level details may not be accessible. Secondly, if the system relies on a certain graphics generation package, then the dataset created only reflects the characteristics of that package, resulting in lack of variety in the dataset created. Last but not least, the system produces synthetic images, which do not contain real-life noise and effects. To alleviate this drawback, a degradation module such as [10] is needed to introduce deformations, distortions and noise to the final images produced.

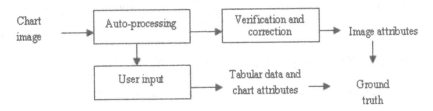

Fig. 1. Semi-automatic ground truthing

3   The Semi-automatic System Revisited

As shown in Figure 1, the system developed based on semi-automatic approach [2] accepts real-life images that were downloaded from the web or scanned in. Basic image processing techniques are performed to automatically extract image attributes of the graphical symbols in the input chart image. As the image processing techniques applied are imperfect dealing with noisy real-life images, the result obtained may be erroneous. Thus in the following step, the user needs to verify the result and make corrections when necessary. Since the structural and semantic information of the input image is not available, the user takes another responsibility which is to input these types of information. Through investigation, we found out that getting direct input from the user is more efficient than an automatic recognition of the structural and semantic information followed by corrections, as keying in high-level information is convenient for the user. The resulting ground truth information is defined to be multi-level, including the pixel level, the text level, the vector level and the chart level. By having multi-level ground truth, the dataset can be used not only by chart recognition systems but also by other systems focusing on different levels, such as text recognition systems or graphical symbol extraction systems. More details of the multi-level ground truth can be referred to in [2], which also suggested the metrics for

Table 1. Statistics of the dataset generated using semi-automatic approach

Image Information	
Chart type	Number of images
Bar chart	80
Pie chart	60
Line chart	60
Ground Truth Information	
Entity	Quantity
Text level	
Text block	4212
Word	5692
Graphics level	
Straight line	10165
Arc	129
Chart level	
Chart title	151
X-axis label	1820
Y-axis label	1308
Bar	1719
Wedge	401
Polyline vertex	681

performance evaluation at individual level as well as the overall performance measure. The dataset reported in [2] initially contained 120 chart images with ground truth. The chart images are of 4 different types: 2D bar chart, 2D pie chart, 3D pie chart and Line Chart. The ground truth data are of two different formats: plain text and XML format. The dataset has since been expanded to 200 chart images, with the same types and ground truth formats. Table 1 summarizes some statistics about the dataset and ground truth generated.

4   The Automatic System

As mentioned in section 2, both the semi-automatic approach and the automatic approach have pros and cons. One obvious problem with the semi-automatic system introduced is its low efficiency. It cost several minutes to process one chart image. Thus the dataset created is fairly small in terms of number of images. To expand the dataset to a reasonably large scale, we also implemented an automatic system. One possible way to achieve automatic ground truthing is to decode the graphics generation software and capture the intermediate data as the ground truth data. However, through investigation of some existing graphics packages, such as Microsoft Excel and PSTricks, we found that this task was not easy to achieve as most graphics generation software reveals only high-level details. Low-level details such as vector information and text bounding boxes cannot be obtained unless reverse engineering is applied. Thus we decided to implement an automatic system on our own. The system should generate chart

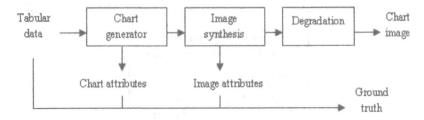

Fig. 2. Automatic ground truthing

images and store detailed ground truth data at all levels that were mentioned in the previous section. The major steps in the automatic system are shown in Figure 2. By comparing this figure with Figure 2, we can see that the automatic ground truthing process focuses on "generation" while the semi-automatic ground truthing process focuses on "extraction".

4.1   The Chart Generator

Randomly generated tabula data (label plus value) is used as the basis for chart generation. The data is passed into the chart generator to create a chart of a certain type chosen by the user. The current version of the system generates four common types of chart: 2D bar chart, 3D bar chart, 2D pie chart and 3D pie chart. Each chart type consists of a set of essential components, which can be further decomposed into text entities and regular graphical entities. Each graphical entity is represented as a combination of graphical primitives following geometric constraints. To draw a generated chart as an image, the drawing functions in the Windows GDI+ library are called to draw the graphical primitives such as line segments and arcs. The thickness of a line or an arc can be specified by user. GDI+ library also provides functions to render text strings in an image and estimate the bounding box of each text string. Figure 3 illustrates how a chart is decomposed and converted into an image. Note that the existence of axis is type-dependent. If a chart type does not require axis, such as a pie chart, then the system does not include it. Drawing 3D charts is more complicated than drawing 2D charts, in our approach the following steps are carried out:

Step 1: Draw a 2D version of the chart.
Step 2: Construct 3D chart based on the 2D version, using geometric transformations. To draw a 3D bar chart from its 2D version, translation is used. To draw a 3D pie chart from the 2D version, perspective distortion and translation are both applied.

4.2   The Degradation Module

For each chart generated, a clean synthetic image is created through rasterization. The degradation module is applied on the clean image to add less-than-ideal effects to simulate real-life image quality. Our degradation module is based

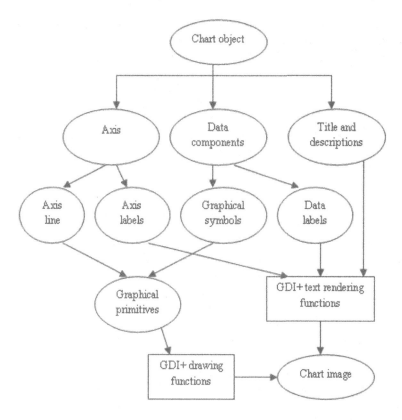

Fig. 3. Drawing chart image using GDI+ functions

on the degradation model proposed by Baird [10]. The original model listed 10 parameters. Considering the problem domain we are dealing with, we only adopt a subset of them. As listed in Table 2, the parameters included in our degradation module are used to perform the following tasks: rotation (skew angle), shearing, edge distortion, Gaussian noise and motion blur.

Table 2. Overview of the parameters in the degradation module

Parameter	Data Type	Range	Meaning
β	Real	$(-\pi, \pi)$	Skew angle, measured in degrees
λ	Real	$[-1, 1]$	Horizontal shearing factor
L	Integer	$[0, 10]$	Degree of edge distortion
v	Integer	$[0, 5]$	Radius of motion blur
θ_b	Real	$(-\pi, \pi)$	Angle of motion blur, measured in degrees
σ	Real	$[0, 50]$	Degree of Gaussian noise

- Rotation. Rotation is a deformation operation. The whole chart is rotated to add a skew angle to the image. For each pixel (x, y) in the image plane, if the skew angle is β, then the new pixel location (x', y') in vector form is:

$$\begin{bmatrix} x' \\ y' \end{bmatrix} = \begin{bmatrix} \cos\beta & -\sin\beta \\ \sin\beta & \cos\beta \end{bmatrix} \begin{bmatrix} x \\ y \end{bmatrix} \qquad (1)$$

- Shearing. Shearing is a common deformation type that changes the shape of a geometric object. The shearing process requires one parameter, the shearing factor $\lambda = \cot\alpha$, and a pixel (x, y) will be mapped to the new location:

$$\begin{bmatrix} x' \\ y' \end{bmatrix} = \begin{bmatrix} 1 & \cot\alpha \\ 0 & 1 \end{bmatrix} \begin{bmatrix} x \\ y \end{bmatrix} \qquad (2)$$

- Edge distortion. In real-life, distortions are very likely to occur along the edges of lines or regions, mainly due to the reproduction process such as scanning or faxing etc. To simulate edge distortion, we adopt a convolution method based on [11], with the modification that besides pixel-adding in the original method, pixel-reduction is also performed. Here pixel-adding means a pixel change from fore-ground color to background color and pixel-reduction means vice versa. A parameter L here is used to controls the degree of edge distortion.
- Motion blur. Motion blur most often occurs during a camera-based capturing process. The modeling of motion blur is based on [12]. Let f(x, y) be the input image, and H(x, y) be the blurring function. With two parameters $v = $ the level of motion blur and $\theta = $ the angle of the motion blur, the blurred image g(x, y) is generated as:

$$g(x,y) = \sum_{n=1}^{width} \sum_{m=1}^{height} f(x-n, y-m)H(n,m) \qquad (3)$$

where

$$H(x,y) = \begin{cases} \frac{1}{2v+1} & , \text{ if } 0 \le |x| \le (2v+1) * \cos\theta \\ & and \ \ 0 \le |y| \le (2v+1) * \sin\theta \\ 0 & , \text{ otherwise} \end{cases} \qquad (4)$$

- Gaussian noise. Gaussian noise models the thermal noise in electronic imaging systems. To generate Gaussian noise, the crucial step is to obtain a Gaussian (normal) distribution, a random variant with its probability density function as:

$$p(X) = \frac{1}{\sigma\sqrt{2\pi}} e^{\frac{-x^2}{2\sigma^2}} \qquad (5)$$

Here we use an algorithm called ran0 [13] to realize the polar method [14] for obtaining a standard normal variable X0. To add Gaussian noise, each pixel Gij in the original image is added with a value σX0. σ is a parameter that controls the level of noise.

4.3   The Ground Truth Generation

The initial tabular data become the semantic level ground truth. The vector information of the lines recorded during drawing process becomes the vector level ground truth. The text strings and their bounding boxes form the text level ground truth. The chart entities created during chart generation are also recorded to form another part of the chart level ground truth. An extra part of the ground truth contains the parameters used by the degradation module. This part of information was not obtainable using the semi-automatic approach.

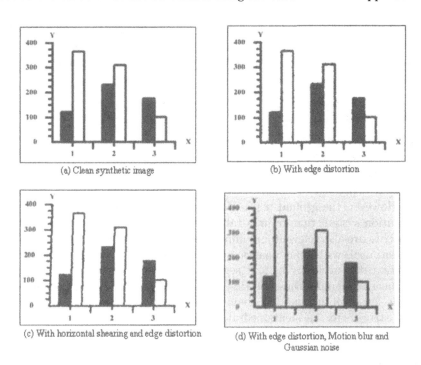

(a) Clean synthetic image

(b) With edge distortion

(c) With horizontal shearing and edge distortion

(d) With edge distortion, Motion blur and Gaussian noise

Fig. 4. Sample synthetic image and degradation effects

5   The Final Ground Truthed Dataset

5.1   Dataset Description

The final dataset contains two subsets from the two works we have done: a collection of real-life images and a collection of synthetic images. For the real-life collection, 200 images were collected and the corresponding ground truth data were also extracted using the system presented in [2]. In the synthetic collection produced using this automatic system, 400 clean images were created for each of the four chart types. For a clean image, one of the eight different combinations of degradation effects was added to create a noisy version. Example of a synthetic image and its corresponding degraded versions are shown in Figure 4. Thus the

Table 3. The final data set

Chart type	Real			Synthetic		
	Scanned	Downloaded	**Total**	Clean	Noisy	**Total**
Bar chart	61	19	**80**	800	800	**1600**
Pie chart	-	60	**60**	800	800	**1600**
Line chart	14	46	**60**	-	-	**-**
Total	**75**	**125**	**200**	**1600**	**1600**	**3200**

final dataset contains 3200 chart images, with ground truth data in XML format. We put them together with the first dataset produced using the semi-automatic system, resulting in the final dataset with a total of 3400 chart images and their corresponding ground truth data. Some statistics about the complete dataset are shown in Table 3.

5.2   Discussions

In automatic ground truth generation, there is a trade off between the complexity of the implementation and the level of details to be kept in the ground truth data. If only tabula data are required, then the generation process is very simple: use a graphical package to create electronic charts and then convert it into image format. However, the ground truth will only be useful when evaluating a chart interpretation system that returns tabula data. Besides the tabula data itself, other metrics are also relevant and important to the performance evaluation of a system that deals with chart images, including the accuracy of graphical symbol construction, the accuracy of text segmentation and recognition etc. Thus to provide measurement for these metrics, the ground truth should be more enriched to include low-level information about graphical symbols, text bounding boxes and text strings etc. As mentioned at the beginning of section 4, the low-level information is not directly obtainable from commercial graphical packages. Thus to obtain such information, we need to implement our own functions for drawing and recording.

The accuracy of the automatically generated ground truth data is relatively higher than those generated using the semi-automatic system. However, some ground truth data may still be slightly erroneous. More specifically, the bounding box returned by the GDI+ function Graphics.MeasureString() does not reflect the true bounding box of a text string, due to the limitation of the way GDI+ computes the width of the text using hinting and anti-aliasing. The bounding box returned by the current implementation is a bit wider than the truth bounding box. The problem may be solved in the new version of the system, using alternative ways of measuring the width of text strings.

The current version of the system only takes the major chart components into consideration, including: chart axes, data components, titles and labels etc. Although these are the essential components for interpreting a chart, there are other important components to be included. For example, legends are very important in a chart with multiple data series. Grid lines may also be included because they are very often used in real-life charts. Besides, the random text generation unit in the current system only generates very simple text strings such as numeric strings etc. Random alphabetic labels, or even sentence based descriptions should be generated. The points mentioned above will be covered as our future work.

6   Issues on Performance Measure

An important issue raised with the ground truthed dataset is how the data can be used to measure the performance of a system. The system to be evaluated does not need to perform all the tasks and generate all the data to match with the ground truth. It can be a line detection system, a text recognition system or an image understanding system. Thus performance score needs to be defined from multiple aspects.

Performance evaluation issue on pixel level and vector level were well described by Liu et al [6]. We also proposed ways to perform evaluation on higher levels in [2]. Below are some of them re-visited:

At a higher level which is the chart level, the detection rate of graphical data components can be obtained by calculating the data component recovery index:

$$DRI = \mu D_d + (1 - \mu)(1 - F_d) \tag{6}$$

where μ is the relative importance of detection and $1-\mu$ is the relative importance of the false alarm. And here:

$$D_d = \frac{\Sigma_{k \in C_g} D_d(k) S(k)}{\Sigma_{k \in C_g} S(k)} \tag{7}$$

where D_d is the overall detection rate, $D_d(k)$ is the detection rate for ground truth component k and S(k) is the size of ground truth component k, C_g is the set of graphical data components in the ground truth.

$$F_d = \frac{\Sigma_{k \in C_d} F_d(k) S(k)}{\Sigma_{k \in C_d} S(k)} \tag{8}$$

where F_d is the overall false alarm rate, $F_d(k)$ is the false alarm rate of the detected component k, C_d is the set of graphical data components detected. $D_d(k)$ and $F_d(k)$ are defined as:

$$D_d(k) = \frac{S(C_d(k) \cap C_g(k))}{S(C_g(k))} \tag{9}$$

$$F_d(k) = 1 - \frac{S(C_d(k) \cap C_g(k))}{S(C_d(k))} \tag{10}$$

where $C_d(k)$ is the detected component and $C_g(k)$ is the ground truth component.

For evaluation of text recognition results, well known IR metrics precision P and recall R are used instead of detection rate and false alarm. Calculation of the precision and recall for character recognition is straightforward:

$$P = \frac{|Ch_g \cap Ch_d|}{|Ch_d|} \tag{11}$$

$$R = \frac{|Ch_g \cap Ch_d|}{|Ch_g|} \tag{12}$$

where Ch_g is the set of characters in the ground truth text and Ch_d is the set of characters recognized. To evaluate the accuracy of text blocks detected, a slight change needs to be made to equation (11) and (12). Instead of the intersection between two sets, the overlap between two corresponding bounding boxes should be calculated.

The overall performance score S may be defined as:

$$S = \Sigma_{i=1}^n w_i S_i \tag{13}$$

where Si is the individual score at a single level i, and w_i is the weight assigned to each $S_i (\Sigma w_i = 1)$. The weights are used to address the aspects that a system emphasizes on. For example, equation (13) is applicable for a system focusing on only one task, by turn off other performance measures (setting all other weights to zero).

7   Conclusion and Future Work

This paper covered our work on constructing a public dataset of chart images and generating multi-level ground truth data for the images. Two approaches were adopted to implement two independent ground truthing systems: the semi-automatic approach and the automatic approach. As the semi-automatic system was reported before, this paper emphasized more on the automatic system which was developed more recently. This paper also discussed the pros and cons of both approaches, and suggested that the ideal way of constructing a large dataset with ground truth is to combine the results of the two approaches. The resulting dataset with ground truth data is publicly accessible, through URL:

http://www.comp.nus.edu.sg/~huangwh/GroundTruth/dataset.html

Acknowledgement. This research is supported by A*STAR grant 0421010085 and NUS URC grant R252-000-202-112.

References

1. Nagy, G.: Twenty years of Document Image Analysis in PAMI. IEEE Transactions on Pattern Analysis and Machine Intelligence 22(1), 38–62 (2000)
2. Yang, L., Huang, W.H., Tan, C.L.: Semi-automatic ground truth generation for chart image recognition. In: Bunke, H., Spitz, A.L. (eds.) DAS 2006. LNCS, vol. 3872, pp. 324–335. Springer, Heidelberg (2006)

3. Haralick, R.M., et al.: UW English document image database I: A database of document images for OCR research. UW CD-ROM

4. Haralick, R. M. et al: UW-II English/Japanese Document Image Database: A Database of Document Images for OCR Research,
http://www.science.uva.nl/research/dlia/datasets/uwash2.html

5. Phillips, I.: Users' reference manual. CD-ROM, UW-III Document Image Database-III (1995)

6. Wang, Y., Haralick, R.M., Phillips, I.T.: Automatic Table Ground Truth Generation and a Background-Analysis-Based Table Structure Extraction Method. In: 6th Int. Conf. on Document Analysis and Recognition, ICDAR 2001, Seattle, pp. 528–532 (2001)

7. Zi, G., Doermann, D.: Document Image Ground Truth Generation from Electronic Text. In: 17th Int. Conf. on Pattern Recognition, ICPR 2004, vol. 2, pp. 663–666 (2004)

8. Yacoub, S., Saxena, V., Sami, S.: PerfectDoc: A Ground Truthing Environment for Complex Documents. In: 8th Int. Conf. on Document Analysis and Recognition, vol. 1, pp. 452–456 (2005)

9. Suzuki, M., Suzuki, S., Nomura, A.: A Ground-Truthed Mathematical Character and Symbol Image Database. In: 8th Int. Conf. on Document Analysis and Recognition, vol. 2, pp. 675–679 (2005)

10. Baird, H.S.: Document Image Defect Models. In: Proceedings of IAPR Workshop on Syntactic and Structural Pattern Recognition, Murray Hill, NJ; Reprinted in: Baird, H.S., Bunke, H., Yamamoto, K.: Structured Document Image Analysis, pp. 546–556. Springer, New York (1990)

11. Zhai, J., Liu, W.Y., Dori, D., Li, Q.: A Line Drawings Degradation Model for Performance Characterization. In: 7th International Conference on Document Analysis and Recognition, Edinburgh, Scotland (2003)

12. Gonzalez, R.C., Wintz, P.: Digital Image Processing, 2nd edn. Addison-Wesley Publishing Company, Reading (1987)

13. William, H.P., Saul, A.T., William, T.V., Brian, P.F.: Numerical recipes in C++: The Art of Scientific Computing. Cambridge University Press, New York (2002)

14. Ross, S.M.: A Course in Simulation. Macmillan Publishing Company, New York (1990)

Performance Characterization of Shape Descriptors for Symbol Representation

Ernest Valveny[1], Salvatore Tabbone[2], Oriol Ramos[1], and Emilie Philippot[2]

[1] Computer Vision Center, Dep. Ciències de la Computació
Universitat Autònoma de Barcelona, Bellaterra, Spain
{oriolrt,ernest}@cvc.uab.es
[2] LORIA, Université Nancy 2, Nancy, France
{Antoine.Tabbone}@loria.fr

Abstract. In this paper we propose a general framework for the characterization of shape descriptors and show its application to graphic symbols. The framework is based on the combination of several performance measures independent of the application. We have applied this framework using a standard set of descriptors and databases. We show how it can be used to characterize the properties of each descriptor for a given database.

1 Introduction

There has been an increasing interest in research in performance evaluation in Graphics Recognition during the last years. Several contests have been organized in past editions of GREC Workshop concerning raster-to-vector conversion [1,2,3], arc segmentation[4] and symbol recognition[5,6]. In the particular domain of symbol recognition, a general framework of evaluation has been proposed[7] with the goal of getting a deeper understanding of the characteristics, pros and cons of various approaches to symbol recognition. The contests aim to analyze the performance of symbol recognition methods with several types of test data, including different number of symbols and several kinds of transformations and degradations. The results have been very positive as they permit to determine the robustness of participant methods under the different kinds of noise included in the test set. However, we cannot get a global understanding of different approaches to symbol recognition as only few methods (those used by the participants in the competition) were evaluated. In addition, as remarked in the conclusions of the last contest (cf. [6]), not always a detailed information about the techniques employed by each participant method is available and, therefore, we cannot have a good understanding of recognition rates according to the different types of methods.

In this paper we propose a different and more general approach for performance evaluation of methods for symbol recognition. If we take a look at them, we can observe that most of them are based on some kind of shape descriptor, as shape is the most characteristic visual feature of symbols. Indeed, the selection of a suitable shape description and representation that permits

W. Liu, J. Lladós, and J.-M. Ogier (Eds.): GREC 2007, LNCS 5046, pp. 278–287, 2008.

to capture the most relevant features of symbols is a key issue in order to obtain good recognition rates. Actually, a large number of shape descriptors have been proposed in the literature [8,9] and most of them have been applied to the problem of symbol recognition.

Then, our main goal is to propose a framework to evaluate the general performance of several shape descriptors and apply this framework to the particular case of symbol representation. This is the first difference with previous approaches to performance evaluation of symbol recognition. Instead of evaluating specific methods for symbol recognition we will evaluate general shape descriptors that can also be used for other problems in pattern recognition. The second difference is in the final goal of the evaluation framework. We do not want to focus on recognition, but we aim at characterizing the behavior of shape descriptors under several circumstances. Then, our evaluation is not only based on the recognition rate, but on the combination of several measures: recognition rate, homogeneity, separability, precision and recall.

In the experiments we have used two databases the symbol database defined for previous contests on symbol recognition, and the standard MPEG shape database. We have taken several standard and well-known shape descriptors grouped in three categories: pixel-based descriptors (Fourier-Mellin, Generic Fourier Descriptor and Zernike moments), contour-based descriptors (shape context, pixel-level constraint and string matching) and structural descriptors (graph-based).

The paper is organized as follows: first, in section 2 we explain the framework for performance characterization, mainly the evaluation measures. Then, in section 3 we describe the experiments using the selected set of descriptors and shapes. Finally, in section 4 we draw the main conclusions of this work.

2   Performance Characterization

As we have said before, our main goal is the characterization of shape descriptors, i.e, to define a kind of *genetic map* of a number of descriptors, i.e, a list of relevant and intrinsic properties for each family of descriptors. Such list of properties can help to choose the most appropriate family of descriptors given a practical pattern recognition problem.

For such a protocol to be of general use, it must be independent of datasets and must evaluate several properties of each descriptor such as the complexity, the robustness to different kinds of transformations and degradations, the power of discrimination as the number of classes and the variability of shapes grows, the genericity with very different datasets or the influence of the setting of the intrinsic parameters of the descriptor.

In this context, we cannot rely only on recognition rate as evaluation measure. Recognition rates can be very dependant on the type of classifier used. In addition, they are largely linked to one kind of practical problem, the classification of unknown shapes and cannot be the best choice to evaluate other kind of applications, such as shape retrieval.

Therefore, we need more general evaluation criteria in order to get a deep understanding of the properties of descriptors. Thus, we have decided to use 5 different measures for that: two of them (separability and homogeneity) try to be independent of the application as they intend to evaluate how well distributed are the shapes in the space of representation provided by the shape descriptor. The other three measures evaluate the performance of the descriptors in two of the most common real problems: recognition rate for the problem of classifying unknown shapes and precision/recall for shape retrieval.

All these measures rely on the computation of the matrix of distances among the representation of all the images in the shape database obtained using a given descriptor. The definition of the distance will assure that all the distances are normalized between 0 and 1 (for instance, using the Pearson correlation coefficient or the normalized euclidean distance).

- **Homogeneity:** A good description of a class of shapes should yield an homogeneous representation in the sense that the representation of all the shapes should be concentrated in a small area of the feature space. In this sense, we have defined a measure so that values close to zero mean that all the feature vectors are close (the descriptor is more homogeneous). This measure, $\mathcal{H}$ is based on the distance between elements belonging to the same class and is defined in the following way:

$$\mathcal{H} = \sum_{c=1}^{N} \frac{H(c)}{N} \tag{1}$$

$$H(c) = \frac{2h(c)}{M_c(M_c - 1)} \tag{2}$$

$$h(c) = \sum_{i=1}^{M_c} \sum_{j=1, j>i}^{M_c} \delta(v_i, v_j) \tag{3}$$

where N is the number of classes, M_c is the number of elements in class c, v_i is the representation of element i using a particular descriptor and $\delta(v_i, v_j)$ is the distance between two elements in the feature space normalized between 0 and 1.

- **Separability:** Another property of a good shape description is that elements belonging to different classes have a dis-similar representation. Thus, we have defined a measure of separability that permits to assess this property. The farther feature vectors of elements belonging to different classes are the more separability of the descriptor. This measure, S, is based on the distance between elements belonging to different classes and defined as:

$$S = \sum_{c=1}^{N} \frac{S(c)}{N} \tag{4}$$

$$S(c) = \frac{s(c)}{M_c \sum_{k=1,k\neq c}^{N}(M_k - 1)} \tag{5}$$

$$s(c) = \sum_{i=1}^{M_c} \sum_{k=1,k\neq c}^{N} \sum_{j=1}^{M_k} \delta(v_i, v_j) \tag{6}$$

- **Recognition rate:** Using the well-known 1-NN classifier, we evaluate the performance of each descriptor for recognition. This is a standard measure that can be used as a benchmark for the performance of the descriptor in recognition tasks.
- **Precision/Recall:** These measures are commonly used in the context of image retrieval and are useful to evaluate the ability of the descriptor to retrieve shapes similar to a given query shape. They can be used to evaluate the performance of the descriptor in retrieval tasks. Precision, P, measures how many retrieved shapes really correspond to the class of the query shape, while recall, R, measures the percentage of the total number of shapes belonging to the query shape actually retrieved by the descriptor. They are defined in the usual way:

$$P = \frac{N_c}{N} \tag{7}$$

$$R = \frac{N_c}{M_c} \tag{8}$$

where c is the class of the query shape, N is the total number of retrieved shapes, N_c is the number of retrieved shapes belonging to class c and M_c is the total number of shapes belonging to class c.

3 Experiments

3.1 Shape Descriptors

All these measures have been applied to evaluate a set of standard and well-known shape descriptors. As it is usually done in the literature we have distinguished between pixel-based, contour-based and structural descriptors.

Pixel-based descriptors are computed directly from the pixels of the whole image. We have used the following descriptors in this category:

- Fourier-Mellin[10]: based on the application of the Mellin and the Fourier transforms to the polar representation of the image. It is invariant to rotation and scaling.
- General Fourier Descriptor (GFD)[11]: based on the Modified Polar Fourier Transform, that applies a 2-D Fourier Transform to the polar representation of the image. The coefficients are conveniently normalized in order to achieve invariance to rotation and scaling.

- Zernike moments[12]: based on computing the projection of the image onto the Zernike polynomials and have been widely used in pattern recognition. They are also invariant to rotation and scaling.

Contour-based descriptors are obtained after extracting the outer contour of the shape. In this category we have used:

- Shape Context[13]: this descriptor is based on taking a sample of points from the contour of the shape and computing the histogram of spatial relations between a reference point and all other sample points in the contour. It is invariant to translation and scaling.
- Pixel-level constraint (PLC)[14]: it is based on the points of the skeleton of the shape. Then, taking any of these points as reference the ratio of angle and length between any other pair of points can be computed and then, the histogram of these ratios is obtained. The histograms obtained taking every point in the skeleton as reference point can be grouped in two matrices, one for angular information and the other one for the length information, that are processed to obtain the final descriptor, that is rotation and scale invariant.
- String matching[15]: based on the representation of the contour as a chain code and applying and edit distance to compute the similarity between chain code of two different shapes.

Structural descriptors are based on representing relationships between components of the shape, normally using graphs or grammars. In our case, we have used a graph representation where nodes correspond to junction points or end points and edges correspond to the lines joining these points. From this graph representation a signature is computed assigning to each node a value based on the number of incident edges and the angle and relative length between them.

3.2   Shape Databases

All these descriptors have been applied to two shape databases: the database of graphic symbols defined for the first contest on symbol recognition at GREC' 2003 [5] and the MPEG-7 contour database.

The GREC database is composed of 50 graphic symbols composed of straight lines and arcs of circumference. The original database generated for the contest contained images with 3 kinds of transformations: geometric transformations (rotation and scaling), binary degradations and vectorial distortions. In our experiments we have used two subsets of images from the original database:

- GREC-50: in this set we have images of the 50 original symbols with rotation, scaling and slight binary degradations (see figure 1(a))
- GREC-Vec: in this set we have included images with vectorial distortion generated by randomly moving junction and end points, but keeping line connectivity as required by the graph-based descriptor. In this case we have only used the 26 symbols composed only of straight lines. In figure 1(b) we can see some examples of the kind of distortions that have been generated.

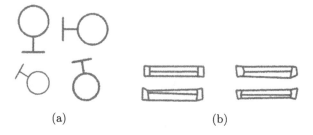

(a)   (b)

Fig. 1. Example of symbols of GREC database. (a) symbols in the set GREC-50. (b) Vectorial distortion applied to symbols included in the set GREC-Vec.

However, in this kind of images, contour-based descriptors do not perform well. In addition, one of the goals of the proposed protocol for the characterization of descriptors was to test the genericity with different datasets. Then, in order to obtain more general results and to be able to better compare pixel-based and contour-based descriptors we have also used images of the MPEG-7 shape database (some examples can be seen in figure 3.2). For this database we have defined 4 subsets:

- MPEG-99: composed of 99 images belonging to 9 different classes.
- MPEG-216: contains images of 18 classes, 12 images per each class.
- MPEG-1045: 1045 images belonging to 42 classes. There are between 3 and 60 images per class.
- MPEG-Occ: this set is the same as MPEG-99, but we have applied a method to generate random partial occlusions of the contour as can be seen in figure 3.2

Fig. 2. Examples of the shapes included in the MPEG-7 shape database

Fig. 3. Examples of the shapes with partial occlusions of the contour

3.3 Analysis of Results

Not all the set of descriptors described in section 3.1 could be applied to all the sets of images explained in section 3.2 due to the properties of each descriptor. Then, in the summary table 1 we can see which descriptors have been applied to each database. Mainly, pixel-based descriptors were applied to both databases, contour-based descriptors only to the MPEG-7 database and structural descriptors only to the GREC database.

Then, once obtained the representation of all the shapes with every descriptor, we computed the distance matrix among all elements and all the evaluation measures: homogeneity, separability, recognition rate and precision/recall. Due to space availability we cannot show the detailed results for all descriptors and databases, but the analysis of these results permit to state some conclusions about the performance of the descriptors, both from a global point of view and from the particular point of view of every descriptor.

If we analyze the results globally, considering all descriptors and databases we can say:

- GFD and Zernike moments have always obtained the best recognition rates. In addition, recognition rates are better for the GREC database than for the MPEG database. This seems logical as shapes in the GREC database have less shape variability.
- In general GFD also gives the best value for homogeneity. For contour-based descriptors, PLC is the descriptor with the best homogeneity. The graph-based descriptor has also a very good homogeneity in the only case where it is used.

Table 1. Summary of the performance of descriptors

	GREC-50	GREC-Vec	MPEG-99	MPEG-216	MPEG-1045	MPEG-Occ
Fourier-Mellin	- - -	+	+	+	-	- - -
GFD	++	++	+++	++	++	+
Zernike	++	++	++	++	+	++
Shape Context			- -	-	- - -	+
PLC			+	++	++	+
String matching			-	-	- -	
Graph		+				

- Zernike moments have always the best separability measure, although they present one of the worst homogeneity values.
- There are some descriptors (GFD, Zernike) with better precision than recall. That means that they prioritize retrieving exact shapes than all meaningful shapes relevant to a given query. On the contrary, other descriptors (Fourier-Mellin, PLC and Shape context) with better recall than precision are better in order to retrieve all relevant shapes although some of them do not correspond to the query. On the other hand, graph-based descriptors and string matching reach a good compromise between precision and recall in the sense that they are able to retrieve a good number of relevant shapes while keeping low the number of non-relevant answers.

Beside these global conclusions we can also state some interesting conclusions about every particular descriptor:

- GFD has a good separability in long and thick shapes and in shapes with curves.
- Fourier-Melllin gives better separability in closed shapes and better homogeneity in long and thick shapes.
- Shape context has better separability in occluded shapes. The recognition rates are, in general, low.
- Pixel-level constraint obtains good recognition rates and good separability in long and thick shapes.
- String matching gives better separability when there are significant changes of direction in the contour of the shape.
- Graph-based descriptors do not really have a good recognition rate. They have difficulty in separating objects with the same structure, but they have good separability for shapes with large number of lines.

We have summarized these conclusions in Table 1 where for every set of images we show how positive $(+,++,+++)$ or negative $(-,--,---)$ are the results obtained for each descriptor. The evaluation of each descriptor is based on the analysis of the results of the recognition rate, the homogeneity and the separability. The precision and recall have not been taken into account. This analysis does not intend to be a rigorous, formal and exact evaluation of the descriptors. However it permits to establish a kind of tendency for each descriptor and can help to choose a descriptor for a given application. For instance, it can be observed that GFD, Zernike and PLC always obtain a positive evaluation, although the "best" descriptor varies depending on the dataset.

4   Conclusions

In this paper we have proposed to use several performance measures for the characterization of shape descriptors. The combination of these measures permits to have a better understanding of the behavior of each descriptor than using single classical indices such as the recognition rate, precision or recall, that are more oriented to specific tasks. We illustrate the usefulness of this approach by analyzing a set of standard shape descriptors using a database of graphic symbols and a database of contour shapes. From the analysis of the results obtained with the proposed evaluation measures we are able to state several conclusions that characterize the performance of the descriptors for each database permitting to summarize it in a table.

References

1. Phillips, I., Chhabra, A.: Empirical performance evaluation of graphics recognition systems. IEEE Transactions on Pattern Analyisis and Machine Intelligence 21(9), 849–870 (1999)
2. Chhabra, A., Phillips, I.: The second international graphics recognition contest - raster to vector conversion: A report. In: Tombre, K., Chhabra, A.K. (eds.) GREC 1997. LNCS, vol. 1389, pp. 390–410. Springer, Heidelberg (1998)
3. Chhabra, A., Philips, I.: Performance evaluation of line drawing recognition systems. In: Proceedings of 15th. International Conference on Pattern Recognition, Barcelona, Spain, September 2000, vol. 4 (2000)
4. Wenyin, L., Zhai, J., Dori, D.: Extended summary of the arc segmentation contest. In: Blostein, D., Kwon, Y. (eds.) GREC 2001. LNCS, vol. 2390, pp. 343–349. Springer, Heidelberg (2002)
5. Valveny, E., Dosch, P.: Symbol recognition contest: A synthesis. In: Lladós, J., Kwon, Y.-B. (eds.) GREC 2003. LNCS, vol. 3088, pp. 368–385. Springer, Heidelberg (2004)
6. Dosch, P., Valveny, E.: Report on the second symbol recognition contest. In: Liu, W., Lladós, J. (eds.) GREC 2005. LNCS, vol. 3926, pp. 381–397. Springer, Heidelberg (2006)
7. Valveny, E., Dosch, P., Winstanley, A., Zhou, Y., Yang, S., Yan, L., Wenyin, L., Elliman, D., Delalandre, M., Trupin, E., Adam, S., Ogier, J.: A general framework for the evaluation of symbol recognition methods. International Journal on Document Analysis and Recognition 9(1), 59–74 (2007)
8. Zhang, D., Lu, G.: Review of shape representation and description techniques. Pattern Recognition 37, 1–19 (2004)
9. da F.R. Costa, L., Cesar Jr., R.M.: Shape analysis and classification. CRC Press, Boca Raton (2001)
10. Adam, S., Ogier, J.M., Cariou, C., Mullot, R., Gardes, J., Lecourturier, Y.: Utilisation de la transformé de fourier-mellin pour la reconnaissance de formes multiorientées et multi-échelles: application l'analyse automatique de documents technique. Traitement du signal 18(1), 17–33 (2001)
11. Zhang, D., Lu, G.: Shape-based image retrieval using general fourier descriptor. Image Communication 17
12. Kim, H., Kim, J., Sim, D., Oh, D.: A modified zernike moment shape descriptor invariant to translation, rotation and scale for similarity-based image retrieval. In: IEEE International Conference On Multimedia and Expo.,

13. Mori, G., Belongie, S., Malik, J.: Efficient shape matching using shape context. IEEE Transactions on Pattern Analyisis and Machine Intelligence 27(11), 1832–1837 (2005)
14. Yang, S.: Symbol recognition via statistical integration of pixel-level constraint histograms: A new descriptor. IEEE Transactions on Pattern Analyisis and Machine Intelligence 27(2), 278–281 (2005)
15. Tsay, Y., Tsai, W.: Model-guided attributed string matching by split-and-merge for shape recognition. International Journal of Pattern Recognition and Artificial Intelligence (1988)

Building Synthetic Graphical Documents for Performance Evaluation

Mathieu Delalandre[1], Tony Pridmore[2], Ernest Valveny[1],
Hervé Locteau[3], and Eric Trupin[3]

[1] CVC, Barcelona, Spain
{mathieu,ernest}@cvc.uab.es
[2] SCSIT, Nottingham, England
tony.pridmore@nottingham.ac.uk
[3] LITIS, Rouen, France
{herve.locteau,eric.trupin}@univ-rouen.fr

Abstract. In this paper we present a system that allows to build synthetic graphical documents for the performance evaluation of symbol recognition systems. The key contribution of this work is the building of whole documents like drawings or maps. We exploit the layer property of graphical documents by positioning symbol sets in different ways from a same background using positioning constraints. Experiments are presented to build two kinds of test document databases : bags of symbol and architectural drawings.

1 Introduction

Performance evaluation of graphics recognition systems goes back to the middle of 90's [1]. At this period the graphics recognition community focussed its researches on the evaluation of vectorization processes for document re-engineering. In recent years there has been a noticeable shift of attention towards the evaluation of symbol recognition [2], especially through the four International Contests on Symbol Recognition at ICPR 2000[1], and GREC 2003, 2005 and 2007[2]. Performance evaluation is divided into two main topics: ground-truthing and performance characterization. The first one is concerned with the production of test document databases and their corresponding ground-truth [3], while the second deals with the matching of system results to that ground-truth [4]. In this paper we are more interested in ground-truthing, focussed on the symbol recognition. Three main approaches exist in the literature: based on paper, CAD[3] and synthetic documents.

The approach based on paper documents is the most common [3]. Representative documents are obtained from paper archives and digital libraries, and ground-truth is created and edited manually using suitable GUI[4]. This kind of ground-truthing results in realistic and unbiased data but raises different problems: how to define the ground-truth, how to deal with the errors introduced by the users, the delay and the cost of

---

[1] http://www.ee.washington.edu/research/isl/IAPR/ICPR00/

[2] http://epeires.loria.fr/

[3] Computer Aided Design

[4] Graphic User Interface

W. Liu, J. Lladós, and J.-M. Ogier (Eds.): GREC 2007, LNCS 5046, pp. 288–298, 2008.
© Springer-Verlag Berlin Heidelberg 2008

the groundtruth acquisition, etc. In many cases these problems render the approach impractical.

A complementary approach that overcomes these problems is to work directly from CAD documents. As these documents are already in a vector graphics form (SVG, CGM, DWG, etc.), it is possible to take advantage of a groundtruth already existing. The CAD documents are next converted into images for the evaluation. Such approach has been used in the past to evaluate the raster to vector conversion processes [5]. It avoids the groundtruthing step required with the scanned images but it still involves collecting the initial documents. This collecting process takes into account several issues [6]: the copyrights, the format registration (to valid, to convert, etc.), the database organization, finding the duplicates, editing the metadata, etc.

A final approach, which avoids all the difficulties, is to create and use synthetic documents. Here, the test documents are built by an automatic system which combines pre-defined models of document components in a pseudo-random way. Test documents and ground-truth can therefore be produced simultaneously. In addition, a large number of document can be generated easily and with limited user involvement. In the past some systems have been proposed to generate synthetic documents to evaluate the vectorization systems [7]. Concerning the symbol recognition this topic is emerging and only the systems described in [8] [9] [2] exist in the literature. Figure 1 gives some examples of document obtained with these systems.

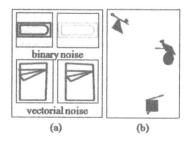

Fig. 1. Examples of synthetic document (a) segmented symbol (b) random symbol set

The systems proposed by [9] and [2] support the generation of degraded images of segmented symbols as shown in the Figure 1 (a). The symbol models are described in a vector graphics format, the vector graphics files are then converted into images. Two kinds of noise are added: binary [9] [2] and vectorial [2]. The system described in [8] employs a complementary approach to build documents composed of multiple unconnected symbols. The Figure 1 (b) gives an example of document generated by this system. Each symbol is composed of a set of primitive (circles, lines, squares, etc.) randomly selected and mildly overlapped. They are next positioned on the image at a random location and without overlapping with the bounding boxes of the other symbols. Finally, noise is added to the generated images using a binary distortion method.

All these systems are interesting, but in order to do a complete evaluation of graphics recognition systems we need whole documents. Indeed, real-life documents (engineering and architectural drawings, electrical diagrams, etc.) are composed of multiple objects constrained by spatial relations (connectivity, adjacency, neighbourhood, etc.). The design of a suitable process to build such documents is a challenging task. Indeed, realistic documents cannot be produced without human know-how into the process. In our work we have considered a shortcut way to solve this problem. Our key idea observes that the graphical documents are composed of two layers : a linear layer (the background) and a symbolic one. We use this property to build several document instances: *ie* symbol sets positioned in different ways using a same background as shown in the Figure 2. In this way, the building process of whole document is made easier and can be considered as a problem of symbol positioning on a document background.

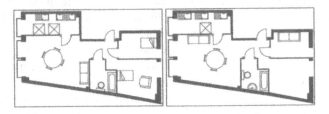

Fig. 2. Two document instances

The main architecture of our system is presented in the Figure 3. It uses as entry data a background image, a database of symbol model and a file containing the positioning constraints. These positioning constraints are edited by a user overlaying the background image and using the models of the symbols to include in the document. Based on these entries two main processes are used to produce the document instances: a symbol factory and a symbol positioning. In what follows we present each of them in the sections 2 and 3. In section 4 we present the building manager supervising these two processes. Section 5 describes some initial experiments and results we are able to produce. Finally, in section 6 we conclude and give our perspectives.

2   Symbol Factory

Following the systems proposed by [9] and [2] we use geometrical primitives (straight lines, arcs and circles) and their associated thickness attributes to describe the symbol models. Each model is stored in an individual file kept inside a database. The user accesses the contents of the database by defining in the file of positioning constraints the models he wants to use. Obviously, in order to produce different document instances, these models are selected at random. The user controls the selection probabilities in the file of positioning constraints. Once selected we load the symbols from their model files, scale them to adapt them to the background size, and compute their bounding boxes. Indeed, the bounding boxes are a common way to handle graphical objects inside a document analysis system. In ours we use them during the positioning process presented in the next section.

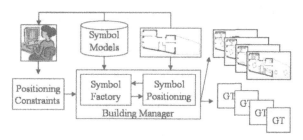

Fig. 3. Our system

3 Symbol Positioning

The goal of our system is to place randomly symbols on a given background. In order to do that, we use positioning constraints that will determine where and how the symbols could be placed. A natural way to define these constraints is to use some of the graphics primitives composing a symbol: these primitives can be exploited next to position the symbol on the background. Some examples could be the two connection points of a resistor, the top line of a bed, the two borders of a frame, etc. This definition makes complex the addition of new models in the system: the user has to define the constraints proper to a model before using. Also in order to position a symbol on the background, primitives corresponding to the constraints must be edited on the background. This process could take lot of time to the user in regard to the number of model and associated constraint. In our work we have considered another approach. We have defined generic constraints fully independent of models. The parameters of the constraints are computed in an automatic way function of models. It is then not necessary to worry about the models to handle during the edition of constraints.

Our constraints are taken at random from the symbols produced by the factory. The links between the constraints and the symbols are defined in the file of positioning constraints edited by the user. Next, the key mechanism of positioning the symbols on the background according to the constraints is detailed in the Figure 4. It raises on the matching between two points: a control point on the symbol and a positioning one defined on the background. The symbol is then positioned in order to fit the control point with the positioning one. To make more flexible our approach we have defined three possibilities to select the positioning points: using a fixed position on the background, or taking a random point in a geometrical shape. In the the first case the constraint defines a fixed value $(x; y)$ where the symbol must be positioned. In the second case, the constraint defines a geometrical shape were points could be selected at random and used for the positioning. We have used two types of shape, either a straight line (the point is selected at random along the line) or either a polygon (the point is selected at random inside the polygon). We will talk next about fixed, sliding and zone constraint to refer these three different positioning processes.

Also, in order to extend the positioning possibilities we also permit on a symbol to define a particular control point and to apply a rotation transformation. The symbol is then positioned in four steps: it is rotated (using a parameter that can be null, a fixed

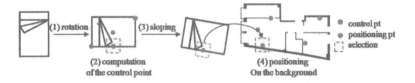

Fig. 4. Constraint mechanism

value or a range), its control point is computed, a slope parameter (between 0 and $2 \times \pi$) is used to incline both the symbol and the control point, and finally it is fixed on the positioning point using the control one. The key step of this process is the computation of the control point. This point is defined for every each constraint using unit polar coordinates (ρ, θ) from the center of the bounding box. These unit polar coordinates are used to compute the values of length and direction (l, α) used to project the center of the bounding box to obtain a control point as explained in the Figure 5 (a). The α value is equal to $\theta \times 2\pi$, l is computed in different ways (1,2,3 and 4) according to the size of the bounding box sides and weighted at last by ρ. The Figure 5 (b) gives some examples of positioning around a point using $\rho = 1$ and $\theta = \{0, \frac{3}{20}, \frac{6}{20}, \frac{9}{20}, \frac{3}{20}, \frac{15}{20}, \frac{18}{20}\}$. The Figures 5 (c) and (d) gives examples using previous rotations of the symbol and with control points defined by $(\rho = 1.0, \theta = 0.25)$ and $(\rho = 1.0, \theta = 0.75)$.

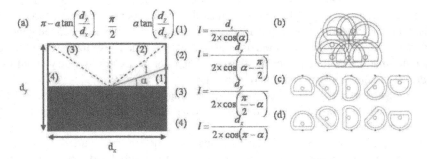

Fig. 5. Control point (a) computation (b) (c) (d) examples of result

4   Building Manager

In the proposed system the factory and the positioning processes are managed by an explicit document building process. It starts with empty documents and fills them with symbols in a pseudo-random way. However, a positioning might fail. These failures appear for example when a symbol is positioned to overlap an existing one, when parts of a symbol overflow a constraint area, etc. The system must be able to identify these failures in order to cancel the positioning. Moreover, users might define constraints that could be hard to satisfy. The system must then detect these cases in order to avoid an infinite building process. To solve these problems our building manager uses five tests, four to check the positioning of symbols and one to stop the building process.

Our first positioning checking concerns the management of the free space of document. Indeed, during the building process several symbols can share the same place. In order to prevent such a case we test the overlapping between the bounding boxes of symbols. This test is computed in three steps as explained in Figure 6: first between a line and a point (a), then between two lines (b) and at last between the two bounding boxes (c). We test then the overlapping between the new symbol we want to position with all the symbols already positioned on the document. Any positive case produces a building failure.

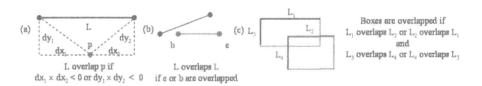

Fig. 6. Overlapping test (a) line-point (b) line-line (c) box-box

Our second positioning checking deals with the sliding constraint as shown in the Figure 7. The positioning process of this constraint could produce overflows of symbols around the line borders (a). In order to limit the positioning to the line areas we have defined an overflow test. This test is based on the covering between two lines (b). It is just a logical adaptation of the overlapping test presented in the Figure 6. A symbol can be considered as overflowing if any of the borders $\{right, up, left, bottom\}$ of its bounding box is not covered by the constraint line L (c). A positive case produces then a building failure.

Our next positioning checking is related to the zone constraint. Indeed, in the same way as in the sliding one, overflows of symbols can appear. The next Figure 8 (a) gives an example of this case. It corresponds to a random fixed point generated too near of the borders of a polygon. In order to detect such a case we exploit the bounding box's corners of the symbol as explained in the Figure 8 (b). We test then if these

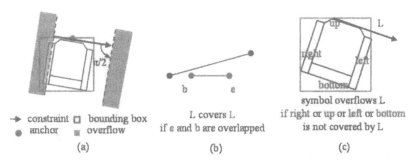

Fig. 7. Sliding checking (a) symbol overflow (b) covering test (c) overflow test

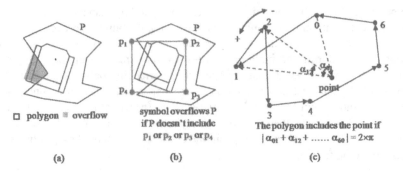

Fig. 8. Zone checking (a) symbol overflow (b) overflow test (c) trigonometric inclusion test

corners are included in the polygon. Any false case will produce a building failure. This inclusion test is based on the method presented in the Figure 8 (c). This method sums the trigonometric angles of successive vectors joining the random point and the polygon ones. A $2 \times \pi$ value corresponds to an inclusion case.

We also check the number of positioned symbol per constraint. Indeed, for every each constraint a maximum number of symbol to position is defined. This number is one for a fixed constraint and can be larger for a sliding and a zone constraint. In this last case, it is defined by the user in the file of positioning constraints. During the building process, the system computes for each constraint the number of symbol already positioned. When this number becomes greater than the maximum a building failure is produced.

In the last test we control the progress of the building process in order to stop it if necessary. Indeed, the system must detect the number of building failure in order to avoid an infinite building process. To do this we use the number of symbol per document as stop criterion. This number corresponds to the sum of the maximum numbers of allowed symbol per constraint. If the number of building failures becomes greater than this number, we stop the process.

5   Experiments and Results

In this section we present some initial experiments and results of our system. The main objective of these experiments is to create databases of test document, with their corresponding ground-truth, for the series of the Symbol Recognition Contests[2]. To do it we have used the symbol model library defined for the previous editions of the Contests[2]. It is composed of 150 models of architectural and electrical symbols. Based on this library we have edited several constraint sets in order to build test document databases of different types. Obviously, the documents produced by our system are in a vector graphics form. For the Contest these documents should therefore be converted into binary images; noise can then be added by the distortion methods used in the past editions of the Contests [2].

We have edited a first set of constraint in order to build "bag of symbol" documents. The Figure 9 presents examples of these bags. In them the symbols are positioned at

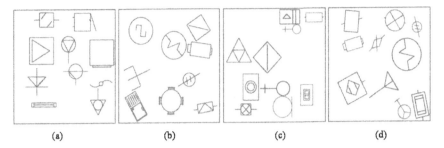

Fig. 9. Examples of bag of symbol (a) none transformation (b) rotated (c) scaled (d) rotated & scaled

random on an empty background, without any connection, and using different rotation or scaling parameters. So these documents look similar to the ones generated by [8] (see Figure 1 (b)). However, they are composed of real-life symbols and not only of geometrical shapes. The key idea of this data set is to create an intermediate level of evaluation between the documents composed of a single segmented symbol (as proposed in the past editions of the Contest[2]) and whole documents (drawings, maps, diagrams, etc.).

To generate these bags we have defined in our setting a single squared zone constraint surrounding an empty background. In order to produce bags of a reasonable size we have resized the original symbol models of the past editions[2] from 512×512 to 256×256 pixels. Based on this initial size we have generated bags of 1024×1024 pixels composed of 10 symbols each. This corresponds to a mean symbol density of 0.625 $(\frac{128^2 \times 10}{512^2})$ which respects a good partitioning between the background and the foreground parts as shown in the Figure 9.

Using these size parameters we have generated 16 databases of 100 bags each. This corresponds to an overall number of 1600 bags composed of around 16000 symbols. These 16 databases have been generated by respecting the protocol used during the previous editions of the Contest[2]. First we have used different model numbers (25,50,100 and 150) in order to test the scalability of the methods. Next we have applied and combined different geometrical operations as illustrated in the Figures 9 (a), (b), (c) and (d). These transformation has been set as follow: from 0 to $2 \times \pi$ for the rotation with a gap of $\frac{2 \times \pi}{1000}$, and from 75 % to 125% for the scaling with a gap of 0.05 % $(\frac{50\%}{1000})$.

Our second set of constraints deals with the building of whole graphical documents using filled backgrounds. For that we have limited our experiments to the building of architectural drawings. The next Figure 10 presents some examples of the drawings we produce. We argue here that the positioning constraints presented in this paper are not domain dependant and could be re-used to build other kinds of document (electrical drawings, geographical maps, etc.). However, the future edition of the Contest[5] will be a kickoff concerning the evaluation of whole documents. It will constitute an important gap for the systems and to limit it to a single domain seems to be fair. We have chosen

---

[5] In 2009 at La Rochelle city (France).

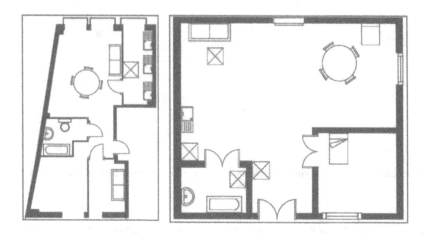

Fig. 10. Examples of built architectural drawing

the architectural drawings in recognition to their interesting properties concerning the connectivity and the orientation of symbols.

To generate these drawings we have retained the size parameter defined for the bags: 256×256 pixels per symbol. Obviously, the use of filled backgrounds makes the images bigger in regard to the one of bags. In order to produce drawings of reasonable dimensions we have fixed a limit of about 4096^2 pixels per image by considering only the backgrounds composed of a small number of rooms (from 4 to 8). We have then selected 10 real-life drawings and created the backgrounds by cleaning their text and symbol parts with an image editor. Using these backgrounds we have defined sets of constraint in order to generate databases of 100 images per background and with 14 to 28 symbols per image. This corresponds to an overall number of 1000 drawings composed of around 18 000 symbols. Obviously, to generate these drawings we have selected only the architectural models of the Contest library[2]. It corresponds to an overall number of 16 models. The Figure 11 gives snapshots of these models with their corresponding labels. For all these models we have also defined resizing parameters, from

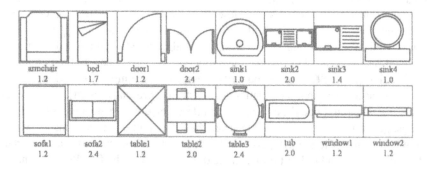

armchair	bed	door1	door2	sink1	sink2	sink3	sink4
1.2	1.7	1.2	2.4	1.0	2.0	1.4	1.0

sofa1	sofa2	table1	table2	table3	tub	window1	window2
1.2	2.4	1.2	2.0	2.4	2.0	1.2	1.2

Fig. 11. Architectural symbols(labels & resizing parameters)

1.0 to 2.4, in order to respect the proportions between the symbols on the drawings. The resizing parameter of 1.0 corresponds then to symbols of 256×256 pixels.

We have then used these models and the resizing parameters in the constraints. The number of constraints per background is about 20. These constraints can be of fixed, sliding or zone type. We have used the fixed constraint to position the door and the window symbols on the drawings. The sliding constraint has allowed us to connect the symbols like the skins, the tubs or the beds along the walls. In each sliding constraint the symbols are positioned in the direction of the line and rotated using a gap of $\frac{\pi}{2}$ in order to respect the wall/symbol alignment. Finally, we have used the zone constraints to define the boundaries of rooms in order to position the other furniture elements like the armchairs, the tables or the sofas. Inside, the symbols have been rotated from 0 to $2 \times \pi$ with a gap of $\frac{2 \times \pi}{1000}$.

6 Conclusion and Perspectives

In this paper we have presented a system for the building of synthetic graphical documents for the performance evaluation of symbol recognition systems. Our main contribution is to extend the past works in this field to the building of whole documents (drawings, maps, diagrams, etc.). To do it we have exploited the layer property of graphical documents in order to position symbol sets in different ways using the same background. Our approach raises on the use of constraint in order to coerce the positioning of symbols. The system that we propose is composed of three components: a symbol factory to select and to load the symbols, a symbol positioning to solve the constraints, and a building manager to supervise the whole process. Experiments show how our system allows to produce large databases of document that look real.

Concerning future perspectives different works are planned. In the short term we plan to develop a GUI to edit the positioning constraints. It will speed up the editing process and help users to build their own databases. Also, based on this GUI we want to use our system to generate other kinds of document like electrical drawings or geographical maps. A more long-term perspective concerns the development of a performance characterization method. Such methods are now required in order to compare the system results with the ground-truth. However, when we work with whole documents the characterization becomes harder because it has to be done between symbol sets. These symbol sets can be of different size, and large gaps can also appear concerning the locations of symbols. Different matching cases can appear and the characterization method should be able to detect and handle them properly.

Acknowledgements

The authors wish to thank Karim Zouba and Murielle Ramangaseheno (LITIS, Rouen University, France) for their contributions to this work. This work was funded by the Spanish Ministry of Education and Science under grant TIN2006-15694-C02-02, and supported by the EPEIRES[2] project of the French Techno-Vision program 2005.

References

1. Kasturi, R., Phillips, I.: The first international graphics recognition contest-dashed-line recognition competition. In: Kasturi, R., Tombre, K. (eds.) Graphics Recognition 1995. LNCS, vol. 1072. Springer, Heidelberg (1996)
2. Valveny, E., et al.: A general framework for the evaluation of symbol recognition methods. International Journal on Document Analysis and Recognition (IJDAR) 1(9), 59–74 (2007)
3. Lopresti, D.P., Nagy, G.: Issues in ground-truthing graphic documents. In: Blostein, D., Kwon, Y.-B. (eds.) GREC 2001. LNCS, vol. 2390, pp. 46–66. Springer, Heidelberg (2002)
4. Wenyin, L., Dori, D.: Principles of constructing a performance evaluation protocol for graphics recognition algorithms. In: Performance Characterization and Evaluation of Computer Vision Algorithms, pp. 97–106. Springer, Heidelberg (1999)
5. Chhabra, A., Phillips, I.: The second international graphics recognition contest - raster to vector conversion: A report. In: Chhabra, A.K., Tombre, K. (eds.) GREC 1997. LNCS, vol. 1389, pp. 390–410. Springer, Heidelberg (1998)
6. Phillips, I., Ha, J., Haralick, R., Dori, D.: The implementation methodology for the cd-rom english document database. In: International Conference on Document Analysis and Recognition (ICDAR), pp. 484–487 (1993)
7. Wenyin, L., Zhai, J., Dori, D.: Extended summary of the arc segmentation contest. In: Blostein, D., Kwon, Y.-B. (eds.) GREC 2001. LNCS, vol. 2390. Springer, Heidelberg (2002)
8. Aksoy, S., et al.: Algorithm performance contest. In: International Conference on Pattern Recognition (ICPR), vol. 4, pp. 870–876 (2000)
9. Zhai, J., Wenyin, L., Dori, D., Li, Q.: A line drawings degradation model for performance characterization. In: International Conference on Document Analysis And Recognition (ICDAR), pp. 1020–1024 (2003)

A Study on the Effects of Noise Level, Cleaning Method, and Vectorization Software on the Quality of Vector Data

Hasan S.M. Al-Khaffaf, Abdullah Zawawi Talib, and Rosalina Abdul Salam

School of Computer Sciences, Universiti Sains Malaysia, 11800 USM Penang, Malaysia
{hasan,azht,rosalina}@cs.usm.my

Abstract. Correct detection of line attributes by line detection algorithms is important and leads to good quality vectors. Line attributes includes: end points, width, line style, line shape, and center (for arcs). In this paper we study different factors that affect detected vector attributes. Noise level, cleaning method, and vectorization software are three factors that may influence the resulting vector data attributes. Real scanned images from GREC'03 and GREC'07 contests are used in the experiment. Three different levels of salt-and-pepper noise (5%, 10%, and 15%) are used. Noisy images are cleaned by six cleaning algorithms and then three different commercial raster to vector software are used to vectorize the cleaned images. Vector Recovery Index (VRI) is the performance evaluation criteria used in this study to judge the quality of the resulting vectors compared to their ground truth data. Statistical analysis on the VRI values shows that vectorization software has the biggest influence on the quality of the resulting vectors.

Keywords: salt-and-pepper, raster-to-vector, performance evaluation, engineering drawings.

1 Introduction

Raster to vector conversion is a hot topic in the field of graphics recognition [1]. Good line detection method could be judged by its ability to recognize line features correctly and thoroughly. Line features include: end points, width, line style, line shape, and center (for arcs). Since line detection usually follows other image analysis stages, its action upon the image would be affected by prior stages that change image content. Among the many factors affecting the quality of detected vector are: all kind of noise, cleaning method used, and vectorization algorithm used. The previous two contests on graphics recognition [2, 3] accompanying GREC'03 and GREC'05 give some insight to the effect of noise on the resulting vector data, but they did not include extensive test on different noise levels or study the effect of different cleaning methods on the quality of the vectors. It also did not reveal the major factor that affects vector quality. Their findings could answer only limited questions regarding the interaction between different factors and treatments.

In the noise factor three treatments (levels) are studied which is 5%, 10%, and 15% noise levels. Uniform salt-and-pepper noise is used in all three treatments. A study on

W. Liu, J. Lladós, and J.-M. Ogier (Eds.): GREC 2007, LNCS 5046, pp. 299–309, 2008.

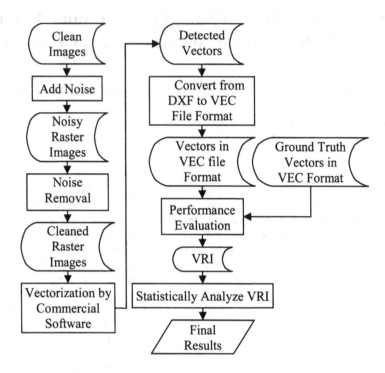

Fig. 1. Flowchart showing the steps of our experiment

the effect of different noise levels on the quality of vector data is carried out. We also studied vectorization performance within different noise levels. Six algorithms are studied for the cleaning factor. The performance of these algorithms within vectorization software is also described. Finally, three commercial vectorization software are used to study vectorization factor. This factor proved to be major player on the quality of vector data. The Vector Recovery Index (VRI) is the performance evaluation criteria used to judge the quality of vector data. Statistical analysis is used to further analyze the data. Analysis of Variance (ANOVA) as well as Estimated Marginal Means (EMM) of the VRI are used. Fig. 1 shows the steps of our experiment.

The rest of the paper is organized as follows: The image data for the experiment is discussed in Section 2. This includes the method used to add the noise. Cleaning methods and vectorization software used in the study are explained in Sections 3 and 4, respectively. Performance evaluation method is described in Section 5. Statistical analysis is explained in Section 6. Finaly, conclusion and future work are presented in Section 7.

2   Image Data

The images from GREC'03 and GREC'07 contests are used since ground truth files are readily available for the performance evaluation task [4]. Another reason is that the graphical elements in GREC'03 images are relatively thin. Noise will affect these thin elements more than other thick elements which make it more challenging for the cleaning method to retain it and the vectorization software to recognize it correctly.

A random noise (Salt-and-Pepper) is added to each image. The algorithm is as follows:

PR = 1 − NL / 100

For each pixel in the image do the following

> Create a uniform random number (R) in the range of -1 to +1
> If R > PR then add Salt noise to the current pixel
> Else if R < -PR then add Pepper noise to the current pixel

NL is the percentage of the noise level to be added to the image and it is between 0 and 100. Mersen Twister random number generator is used to obtain a sequence of uniform random numbers with good randomness and long repetition cycle. Uniform distribution is selected to give all pixels the same chance to be distorted by noise.

Using the above algorithm we create three distorted images with 5%, 10%, and 15% noise levels for each original image.

3 Cleaning Methods

Each distorted image is then cleaned by three Salt-and-Pepper cleaning methods namely: kFill [5, 6], Enhanced kFill [7], Activity Detector [8]; and their enhanced counterparts named as Algorithm A (Alg A), Algorithm B (Alg B), and Algorithm C (Alg C), respectively [9] totaling to six cleaning methods. kFill is a multi-pass two iteration filter capable of removing salt-and-pepper noise. Enhanced kFill (Enh. kFill) cleans the image in a single pass. Activity Detector (Act. Detec.) studies the activity around each connected component (CC) and classifies CC's into three categories. The cleaning is performed by removing selected CC's based on specified criteria. A procedure named TAMD is developed to enhance noise cleaning by protecting weak features such as one-pixel-wide graphical element (GE) while removing small spurious limbs attached to the GE's. Alg A and Alg B are created by integrating TAMD into kFill and Enhanced kFill logic. TAMD is performed as a post processing step in Alg C. The parameters for the methods are set as in our previous study [9].

4 Vectorization

Three commercial software (Vectory [10], VPstudio [11], and Scan2CAD [12]) are used to vectorize cleaned images and detected vectors are saved as DXF files. These files are then converted to VEC files which have a simple format and are easier to deal with using the performance evaluation tool. Software selections are based on available features. Having the feature of detecting arcs and circles is the most important. So is the ability to output in DXF format. It would also be advantageous to use software that have been used by other researchers for performance evaluation since they may provide us with some information and clue about its performance. The above three software were used in [13, 14].

Note that vectorization software include many features that could be utilized to enhance the detection of graphical elements thus enhancing the vector quality. Our interest is in the automatic conversion process, thus most of these features are not used.

5 Performance Evaluation

Vector Recovery Index [15] of the detected vectors is the criteria used to judge the quality of the resulting vectors. Performance evaluation tool (ArcEval2005.exe) compares the detected vector file with the ground truth file and output the VRI score. The version of the tool used carries out performance evaluation based on arcs only. All straight lines in the detected vectors file are skipped. For real scanned images the ground truth data may be prepared manually.

VRI is an objective performance evaluation of line detection algorithms (vectorization software in our case) that works at vector level. The VRI index is a combination of two matrices which are vector detection rate (D_v) and vector false alarm rate (F_v). The VRI is calculated as follows:

$$VRI = \beta D_v + (1 - \beta)(1 - F_v) \cdot \tag{1}$$

where β is taken as 0.5 in this work to give similar weight to vector detection rate and vector false alarm rate.

Vector detection rate is defined by two terms which is line basic quality and fragmentation quality. Line basic quality represents the accuracy of the detection of line attributes which include end points, width, line style, line shape, and center (for arcs) compared with the attributes of ground truth data. Fragmentation quality measures the fragmentation of the detected line compared to the ground truth line. The False alarm rate measures the degree of a detected line being a false alarm. VRI value is in the range of 0 to 1, the higher the better in detection.

6 Statistical Analysis

SPSS software is used to analyze the resulting VRI values. We have three factors: noise level, cleaning method, and vectorization. Hence three independent variables (IV) are created in SPSS: noise [three levels: 5%, 10%, 15%], clean [six levels: kFill, Enh. kFill, Act. Detec., Alg A, Alg B, Alg C], and vectorization [three levels: VPstudio, Vectory, Scan2CAD]. One dependent variable (DV) is created (VRI).

Since we have three different factors to study, Three-Way ANOVA is used in our analysis. The analysis are used to show the main effects and interaction (combination) effects of the IV's on the DV. The interaction effects show combination effects of two or more IV's on the DV. The description of the Three-Way ANOVA is complicated because of the three factors involved. So, an explanation of One-Way ANOVA which has only one IV (vectorization) and one DV (VRI) is illustrated below.

We start our analysis by formulating a hypothesis on our data. Our hypothesis (called null hypothesis) assume that the means of the VRI for the different levels of vectorization are equal as shown below:

$$H_0 : \mu_{VPstudio} = \mu_{Vectory} = \mu_{Scan2CAD} \tag{2}$$

Our alternative hypothesis is mutually exclusive compared to the null hypothesis and it should be exhaustive. The alternative hypothesis is shown below:

$$H_1 : \text{Not all the means are equal} \tag{3}$$

It is the significance of the F-test that shows if the group means differ. The F-test insures that any difference in group means does not happen by chance. If the change in the group means is not significant then we will assume that the IV (vectorization in this example) has no effect on DV (VRI in our case). If the significance of F-test is equal or less than 0.05, then the change in mean is considered as significant and we will reject the null hypothesis formulated above and accept the alternative hypothesis. The value 0.05 is called α and it represents the probability of rejecting the null hypothesis when it is true.

6.1 Setting-Up the Experiment

Some parameters for the three vectorization software need to be preset prior to applying vectorization. That is to ensure consistency between different software such as: same measuring units are used and Mechanical Engineering Drawing is used as drawing type. Other parameters and thresholds are left unchanged.
    For each vectorization software used, we:

0. Preset software parameters.
1. Load and convert the cleaned image into vector form and save the result as a DXF file.
2. Convert DXF file into VEC file.
3. Use the performance evaluation tool to get the VRI of the detected vectors.

These are the typical steps for the experiment, but in VPstudio one parameter needs to be preset after loading the image.

6.2 Experimental Results and Discussions

Eleven raster images are distorted with the three different levels of noise and then cleaned by the six cleaning methods. The cleaned images are then vectorized by the three commercial raster to vector software. One VRI value is computed from each detected vector and the ground truth vector files. A total of 594 separate VRI values are to be generated, but some values could not be generated and thus reducing the number of VRI values to 588. The VRI values are then analyzed by SPSS. The values that could not be generated are related to Act. Detec. and Alg C when the noise level

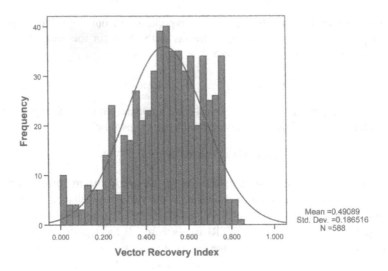

Fig. 2. Frequency table for VRI

is set to 15%. This is due to the number of CC's generated become larger than the space allocated to it in the implementation. Fig. 2 shows the frequency histogram for the VRI.

The minimum value of VRI is 0 which indicates no vector is detected. The mean value of VRI is 0.491 which is below the satisfactory value of 0.8 as suggested by [2]. The low value of VRI is due to the setting of the detected vectors width to 1 as we are not able to obtain the actual width of the detected vectors. The low value of VRI is also due to the weak features of some of the original images as well as the amount of noise added to the image. Another reason for the low value of VRI is that vectorization parameters for the three software are not modified to give better quality since we are focusing on the automatic conversion capabilities of the vectorization software. The mean value (.491) is close to the median (.508), suggesting normal distribution of the data. Small negative value of skewness (-.516) indicates that the distribution has tiny tail to the left. Negative value of the kurtosis (-.273) suggests that small proportions of the data are located in the tails of the distribution.

First we need to know factors that have major impact on the quality of vector data. Three-Way ANOVA is used to analyze the effects of different independent variables (noise, clean, and vectorization) on the dependent variable (VRI). Table 1 shows the significance of each separate factor and the combinational effect of different factors on VRI.

As shown in Table 1, the significant value (Sig.) of vectorization factor (.000) and the interaction between the two factors vectorization*noise (.012) is less than the threshold value 0.05 leading to the conclusion that vectorization and the combination of vectorization and noise do affect VRI values.

Other factors (clean and noise) and combination of factors (vectorization * clean, clean * noise, and vectorization * clean * noise) have significant values of more than 0.05 which lead to the conclusion that it does not effect VRI.

Table 1. Tests of Between-Subjects Effects

Effect	Source	F	Sig.
	vectorization	33.413	.000
Main effect	clean	1.433	.211
	noise	1.981	.139
Two-way interaction	vectorization * clean	1.341	.205
	vectorization * noise	3.227	.012
	clean * noise	.215	.995
Three-way interaction	vectorization * clean * noise	.296	.999

6.2.1 Vectorization

As shown in Fig. 3, VPstudio produces better quality of vector data compared to the other software. It also performs better with increased amount of noise when the noise level is moderate and the performance drops with increase amount of noise when the noise level is high. In fact, we have also carried out further investigation regarding performance of VPstudio by running an experiment with 20% noise for images of GREC'03 only. The result as shown in Fig. 4 confirms further that performance will drop as for other software when the noise increases. The other two software show a drop in performance with an increase amount of noise regardless of noise levels. Contrast test also shows that VPstudio has significant difference over the other two vectorization software.

VPstudio which has the best performance in VRI has the least sensitivity with any cleaning method as shown in Fig. 5. However, its best performance is when it works with Enh. kFill (Estimated Marginal Means of VRI = 0.589) and lowest result when it works with Alg B (0.546).

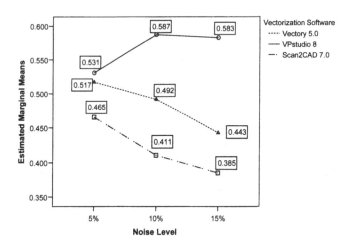

Fig. 3. Software efficiency with different noise levels

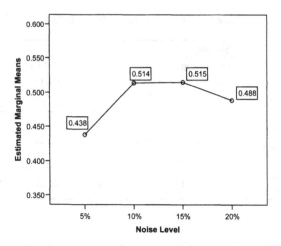

Fig. 4. VPstudio efficiency with different noise levels

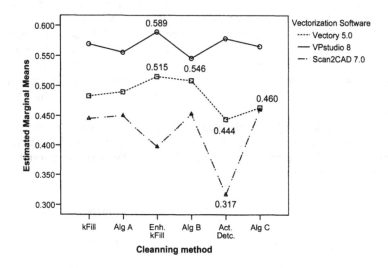

Fig. 5. Efficiency of vectorization software with many cleaning methods

Fig. 5 also shows that Vectory exhibit moderate sensitivity to cleaning methods and has better quality with images cleaned by Enh. kFill (0.515) and lowest VRI with images cleaned by Act. Detec. (0.444).

Based on Fig. 5, Scan2CAD shows highest sensitivity to cleaning methods. It has best performance when working with Alg C (0.460) and global lowest value of 0.317 with Act. Detec.

6.2.2 Noise Levels

The EMM of VRI show a slight drop in performance with increase amount of noise as shown in Table 2. The three levels of noise used in this study show little impact on

Table 2. Estimated Marginal Means of VRI with different noise levels

Noise Level	EMM of VRI
5%	0.505
10%	0.496
15%	0.471

the result of VRI. More levels of noise are required in order to show the real impact of noise levels on VRI.

6.2.3 Cleaning Methods

All cleaning methods (except Act. Detec.) show close performance (see Table 3). Act. Detec. has the lowest performance compared to others because it could not remove noise that touches GE and may lead to difficulties during the recognition process. Alg C which is an enhanced version of Act. Detec. performs better than Act. Detec. since it did not suffer the aforementioned drawback, but its performance is close to the other four algorithms.

Table 3. Estimated Marginal Means of VRI for cleaning methods

Cleaning method	EMM of VRI
kFill	.499
Alg A	.498
Enh. kFill	.500
Alg B	.502
Act. Detec.	.448
Alg C	.496

We have observed that even if some noise still exist in most image area (especially in 15% noise level) such as in Enh. kFill and Alg B due to its single pass nature these methods perform close to multi pass filters, with respect to EMM of VRI.

For cleaning algorithms, EMM of VRI shows that Alg A and Alg B have similar performance compared to their original counterparts.

7 Conclusions and Future Work

Many factors that may affect the quality of the vector data are studied in this paper including noise, cleaning methods and vectorization software. An experiment on a scanned drawings shows that vectorization software has the biggest impact on the quality of the vector data. Investigation on the interactions between vectorization and cleaning methods is also carried out.

We believe that the experiment in this paper should be extended into different directions in order to make it more general. Ongoing investigations include using Gaussian noise (more common in document images), and for the cleaning methods some state of the art filters are being used such as median filter and its variants.

Morphological operators should also be investigated. The set of test images is to be expanded to include more images. There are many other raster to vector software available hence the need to study their performance.

We also suggest adding more factors to the experiment. For example, if the images are classified into (simple, moderate and complex) using some criteria then we could add image complexity as a factor. The analysis may reveal new information about the interaction of image complexity with other factors. Other factors could further be classified into more specific types such as using Gaussian vs. uniform noise, and single-pass vs. multi-pass filters.

Acknowledgment

We would like to thank Low Heng Chin and Ataharul Islam of the School of Mathematical Sciences – USM for their helps on statistical analysis. Thanks also go to Faisal Shafait of University of Kaiserslautern who provides the ground truth data for GREC'07 contest's test images.

This work is fully supported by a Science Fund grant from the Ministry of Science, Technology, and Innovation (MOSTI), Malaysia under project number 01-01-05-SF0147.

References

1. Tombre, K.: Graphics recognition: The last ten years and the next ten years. In: Liu, W., Lladós, J. (eds.) GREC 2005. LNCS, vol. 3926, pp. 422–426. Springer, Heidelberg (2006)
2. Liu, W.: Report of the Arc Segmentation Contest. In: Graphics Recognition: Lecture Notes in Computer Science: Recent Advances and Perspectives, pp. 363–366. Springer, Heidelberg (2004)
3. Wenyin, L.: The third report of the arc segmentation contest. In: Liu, W., Lladós, J. (eds.) GREC 2005. LNCS, vol. 3926, pp. 358–361. Springer, Heidelberg (2006)
4. Shafait, F., Keysers, D., Breuel, T.M.: GREC 2007 Arc Segmentation Contest: Evaluation of Four Participating Algorithms, vol. 5046. Springer, Heidelberg (2007)
5. O'Gorman, L.: Image and document processing techniques for the RightPages electronic library system. In: Proc. 11th IAPR International Conference on Pattern Recognition. Conference B: Pattern Recognition Methodology and Systems, The Hague, pp. 260–263 (1992)
6. Story, G.A., O'Gorman, L., Fox, D., Schaper, L.L., Jagadish, H.V.: The RightPages image-based electronic library for alerting and browsing. Computer 25(9), 17–26 (1992)
7. Chinnasarn, K., Rangsanseri, Y., Thitimajshima, P.: Removing salt-and-pepper noise in text/graphics images. In: The 1998 IEEE Asia-Pacific Conference on Circuits and Systems, Chiangmai, pp. 459–462 (1998)
8. Simard, P.Y., Malvar, H.S.: An efficient binary image activity detector based on connected components. In: Proc. IEEE International Conference on Acoustics, Speech, and Signal Processing, pp. 229–233 (2004)
9. Al-Khaffaf, H.S.M., Talib, A.Z., Abdul Salam, R.: Internal Report, Artificial Intelligence Research Group, School of Computer Sciences, Universiti Sains Malaysia (2006)
10. Vectory 5.0. Raster to Vector Conversion Software. Graphikon GmbH, Berlin, Germany, http://www.graphikon.de

11. VPstudio ver 8.02 C6. Raster to Vector Conversion Software, Softelec, Munich, Germany, http://www.softelec.com, http://www.hybridcad.com
12. Scan2CAD 7.5d. Raster to Vector Conversion Software, Softcover International Limited, Cambridge, England, http://www.softcover.com
13. Phillips, I.T., Chhabra, A.K.: Empirical performance evaluation of graphics recognition systems. IEEE Transactions on Pattern Analysis and Machine Intelligence 21(9), 849–870 (1999)
14. Chhabra, A.K., Phillips, I.T.: Performance evaluation of line drawing recognition systems. In: Proc. 15th International Conference on Pattern Recognition, Barcelona, pp. 864–869 (2000)
15. Liu, W.Y., Dori, D.: A protocol for performance evaluation of line detection algorithms. Machine Vision and Applications 9(5-6), 240–250 (1997)

GREC 2007 Arc Segmentation Contest: Evaluation of Four Participating Algorithms

Faisal Shafait[1], Daniel Keysers[1], and Thomas M. Breuel[2]

[1] Image Understanding and Pattern Recognition (IUPR) research group
German Research Center for Artificial Intelligence (DFKI) GmbH
D-67663 Kaiserslautern, Germany
faisal@iupr.dfki.de, keysers@iupr.dfki.de
[2] Department of Computer Science, Technical University of Kaiserslautern
D-67663 Kaiserslautern, Germany
tmb@informatik.uni-kl.de

Abstract. Automatic conversion of line drawings from paper to electronic form requires the recognition of geometric primitives like lines, arcs, circles etc. in scanned documents. Many algorithms have been proposed over the years to extract lines and arcs from document images. To compare different state-of-the-art systems, an arc segmentation contest was held in the seventh IAPR International Workshop on Graphics Recognition - GREC 2007. Four methods participated in the contest, three of which were commercial systems and one was a research algorithm. This paper presents the results of the contest by giving an overview of the dataset used in the contest, evaluation methodology, participating methods and the segmentation accuracy achieved by the participating methods.

Keywords: Graphics Recognition, Line Drawings, Technical Drawings, Arc Segmentation Contest.

1 Introduction

Reliable detection of geometric primitives like lines, arcs, circles etc. in document images is one of the key problems in graphics recognition. Due to the importance of this task, the International Association for Pattern Recognition's Technical Committee on Graphics Recognition (IAPR TC10) has been organizing biennial arc segmentation contests since 2001 [1,2,3]. The purpose of these contests was to provide a platform for comparative evaluation of state-of-the-art research and commercial graphics recognition algorithms. The benchmarking of algorithms in this way helps in objectively evaluating the performance of participating systems and highlights the strengths and weaknesses of these systems. Therefore the contest-based approach for comparing algorithms is also used in other domains of document analysis research, like page segmentation [4], handwriting recognition [5], and document image dewarping [6].

This contest is fourth in the series of arc segmentation contests and was held at the seventh IAPR International Workshop on Graphics Recognition (GREC

W. Liu, J. Lladós, and J.-M. Ogier (Eds.): GREC 2007, LNCS 5046, pp. 310–320, 2008.
© Springer-Verlag Berlin Heidelberg 2008

2007), in Curitiba, Brazil, September 20-21, 2007. A dataset of five training and five test images was used in the contest. This contest was different from the previous three contests from the view point of ground-truth representation and performance evaluation protocol. Previous arc segmentation contests used the VEC format for representing arcs, and used the VRI score [7] as a measure of performance of arc segmentation. In this contest, we have used a color-based representation and evaluation scheme [8] discussed in Section 2 and Section 3. Four methods participated in the contest. A brief description of the methods is given in Section 4. The dataset used in the contest and the results of the contest are given in Section 5 followed by a conclusion in Section 6.

2   Representation of Geometric Primitives

The traditional way of representing geometric primitives like lines, arcs, or circles in a drawing is to use their parametric representation. The VEC format uses this representation in plain text form, using one line of parameters for each arc. Other drawing tools can then read this format and reproduce an image containing exactly the arcs given in the VEC-format text file. However, if some of the arcs are incorrect, it is hard to find the source of error, since the correspondence of the arcs in VEC-format to pixels in the original image can not be easily established. In addition, the performance of the algorithm can not be judged by looking only at the VEC-format text file, and specialized software is needed to view the detected arcs and analyze the segmentation errors.

To overcome these problems, we propose a new representation of arc segmentation. This representation is based on pixel-accurate color-coding of page segmentation as proposed in [8]. Arc segments in an image are represented within the image such that each pixel belonging to an arc is assigned as its value the index of the arc. A particular color can be assigned to the page background (e.g. 0xffffff) and to all pixels not belonging to any arc (e.g. 0x000000). This representation of arc segmentation is particularly convenient because it can be used

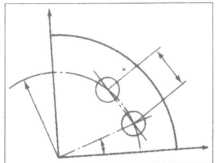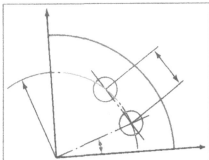

Fig. 1. An example image to demonstrate color encoding of arc segments. Each arc found in the image is labeled with a unique color.

to accurately represent different arcs in the same image as shown in Figure 1. Secondly, it can be saved and exchanged using any lossless color image format, thereby avoiding the need for specialized software for viewing the arcs. The assignment of colors to arcs is arbitrary, so any colors can be chosen for representing different arcs. Pixels belonging to more than one arc can be assigned a unique color if needed.

3 Vectorial Score for Performance Evaluation

To evaluate the performance of an arc segmentation algorithm, we use the vectorial score proposed in [8]. According to this vectorial score, different errors that are measured are:

Oversegmented arcs: The number of arcs that are either split into more than one arc, or are partially detected.

Undersegmented arcs: The number of arcs merged with some other arc.

Total Oversegmentations: The total number of segmentations that ground-truth arcs were split into.

Total Undersegmentations: The total number of segmentations that would be needed to split all merged arcs.

Missed arcs: The number of arcs that were not found by the algorithm.

False alarms: The number of detecting arcs originating from noise or nongraphics elements.

4 Participating Methods

Results of four methods were presented for participation in the contest:

1. Liu Wenyin's method [9]
2. Vectory software ver. 5.0 (http://www.graphikon.de)
3. Scan2CAD software ver. 7.5d (http://www.softcover.com)
4. VPstudio software ver. 8.02 (http://www.softelec.com)

Liu Wenyin from City University of Hong Kong provided the results of his method, whereas Hasan Al-Khaffaf from Universiti Sains Malaysia presented the results of the other three commercial systems.

5 Results

The results of all participating methods on each test image are shown in Figures 2 to 6. The test images were obtained by scanning selected engineering drawings from different books. The ground-truth was then generated manually by coloring all the pixels belonging to an arc with a unique color using an off-the-shelf image manipulation program.

Table 1 shows the vectorial score obtained by all participating methods on the test images. The results were obtained by using a relative threshold of 0.1

Table 1. Different types of errors made by each algorithm on the test images. The column labels are: total oversegmentations (T_o), total undersegmentations (T_u), oversegmented components (C_o), undersegmented components (C_u), missed components (C_m), false alarms (C_f).

Algorithm	T_o	T_u	C_o	C_u	C_m	C_f
Wenyin's method	21	8	13	6	1	93
Scan2CAD	72	9	48	7	9	64
Vectory	54	9	43	9	14	0
VPstudio	55	4	49	3	8	64

Table 2. Different types of errors made by each algorithm on the test images when arc width was set to a constant value of 10 pixels in the output of all algorithms

Algorithm	T_o	T_u	C_o	C_u	C_m	C_f
Wenyin's method	17	9	9	6	1	94
Scan2CAD	36	10	20	6	9	66
Vectory	35	13	26	9	13	1
VPstudio	7	5	7	4	8	62

and an absolute threshold of 100 pixels. This implies that a segmentation error was considered significant only if the number of in-correctly segmented pixels was either larger than 10% of the pixels belonging to an arc or was larger than 100 pixels in total. The results show that all the algorithms over-segmented the arcs. This happened when all pixels belonging to an arc were not assigned to the arc by the algorithm. One major reason for this was that the VEC-files for the commercial systems were supplied with a constant line width of one pixel for all the arcs. An example showing the segmentation results of the VPstudio software on a test image is shown in Figure 7(a). Evaluation result for this image reported that all 13 arcs were over-segmented. To see the influence of this problem, we re-ran the evaluation using a constant line width of 10 pixels for all systems. Since we use only the foreground pixels while ignoring the background pixels, setting the arc width to 10 pixels has the effect of actually ignoring the arc width. The effect of setting the arc width of all arcs to 10 pixels for the example image of Figure 7(a) is shown in Figure 7(b). It can be seen that all the pixels belonging to an arc are now correctly assigned to that arc. Table 2 shows the evaluation results by ignoring the arc width. This table shows that most of the over-segmentations were due to the small line thickness supplied by the systems.

The results show that Wenyin's method and VPstudio software worked very well in segmenting arcs from the images and did uniformly better than the other two systems on most of the performance measures. The number of false alarms were high for these systems because they did not remove text parts in the images prior to arc recognition. From that aspect Vectory software performed the best by removing all textual components from the image, thereby resulting in no

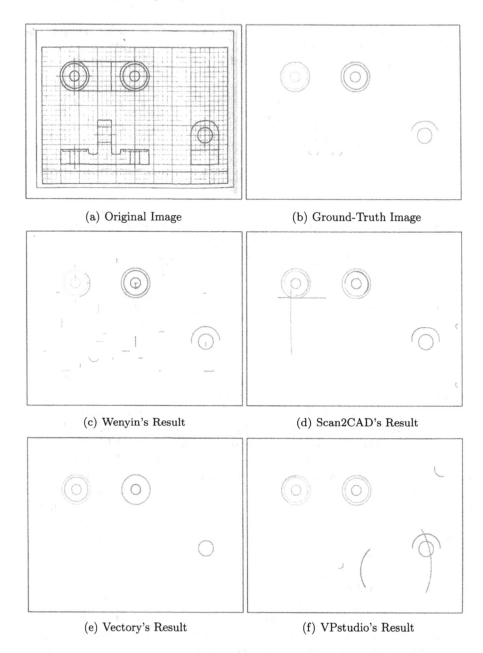

(a) Original Image

(b) Ground-Truth Image

(c) Wenyin's Result

(d) Scan2CAD's Result

(e) Vectory's Result

(f) VPstudio's Result

Fig. 2. Results of the four participating methods on the first test image. For clarity in the ground-truth image, only those foreground pixels that belong to an arc are shown. Due to the presence of the background grid, none of the participating methods could correctly segment all arcs from the image.

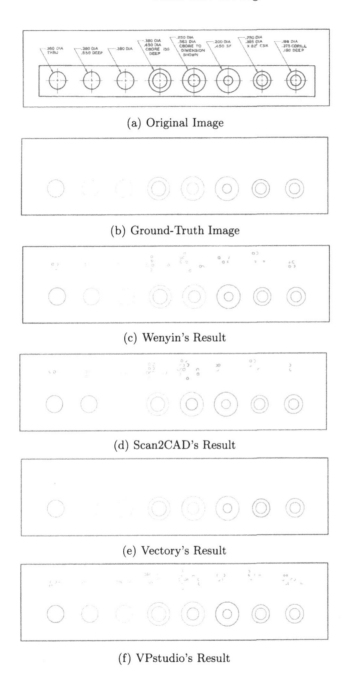

(a) Original Image

(b) Ground-Truth Image

(c) Wenyin's Result

(d) Scan2CAD's Result

(e) Vectory's Result

(f) VPstudio's Result

Fig. 3. Results of the four participating methods on the second test image. All methods correctly found the circles, but also produced many false alarms originating from the text in the annotations, except the Vectory software which seems to have removed the text parts prior to arc segmentation.

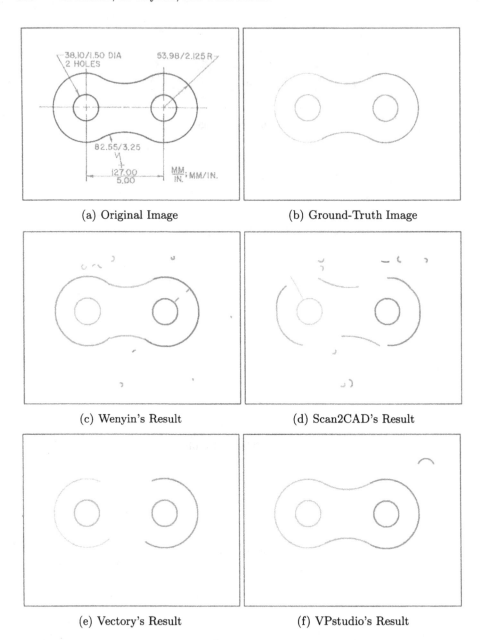

(a) Original Image

(b) Ground-Truth Image

(c) Wenyin's Result

(d) Scan2CAD's Result

(e) Vectory's Result

(f) VPstudio's Result

Fig. 4. Results of the four participating methods on the third test image. VPstudio software had the best results in this case, since it was the only method that correctly segmented the two concave curves.

false alarms. Interestingly, for the test image shown in Figure 3, evaluation of the output of Vectory software reported all 13 arcs as over-segmented. A closer look revealed that the Vectory software also did a skew correction of the image,

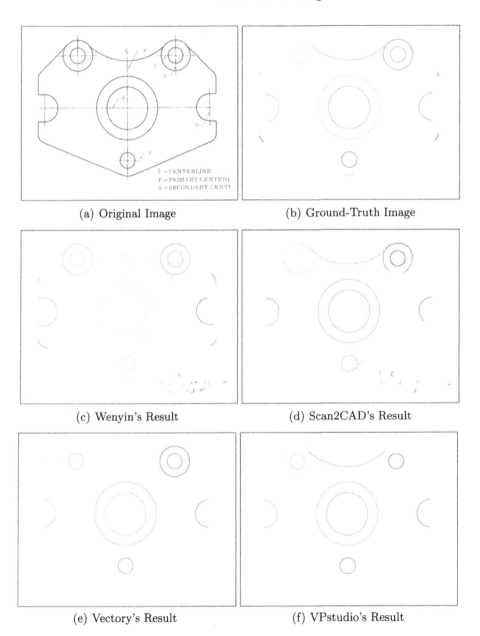

(a) Original Image

(b) Ground-Truth Image

(c) Wenyin's Result

(d) Scan2CAD's Result

(e) Vectory's Result

(f) VPstudio's Result

Fig. 5. Results of the four participating methods on the fourth test image. Liu Wenyin's method had the best results for this image since it was the only method that correctly found the curved corners in the image.

thereby slightly moving all circles from their original position. This resulted in all circles reported as over-segmented. Liu Wenyin's method had the least number of missed errors. Inspection of the results revealed that most of the missed error

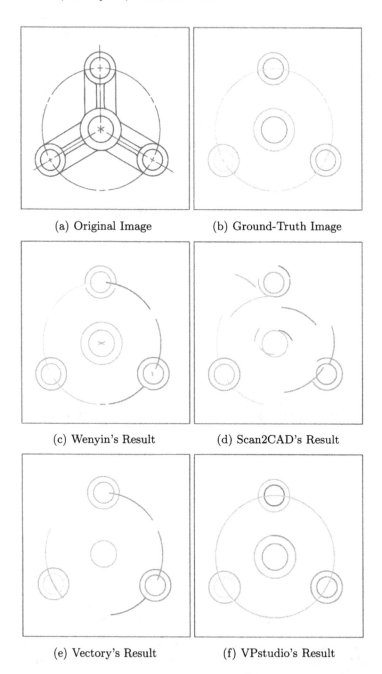

(a) Original Image   (b) Ground-Truth Image

(c) Wenyin's Result   (d) Scan2CAD's Result

(e) Vectory's Result   (f) VPstudio's Result

Fig. 6. Results of the four participating methods on the fifth test image. Both VPstudio and Liu Wenyin's method had comparable results in this case that were better than those of Vectory and Scan2CAD.

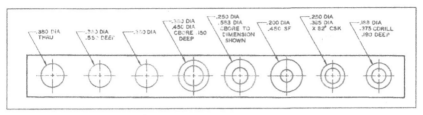

(a) Original segmented image

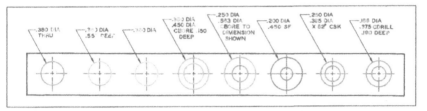

(b) Segmented image after setting arc width to 10 pixels

Fig. 7. (a) The segmentation result of VPstudio on a test image. Despite the algorithm working very well in segmenting the arcs, the evaluation result reported that all 13 arcs were oversegmented in this image, since the results were supplied with a constant line width of one pixel. (b) The segmentation result after setting arc width to 10 pixels. The evaluation result for this image reported no segmentation errors.

in commercial systems originated from ground-truth arcs consisting of round corners as in Figure 5.

6   Conclusion

This paper presented a summary of the GREC 2007 arc segmentation contest. We described the pixel-accurate color-based representation of arc segmentation that was used in the competition along with a vectorial score for measuring arc segmentation accuracy. The vectorial score enables us to evaluate different aspects of an arc segmentation algorithm. One research algorithm by Liu Wenyin and three commercial systems namely Scan2CAD, Vectory, and VPstudio were presented for participation in the contest. Results showed that Wenyin's method and VPstudio out-performed the other two systems, whereas the performance of Wenyin's method and that of VPstudio software was not significantly different from each other.

Acknowledgments

This work was partially funded by the BMBF (German Federal Ministry of Education and Research), project IPeT (01 IW D03).

References

1. Wenyin, L.: The third report of the arc segmentation contest. In: Liu, W., Lladós, J. (eds.) GREC 2005. LNCS, vol. 3926, pp. 358–361. Springer, Heidelberg (2006)
2. Wenyin, L.: Report of the arc segmentation contest. In: Lladós, J., Kwon, Y.-B. (eds.) GREC 2003. LNCS, vol. 3088, pp. 364–367. Springer, Heidelberg (2004)
3. Wenyin, L., Zhai, J., Dori, D.: Extended summary of the arc segmentation contest. In: Blostein, D., Kwon, Y.-B. (eds.) GREC 2001. LNCS, vol. 2390, pp. 343–349. Springer, Heidelberg (2002)
4. Antonacopoulos, A., Gatos, B., Bridson, D.: ICDAR 2007 page segmentation competition. In: Proc. 9th Intl. Conf. on Document Analysis and Recognition, Curitiba, Brazil, pp. 1279–1283 (2007)
5. Maergner, V., Abed, H.E.: ICDAR 2007 - arabic handwriting recognition competition. In: Proc. 9th Intl. Conf. on Document Analysis and Recognition, Curitiba, Brazil, pp. 1274–1278 (2007)
6. Shafait, F., Breuel, T.M.: Document image dewarping contest. In: 2nd Int. Workshop on Camera-Based Document Analysis and Recognition, Curitiba, Brazil, pp. 181–188 (2007)
7. Wenyin, L., Dori, D.: A protocol for performance evaluation of line detection algorithms. Machine Vision and Applications: Special Issue on Performance Characteristics of Vision Algorithms 9, 240–250 (1997)
8. Shafait, F., Keysers, D., Breuel, T.M.: Pixel-accurate representation and evaluation of page segmentation in document images. In: 18th Int. Conf. on Pattern Recognition, Hong Kong, China, pp. 872–875 (2006)
9. Wenyin, L., Dori, D.: Incremental arc segmentation algorithm and its evaluation. IEEE Trans. on Pattern Analysis and Machine Intelligence 20, 424–431 (1998)

Report on the Third Contest on Symbol Recognition

Ernest Valveny[1], Philippe Dosch[2], Alicia Fornes[1], and Sergio Escalera[1]

[1] Computer Vision Center, Dep. Ciències de la Computació
Universitat Autònoma de Barcelona, Bellaterra, Spain
{ernest,afornes,sergio.escalera}@cvc.uab.es
[2] LORIA, Université Nancy 2, Nancy, France
{Philippe.Dosch}@loria.fr

Abstract. In this paper we make a brief report of the third edition of the International Symbol Recognition Contest, organized in the context of GREC'07. This contest follows the series started at the GREC'03 workshop. In this report we describe the main changes introduced in the test data according to the conclusions of the past edition of the contest. We also summarize the results obtained by the only participant method. Finally, we point out some conclusions and open issues to be addressed in the next editions of the contest.

1 Introduction

The performance evaluation of symbol recognition has been a focus of research interest in the last years. Several surveys on symbol recognition[1,2,3,4] pointed out the need of standard evaluation tools in order to compare the large number of symbol recognition methods. As a result, a generic framework for the evaluation of symbol recognition has been proposed [5]. In this framework, the main issues to be addressed by any performance evaluation system are identified (mainly, the generation of datasets and groundtruth, the definition of metrics, and the protocol of evaluation) and several alternatives are proposed and discussed in the special case of symbol recognition.

Following this generic framework, and from a practical point of view, several contests have been organized. Actually, the first effort on the evaluation of symbol recognition was undertaken at ICPR'00 [6] where a contest was proposed using a dataset consisting of 25 electrical symbols, which were scaled and degraded with a small amount of binary noise. Afterwards, the series of contests on symbol recognition in the context of the GREC workshop started in 2003. In the first edition [7], the dataset was composed of 50 architectural and electrical symbols. These symbols were rotated, scaled, degraded with binary noise and deformed through vectorial distortion in order to generate up to 72 different tests with increasing levels of difficulty and number of symbols. There were five methods participating in the contest. Then, in the second edition [8] some modifications were introduced according to the conclusions of the first contest. The set of symbols was increased up to 150 different symbols, allowing the

W. Liu, J. Lladós, and J.-M. Ogier (Eds.): GREC 2007, LNCS 5046, pp. 321–328, 2008.
© Springer-Verlag Berlin Heidelberg 2008

definition of more pertinent tests for the evaluation of the scalability. In addition, four new degradation models were added to the framework for the generation of more noisy data. These new degradation models constituted a kind of "torture models". In this way, the robustness of the methods could be tested under very extreme conditions. Four methods participated in the contest.

Among the main conclusions stated in the report of the last contest [8] we can remark some issues that have been taken into account, not only in the design of the third edition of the contest, but also in the work undertaken in the last two years. Firstly, it was stated that evaluation should be a continuous task, not concentrated every two years at specific contests. Therefore, tools for the analysis of the results of recognition methods should be provided. In this sense, the work on the French project ÉPEIRES[1] has set up a web-based framework for the evaluation of symbol recognition where new tests can be easily created and the results obtained by a given method can be uploaded and automatically analyzed. Secondly, it was stressed the need of extending the evaluation to symbol localization and segmentation. Some work on this topic has been undertaken under the framework of the ÉPEIRES project too. As a result, a first approach to the generation of synthetic complete architectural drawings has been developed[9]. This is the first step in order to be able to generate large amounts of data for the evaluation of segmentation. Work has still to be done concerning the metrics to compare the results with the ground-truth. So, in the third edition of the contest we have not considered localization and segmentation and we have constrained the contest only to pre-segmented symbols as in past editions. Thirdly, it was claimed that more heterogeneous data should be included in the framework. In order to give an answer to this demand, we have included in this edition of the contest a dataset composed of logos. Logos are also graphic symbols, but with very different properties (regarding shape, primitives, appearance, etc) with respect to the technical symbols used in the previous contests. In this way, the range and variability of symbols is extended. Finally, it was remarked the need of defining blind tests in order to ensure that participant methods are not adapted to the particular data of the contest. In this edition this remark has been taken into account by including different types of randomly selected degradations in the same test. The goal is to be sure that participants design generic symbol recognition methods, able to work with all kind of (noisy) symbols.

In the next sections, we describe more in details the data provided in this edition of the contest as well as the results obtained by the only participant method. But before, we would like to recall the original purpose of this series of contests as stated in the call for participation: the main goal is not to give a single performance measure for each method, but to provide a tool to compare various symbol recognition methods under several different criteria. The question consists of determining the performance of symbol recognition methods when working on various kinds of symbols, extracted from diverse application domains, under several constraints, with different levels of noise and degradation. Whatever the performance measures are, we strongly believe that the main

---

[1] http://www.epeires.org/

Fig. 1. Some examples of technical symbols

objective of this evaluation framework must be the scientific analysis of the results. This analysis must be intended to determine the different qualities expected for recognition methods: robustness, genericity, precision, computational efficiency.

The paper is organized as follows. In section 2 we describe the datasets that were generated for this edition of the contest. Then, in section 3 we briefly describe the main features of the only participant method and analyze the results of its application to the dataset. Finally, in section 4 we state the main conclusions of the evaluation and some actions to be undertaken in the future.

2 Dataset

As explained in the previous section, we have considered two different kinds of symbols in this edition of the contest: technical symbols and logos. For technical symbols, we used the same dataset as in the last edition, that is, a set of 150 symbols, mainly originally from the domains of architecture and electronics. We can see in figure 1 some examples of this dataset where symbols are composed of linear primitives (straight lines and arcs). Logos are the main novelty in the dataset. We have included them in order to extend the spectrum of symbols. Logos are different of technical symbols in the sense that they are not composed only of linear primitives. They can include solid regions, texture, characters, more than one graphic component, etc. Thus, it is a completely different kind of symbol representation and can be useful to test whether recognition methods are generic enough. This dataset is composed of 105 different logos and some examples can be seen in figure 2.

We have used the same kind of transformations and degradations as in the last contest to generate the final tests for evaluation. Thus, rotation, scaling

Fig. 2. Some examples of logos

Fig. 3. Some examples of degraded images

and binary degradation using the Kanungo's method [10] have been applied to the ideal models of the symbols. In figure 3, we can see some examples of the degraded images. We have considered the same six models of degradation defined in the last contest as it was concluded that no new models were needed. As explained in the previous section, some of these models introduce heavy distortions in the images and thus, the level of difficulty is high.

The final tests for the evaluation have been generated combining all these elements. In table 1, there is a summary of all the tests with their main features. We can see that we have designed tests for two different sizes of the database for technical symbols. A first set of tests with 50 symbols and a second set with 150 symbols. In this way, we can evaluate the robustness to the scalability in the number of symbols. For both sets, all the possible combinations of rotation and scaling have been considered. Moreover, all the tests include binary degradation. Degradation is always randomly selected among the six possible models. Thus, we achieve the goal of generating blind tests, as explained in the introduction.

For logos, all the tests include the whole database of 105 symbols. In this case, several combinations of rotation, scaling and degradation have been considered. Two tests including specific models of degradation have been defined but, for

Table 1. Description of all the tests

Test	Dataset	No. of Models	No. of Images	Rotation	Scaling	Degradation
1	Technical	150	500	Random	None	Random among 6 GREC'05 models
2	Technical	150	500	None	Random	Random among 6 GREC'05 models
3	Technical	150	500	Random	Random	Random among 6 GREC'05 models
4	Technical	50	200	Random	None	Random among 6 GREC'05 models
5	Technical	50	200	None	Random	Random among 6 GREC'05 models
6	Technical	50	200	Random	Random	Random among 6 GREC'05 models
7	Logos	105	300	Random	None	None
8	Logos	105	300	None	Random	None
9	Logos	105	300	Random	Random	None
10	Logos	105	300	None	None	Second GREC'05 model
11	Logos	105	200	None	None	Fourth GREC'05 model
12	Logos	105	300	None	None	Random among 6 GREC'05 models
13	Logos	105	300	Random	None	Random among 6 GREC'05 models
15	Logos	105	200	Random	Random	Random among 6 GREC'05 models

the rest of the tests, degradation is randomly selected in order to generate blind tests.

All the information and data related to the tests can be found on the webpage of the ÉPEIRES project at http://www.epeires.org/.

3 Results

In this edition, only one method participated in the evaluation of the proposed tests. The method has been developed by Alicia Fornes and Sergio Escalera, from the Computer Vision Center, in Spain. A paper describing this method appears in the current LNCS volume. Nevertheless, we give an overview of the method in the next section in order to facilitate the understanding of the results.

3.1 Description of the Method

The method works on the skeleton or the contour of the original image. The choice use of skeletons or contours is decided depending on the shape database. Skeletons are preferred for line-based symbols while contours are dedicated for silhouette-based shapes. Images are aligned using the Hotelling transform that is based on principal components to find the main axis of the object. Then, the shape is represented using the Blurred shape model descriptor (BSM) that makes the technique robust against elastic deformations. Afterwards, Adaboost is applied to each pair of classes to train a set of binary classifiers. Finally, the set of binary classifiers is embedded in the framework of Error Correcting Output Codes (ECOC) to improve the final classification.

The main core of this method is the BSM descriptor. With this descriptor, the symbol is described by a probability density function that encodes the probability of pixel densities of image regions: The image is divided in a grid of n x n equal-sized subregions. Every bin receives votes from the pixels in its region but also from the pixels in the neighboring bins. The weight of the vote is set according to the distance to the center of the bin. The output descriptor is a vector histogram where every position corresponds to the weight of the pixels in the context of every sub-region. This vector is normalized in the range [0..1] to obtain the probability density function (pdf) of the n x n bins. In this way, the output descriptor represents a distribution of probabilities of the object shape considering spatial distortions. For further details, see [11].

3.2 Analysis of Results

Unfortunately we cannot present results for all the tests. The participant method was only evaluated using 5 of the proposed tests. In table 2, we show the recognition rates of the method for these 5 tests.

If we try to analyze these results we can draw several conclusions. Only one test with technical symbols was evaluated. This test contains images of 50 symbols with scaling and binary degradation. The recognition rate, 91%, can be

Table 2. Results of the method

Test	Dataset	Rotation	Scaling	Degradation	Recognition rate
5	Technical	None	Random	Random	**91%**
8	Logos	None	Random	None	**95%**
10	Logos	None	None	Second model	**82%**
11	Logos	None	None	Fourth model	**46%**
12	Logos	None	None	Random	**55%**

considered as a good result if we compare it with the recognition rates obtained for similar tests in the past contest. In it, the average of the recognition rates for all the methods, all degradation models and scaling was only 74.25%.

Concerning logos, the recognition rate for images without degradations remains at a high level, 95%. However, it decreases rapidly when degradations are applied. Although we have no other methods to compare these results, we can try to establish some relations with the results obtained in the most similar kind of tests in the last contest. In that case, for tests with 100 symbols (approximately the same number of logos), no scaling and binary degradation, the average of all the methods over all models of degradation was 90%, clearly greater than the recognition rate obtained in this case for the test 12 with logos. It is difficult to draw exact conclusions from these results as we have no other results with the logo database. We cannot state whether the low results for the logos are due to the fact that logos are intrinsically more difficult to recognize than technical symbol or whether they are a consequence that this method is better adapted to linear shapes than to solid shapes.

4 Conclusions and Future Work

In this edition, we have extended the contest with two of the considerations arising from the conclusions of the last contest: we have included a new kind of symbols, logos, and we have generated blind tests combining all the models of degradations. However, no relevant conclusions can be drawn from the experimentation with the logo dataset as we only have results from one method, and not for all the tests.

Nevertheless, after three editions of the contest, the framework for the evaluation of the recognition of pre-segmented symbols recognition seems mature enough. In this sense, this framework can be converted in a tool for continuous evaluation through the web platform of the ÉPEIRES project. This way, any researcher can contribute with new results to the database of the platform and we can have a good overview of the performance of a large number of methods. In this context, many tests have been generated along the three editions of the contest. Maybe it would be interesting to define a set of standard validation tests taking into account all the kinds of transformations and degradations. This set would constitute a kind of standard evaluation that every method should pass. Thus, we would have a generic global evaluation of all the methods. In addition, it would be also interesting to add new symbols to the framework in order to create a really large database of symbols, representative enough of all kinds of graphic symbols.

The big challenge that is still to be addressed is the evaluation of localization/segmentation in complete drawings with non-segmented symbols. In this sense, some advances have been described in the field of ground-truthing with the generation of synthetic documents. The next step should be the definition of metrics to compare the results with the ground-truth, and the definition of the evaluation protocol. We plan to advance in this direction and we hope to be able to propose early a contest on symbol localization.

Finally, we want to make a note on the low participation in this edition of the contest. For next editions, we should increase the efforts in order to promote the participation in the contest. However, this could be another point for providing a continuous framework for the evaluation of the recognition of pre-segmented symbols. We hope that new researchers will be interested by the contest when it will include symbol localization.

Acknowledgment

The authors would like to acknowledge the French Ministry of Research for the funding of the ÉPEIRES project as a part of the Techno-Vision campaign.

This work has also been partially supported by the Spanish project TIN2006-15694-C02-02, and by the Spanish research programme Consolider Ingenio 2010: MIPRCV (CSD2007-00018)

References

1. Chhabra, A.K.: Graphic Symbol Recognition: An Overview. In: Tombre, K., Chhabra, A.K. (eds.) GREC 1997. LNCS, vol. 1389, pp. 68–79. Springer, Heidelberg (1998)
2. Cordella, L., Vento, M.: Symbol recognition in documents: a collection of techniques. International Journal on Document Analysis and Recognition (IJDAR) 3, 73–88 (2000)
3. Lladós, J., Valveny, E., Sánchez, G., Martí, E.: Symbol recognition: Current advances and perspectives. In: Blostein, D., Kwon, Y.-B. (eds.) GREC 2001. LNCS, vol. 2390, pp. 104–127. Springer, Heidelberg (2002)
4. Tombre, K., Tabbone, S., Dosch, P.: Musings on symbol recognition. In: Liu, W., Lladós, J. (eds.) GREC 2005. LNCS, vol. 3926, pp. 23–34. Springer, Heidelberg (2006)
5. Valveny, E., et al.: A general framework for the evaluation of symbol recognition methods. International Journal on Document Analysis and Recognition (IJDAR) 1, 59–74 (2007)
6. Aksoy, S., Ye, M., Schauf, M., Song, M., Wang, Y., Haralick, R., Parker, J., Pivovarov, J., Royko, D., Sun, C., Farneboock, G.: Algorithm performance contest. In: Proceedings of 15th. International Conference on Pattern Recognition, Barcelona, Spain, vol. 4, pp. 870–876 (2000)

7. Valveny, E., Dosch, P.: Symbol recognition contest: a synthesis. In: Lladós, J., Kwon, Y.B. (eds.) GREC 2003. LNCS, vol. 3088, pp. 368–385. Springer, Heidelberg (2004)
8. Dosch, P., Valveny, E.: Report on the second symbol recognition contest. In: Liu, W., Lladós, J. (eds.) GREC 2005. LNCS, vol. 3926, pp. 381–397. Springer, Heidelberg (2006)
9. Delalandre, M., Pridmore, T., Valveny, E., Trupin, E., Locteau, H.: Building synthetic graphical documents for performance evaluation. In: Workshop on Graphics Recognition (GREC), pp. 84–87 (2007)
10. Kanungo, T., Haralick, R.M., Baird, H.S., Stuetzle, W., Madigan, D.: Document Degradation Models: Parameter Estimation and Model Validation. In: Proceedings of IAPR Workshop on Machine Vision Applications, Kawasaki, Japan, pp. 552–557 (1994)
11. Fornés, A., Escalera, S., LLadós, J., Sánchez, G., Radeva, P., Pujol, O.: Handwritten symbol recognition by a boosted blurred shape model with error correction. In: Martí, J., Benedí, J.M., Mendonça, A.M., Serrat, J. (eds.) IbPRIA 2007. LNCS, vol. 4477, pp. 13–21. Springer, Heidelberg (2007)

Is Graphics Recognition an Unidentified Scientific Object?

Karl Tombre

LORIA-INRIA, B.P. 239, 54506 Vandœuvre-lès-Nancy CEDEX, France
Karl.Tombre@loria.fr

Abstract. This paper summarizes the presentation and discussions at the panel session held at the conclusion of the GREC'07 workshop. After making a short review of where the graphics recognition stands, we raise some questions (hopefully) of interest for the future of this community.

1 Introduction: A Short History of Graphics Recognition as an Identified Community

As the name of a scientific community and of a workshop, "Graphics Recognition" has not been used for such a long time. In 1988, the "Structural and Syntactical Pattern Recognition" workshop was organized in Pont-à-Mousson, France, by Prof. Roger Mohr. At that event, such a large number of papers were presented on document image analysis applications that it was decided that the next workshop, held in New Jersey in 1990 and organized by Dr. Henry Baird, would be completely focused on this field. Revised versions of the papers presented at the SSPR'90 workshop were published in what was probably one of the first scientific volumes focused on document image analysis [1]. At this workshop, it was discussed and decided to create a new series of international conferences, the ICDAR series whose first instance was held in 1991 in Saint-Malo, France; this opened the way to the visibility of the document analysis and recognition community as such—although work on these topics had of course existed for a long time before that, within the general pattern recognition community.

Within this general positioning of the community as such, Prof. Rangachar Kasturi moved in 1992 to give a new start to a Technical Committee of the International Association for Pattern Recognition, namely TC10 on "Line Drawing Interpretation", by focusing it on the subfield of document analysis and recognition devoted to graphics-rich documents and the specific problems raised by these documents (raster-to-graphics conversion, text-graphics separation, symbol recognition...) To emphasize the new focus, TC10 was renamed as the technical committee for *Graphics Recognition*. This gave birth in 1995 to the GREC series of workshops, which themselves have led to the publication of reference LNCS volumes for the community [2,3,4,5,6,7]. So one may say that it is only in the last 15 years that there has been a set of researchers identifying themselves as the "Graphics Recognition Community".

W. Liu, J. Lladós, and J.-M. Ogier (Eds.): GREC 2007, LNCS 5046, pp. 329–334, 2008.

This paper, stemming from a panel discussion held at the end of the GREC'07 workshop in Curitiba, Brazil, tries to take a step back and look at the field, its achievements, its remaining challenges, and the homogeneity of its composition.

2 Looking Back: What Did We Conclude Two Years Ago?

GREC'05 marked the 10th anniversary of the workshop and was an opportunity for reviewing achievements, topics which we had drifted away from, others which were still relevant, and emerging new themes.

Our main conclusions at that time were the following [8]:

- There are still open problems in *building complete systems*, so scientists should not drift away too quickly from this question, despite the successes which have been achieved;
- *Symbol recognition* matures but there are remaining open problems, including the recognition of non-segmentable symbols, the issue of scalability, and the recognition of complex symbols made of the assembly of smaller symbols;
- *Cultural heritage documents* emerge as a new theme;
- The issue of *symbol spotting*, i.e. localizing a symbol without necessarily recognizing it explicitly, is getting more focus, including the need to take into account relevance feedback from the user;
- The community has failed to gather around a common base of *software*, largely due to the "not-invented-here" syndrome.

Some hot topics were identified:

- To achieve progress in close-to-optimal, automatic, and non-contextual *vectorization* on black-and-white images, progress will have to include the processing of gray-level images, sub-pixel precision in the segmentation tools, better curve segmentation algorithms, and a seamless integration of user input and of contextual knowledge;
- The problems to be solved for complete document analysis include low-level questions such as the digitizing resolution, and the analysis of digital documents with little or no structure;
- Performance analysis campaigns have been a success, but the contests' use of degradation models can be controversial, there are ususally too few participants, and we need to perpetuate the access to data and evaluation tools, also outside the contests.

It is interesting to look at the community's most recent achievement with these questions in mind, to see how fast and in which direction we are moving.

3 Some Topics Discussed during the GREC'07 Workshop

Let us now try to review a few topics addressed by our community these last years, and discuss to which extent they characterize graphics recognition as a field.

3.1 Features

A lot of work has dealt with defining and using appropriate features for recognition tasks, especially in graphics recognition. At GREC'07, work was presented on blurred shape models, ridgelets, graph representations, region-based signatures, to name a few. We also got a presentation of an interesting attempt at characterizing the performance and usability of various shape descriptors. In recognition as well as in information spotting applications, it is indeed necessary to work on the most appropriate features, and also on the right feature combination methods for optimal recognition.

But it becomes increasingly difficult to answer the question: which features distinguish graphics recognition from general pattern recognition problems. Comparing our contributions with those in content-based image retrieval, for instance, show a real convergence, where the fact that we are dealing with black-and-white, graphical information tends to become a detail.

One contribution our community can make, as it deals with a subset of all possible imaging applications, is to work on the characterization of various features for shape representation and recognition, thus contributing to building up a professionnal repository of features with their properties, and maybe avoiding the recurring appearance of "new" features which are only minor variations on old themes.

3.2 User Interaction

It is increasingly necessary to design analysis and recognition methods which do not work in stand-alone mode, but take into account the user's interaction, so as to be able to perform incremental learning, relevance feedback in recognition and retrieval applications, interactive recognition in sketching mode, etc. But little work has been done on modelling the user, who is mostly considered as some kind of ill-defined, external entity.

The fact is that there is nothing in common between a "vanilla plain" user who may be your uncle or grandma, browsing a collection of images and giving relevance feedback without really knowing anything about the application, and a highly specialized user able to input syntactical rules to represent the knowledge in a specific document analysis application.

If the purpose is indeed to build a highly specialized system, this may not be a problem, but when the application is potentially very general whereas the user interaction paradigm requires the user to have a PhD in pattern recognition or to have trained for months, there is a contradiction in the whole setup which limits the applicability of the method.

3.3 Building Large-Scale Systems

A number of large-scale applications are being dealt with, especially for sketching, and for the analysis, characterization and indexing of large databases of historical documents. However, in order to go beyond the proof-of-concept system, we need to put more emphasis on software, including reusable software which has not been developed in our own group. The community still suffers of the "not invented here" syndrome, despite the availability of a significant number of larger image analysis platforms, some of them specialized in document analysis, others more generic. It will be difficult to have a significant impact without either building oneself production-quality code and putting in enough human resources for that purpose, or using others' code and pushing one's students to use existing code.

It was mentioned that graphics recognition was still looking for its "killer application" and some participants felt that a general sketching interface could be such an application. But in that case, efforts must be coordinated, within academia or in partnership with companies.

The dream of building completely automated systems for converting drawings, maps and diagrams into high-level representations seems to have vanished, as the methods we design reach their limits at a level where there is still a lot of user editing to be done. But as one participant mentioned, there is still a very interesting opportunity to build combined sketching/retrieval/recognition systems, making it possible to navigate in a large document base by sketching simple examples of what is being searched for.

Suprisingly, whereas there are a lot of high-interest applications in dealing with digital documents (e.g. PDF documents or web graphics, in which to search for and recognize various entities), a workshop like GREC seems not to be deemed the right place to present this, as we have not seen any work in this area.

3.4   Performance Evaluation and Contests

This is an area where the field of document analysis in general, and the subfield of graphics recognition in particular, have often been showing the path to the whole image analysis community. This may stem from the fact that the data and problems we work on are more easily circumscribed. But we have also been pro-active in organizing contests, gathering ground-truthed data, making available performance evaluation metrics and tools.

But there are still concerns. We have few participants in the contests, despite the very hard work by the organizers. This is a little bit disappointing. Our community has also started referring to the data and evaluation tools continuously, in day-to-day work without any contest. This effort must be continued, but for that we need to have open-source, robust benchmarking tools available online, with a sufficient amount of ground-truthed data. How do we capitalize on contributions by various teams?

Another question is to assess whether we have benchmarking data covering all the needs. Finally, it is still necessary to ask ourselves what we want to evaluate, so that our benchmarks are not purely academic but that they model real-world challenges.

Some future directions in that area could be to challenge our existing methods by making them available as a web service and letting colleagues "attack" them to test their limits. Another idea would be to announce in advance a grand challenge for the commmunity to work on, from one workshop to the other, for instance.

4   Is Graphics Recognition Still an Appropriate Scientific Area?

One final and somewhat controversial topic discussed at length in our closing panel was whether it still makes sense to gather a community around the theme of "Graphics Recognition". In the first GREC workshops, most people contributed with pure graphics recognition problems, such as vectorization or text-graphics segmentation. There are very few new contributions to these topics, as methods have reached maturity and are assimilated by people as state of the art.

Now, we see a variety of work related to topics such as visual languages, document image layout analysis, shape descriptors, information retrieval and information spotting, biometrics, etc. But in each case, we only have a subset of activities related to each of these areas. Is the subset representative of some specificities, or does it just happen to be the contributions to these topics of the groups used to define themselves as belonging to the "graphics recognition community" (and used to attend GREC)? We should have the courage to ask ourselves these questions...

5   Conclusion

The graphics recognition community has contributed significantly to the field of pattern recognition, setting the path for others in some areas, such as performance evaluation. It is now at crossroads where it has to ask itself whether its scientific interests need to be kept together or if it is time to reconsider the frontiers and join the most appropriate scientific communities.

Looking back at history, the concept of "Graphics Recognition" has been in use for around 15 years only, at least to define a scientific community. Science is a living phenomenon, where one of the biggest dangers is to bury oneself (and one's students) in a narrow pit where one looses the view of evolutions and new advances in neighboring areas. I have had the pleasure of being strongly involved in the adventure of this community for these 15 years, and am personally confident that the groups now active in graphics recognition have the ability to make the most appropriate moves. I would recommend to make them without any tabu on which structures to keep, which structures to throw away, and which structures to change.

References

1. Baird, H.S., Bunke, H., Yamamoto, K. (eds.): Structured Document Image Analysis. Springer, Heidelberg (1992)
2. Kasturi, R., Tombre, K. (eds.): Graphics Recognition 1995. LNCS, vol. 1072. Springer, Heidelberg (1996)

3. Chhabra, A.K., Tombre, K. (eds.): GREC 1997. LNCS, vol. 1389. Springer, Heidelberg (1998)
4. Chhabra, A.K., Dori, D. (eds.): GREC 1999. LNCS, vol. 1941. Springer, Heidelberg (2000)
5. Blostein, D., Kwon, Y.-B. (eds.): GREC 2001. LNCS, vol. 2390. Springer, Heidelberg (2002)
6. Lladós, J., Kwon, Y.-B. (eds.): GREC 2003. LNCS, vol. 3088. Springer, Heidelberg (2004)
7. Liu, W., Lladós, J. (eds.): GREC 2005. LNCS, vol. 3926. Springer, Heidelberg (2006)
8. Tombre, K.: Graphics Recognition: The Last Ten Years and the Next Ten Years. In: Liu, W., Lladós, J. (eds.) GREC 2005. LNCS, vol. 3926, pp. 422–426. Springer, Heidelberg (2006)

Author Index